Mechanical Engineering and Thermodynamics

Mechanical Engineering and Thermodynamics

Edited by **Nathan Rice**

Willford Press
New York

Published by Willford Press,
118-35 Queens Blvd., Suite 400,
Forest Hills, NY 11375, USA
www.willfordpress.com

Mechanical Engineering and Thermodynamics
Edited by Nathan Rice

International Standard Book Number: 978-1-68285-109-8 (Hardback)

Printed in the United States of America.

Contents

Preface

Over the recent decade, advancements and applications have progressed exponentially. This has led to the increased interest in this field and projects are being conducted to enhance knowledge. The main objective of this book is to present some of the critical challenges and provide insights into possible solutions. This book will answer the varied questions that arise in the field and also provide an increased scope for furthering studies.

Mechanical engineering is an established and prominent field of engineering that is concerned primarily with designing, manufacturing and maintaining various machines, equipment and mechanical systems. Thermodynamics plays a significant role in conceptualizing, designing and manufacturing of different machines and mechanical structures. The principles of heat exchange, relation between work and energy output, temperature measurement are crucial to mechanics and engineering. This book encompasses different principles, concepts, tools and techniques involved in mechanical engineering and the role of thermodynamics in this field. Students, researchers and engineers will find this book a wonderful reference source.

I hope that this book, with its visionary approach, will be a valuable addition and will promote interest among readers. Each of the authors has provided their extraordinary competence in their specific fields by providing different perspectives as they come from diverse nations and regions. I thank them for their contributions.

Editor

Functional parameters modelling of transformer

Marius-Constantin Popescu* and Cristinel Popescu

Faculty of Electromechanical and Environmental Engineering, University of Craiova, B-dul Decebal, nr.107, 200440-Craiova, Dolj, România.

This paper analyses transformer loss of life, attending to the realistic variability of both structural and functional parameters. The article begins with the modelling of load and ambient temperature profiles by means of chronological series theory. A simple additive model is proposed and validated with realistic data. Loss of life resulting from probabilistic functional and structural parameters is analysed through a sensitivity study.

Key words: Transformer thermal, estimate thermal parameters.

INTRODUCTION

This paper is devoted to load and ambient temperature (functional parameters) profiles modelling and to the sensitivity study of thermal and loss of life models relatively to functional and structural parameters. In this research, the functional parameters modelling was based on realistic data. It is not objective of this section to exhaustively apply time series theory in modelling realistic load and temperature profiles. Such complete analysis is out of the scope of this work. The objective is based on data representing the load profiles of realistic distribution transformers and ambient temperature profiles, to obtain sufficient accurate models that will give physical support to the probabilistic models. Loads modelling and forecasting play a fundamental role in power systems planning and management. Due to its connection with weather characteristics, loads and weather modelling are joined subjects of some works (Asbury, 1975; Chong and Malhame, 1984; Sachedev et al., 1977; Srinivasan and Pronovost, 1975). Provided a transformer thermal model is chosen, deterministic hot-spot temperature can easily be computed, given the input profiles of load and ambient temperature. When analysing the time series representative of a given transformer load (or the time series of a localised ambient temperature), one can visualise a cycling (deterministic) behaviour (daily, weekly, monthly, seasonally) to which is superposed a random behaviour.

*Corresponding author. E-mail: mrpopescu@em.ucv.ro.

Such input profile structure will be reflected on consequent hot-spot temperature profile: deterministic and random components. Apart from specific characteristics and improvements that transformer thermal model may reflect, the validity of deterministic input profiles is questionable, due to unpredictable (random) changes that realistic profiles do present. This fact determines a probabilistic analysis of the system, which is being discussed in this research. The objective of this section is to study loss of life sensitivity over given statistics of the inputs. Simulation results using thermal and loss of life models linearisation and direct Monte Carlo methods will be presented for Normal and Uniform distributions of both input variables. Thermal and loss of life models sensitivity, relatively to structural parameters, is also developed. Usually, it is assumed that model structural parameters are known without error. Some of these parameters are transformer specific ($\Delta\Theta_{0R}$, $\Delta\Theta_{ohR}$, *R*), determined either from tests or from manufacturer's catalogue data. And they do present some variability for a given transformer rated power, depending on the manufacturers, as presented on by Popescu (2008). Others, like *n* and *m* are difficult to determine with precision from tests, since they are closely related to transformer cooling conditions and geometry. The variability that these parameters do present in practice, is the basis of this study which objective is under a reference scenario of functional inputs to analyse models output (*LOL*) sensitivity, and models structural parameters, namely, $\Delta\Theta_{0R}$, $\Delta\Theta_{ohR}$, *R, n,* and *m.*

FUNCTIONAL PARAMETERS MODELLING

Time series descriptive techniques

Despite the diversity of approaches, methodologies and end-use applications, when studying realistic load and weather data profiles, one can always identify trends (deterministic) components, to which are superposed irregular (random) behaviours. Deterministic data can be described by an explicit mathematical relationship (a mathematical model). In as much as no unforeseen event in the future will influence the phenomenon, producing the data set under consideration for identical experience conditions, the mathematical model will reproduce the same exact data set, no matter how many times the experiment is repeated. Random data values are unpredictable in a future instant in time, and therefore must be described in terms of probability statements and statistical averages, rather than by explicit mathematical relationships. In practical problems, involving random variables, one must not expect to obtain theoretical results, namely, a purely random variable. The main reason is that a random variable is a theoretical concept, which can not be reproduced (simulated) in practice; only samples of random variables can numerically be simulated. The statistics of samples only asymptotically (with the increasing length of the sample) tend to be random variable statistics. A random variable can be viewed as a sample of infinite length. Time series representative of load and ambient temperature profiles do present deterministic and random characteristics simultaneously (Figure 1). Such a data set, presenting concomitant time and random characteristics is referred to as a stochastic process. Time plot will often show the most important properties of a time series (Chatfield, 1975; Popescu et al., 2009a). It was predictable and can be visually confirmed that ambient temperature time series do exhibit a seasonal effect that, although not representative of the sample, is expected to be cyclic. Possible long-term trends will not be considered since, although might be present, the sample length is insufficient to allow this kind of analysis. A common model to describe time series as the one represented on Figure 1 is the additive model of the form (Chatfield, 1975; Friedlander and Francos, 1996; Mastorakis et al., 2009b):

$$x_t = x_{\det t} + x_{rant}\,, \tag{1}$$

Where; x_{dett} represents the deterministic cyclic component and x_{rant} the random component.

Most of the time series theory concerns stationary time series, which, intuitively, is a time series where no systematic temporal variations in mean and variance occur. From the analysis of series residuals, after removing the seasonal effects (and trends when existent), one may conclude that it is possible to model residuals by means of a stationary stochastic process. Several approaches, methodologies and tests can be used to detect time series characteristics such as cyclic variations, stationary, randomness (Gutmann and Wilks, 1982; Popescu, 2006; Popescu et al., 2009; Ross, 1987). However, a complete and powerful tool is provided by the autocorrelation function. If x_t and y_t are two samples, length N of two stationary ergodic processes, an estimator of their correlation function, $\hat{\rho}_{xy}(k)$ is, according to Chatfield (1975) and Popescu (2006):

$$\hat{\rho}_{xy}(k) = \frac{C\hat{O}V_{xy}(k)}{C\hat{O}V_{xy}(0)}\,, \tag{2}$$

Where; $C\hat{O}V_{xy}(k)$ denotes an estimator of the covariance function

$$C\hat{O}V_{xy}(k) = \frac{1}{N}\sum_{t=1}^{N-k}(x_t - \overline{x})(y_{t+k} - \overline{y})\,, \tag{3}$$

with $\overline{x}$ and $\overline{y}$ representing the samples averages given by:

$$\overline{x} = \frac{1}{N}\sum_{t=1}^{N} x_t \ and\ \overline{y} = \sum_{t=1}^{N} y_t\,, \tag{4}$$

When $x_t \equiv y_t$, expression (4) represents the autocorrelation function and expression (3), the autocovariation function. When the variable it refers to is clear, the estimator of the autocorrelation function will be denoted by $\hat{\rho}_x(k)$ or simply by $\hat{\rho}(k)$, and the estimator of the autocovariation by $C\hat{O}V_x(k)$ or simply by $C\hat{O}V(k)$. The analysis of the corresponding sample autocorrelogram (plot of the autocorrelation coefficients as a function of time lag k) often provides fully insight into the probabilistic model that describes the data. The autocorrelation function is the measurement of correlation (link) between series data values at different time and distances apart.

For a random variable, correlation coefficients must be null for any lag k, but $k = 0$. It should be remarked that, mathematically, the maximal time lag k in (2) is limited to N/2, although (Chatfield, 1975) states that N/4 is the usual limit. Information contained in the sample time series may not always be sufficient to completely characterise it. Figure 2a represents the autocorrelogram of the one-year time series of data represented on Figure 1. And although yearly cyclic variations are expected to occur, the autocorrelogram does not evidence them. However, by increasing the sample size to two years length, the respective autocorrelogram being represented on Figure 2b, clearly evidences an almost sinusoidal variation, which, although expected, should be confirmed with a longer size sample. If a time series could be

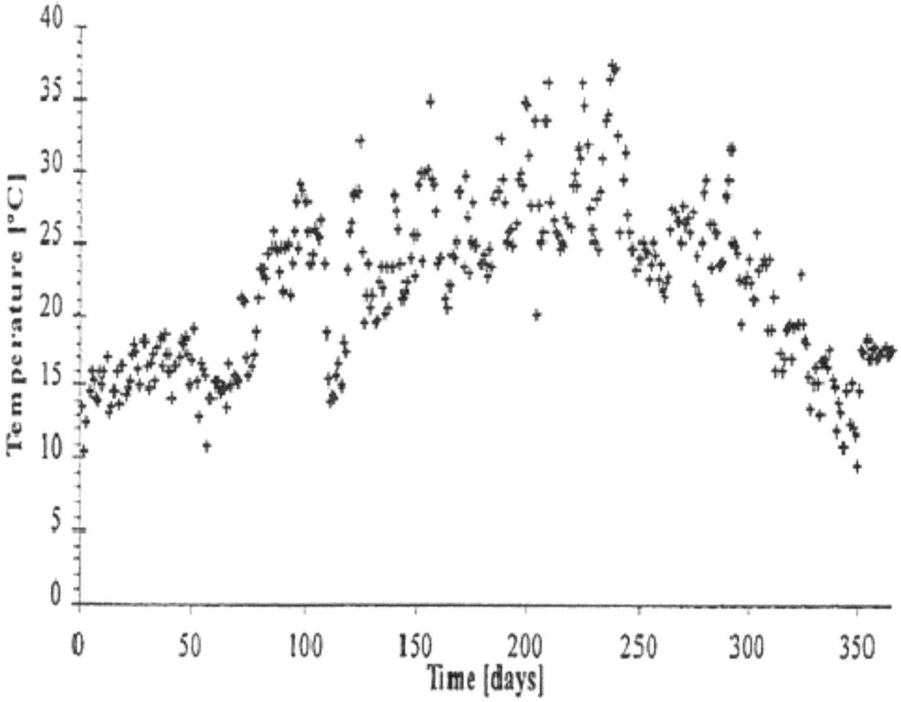

Figure 1. Annual time series representing daily maximal ambient temperatures. Data from 2005.

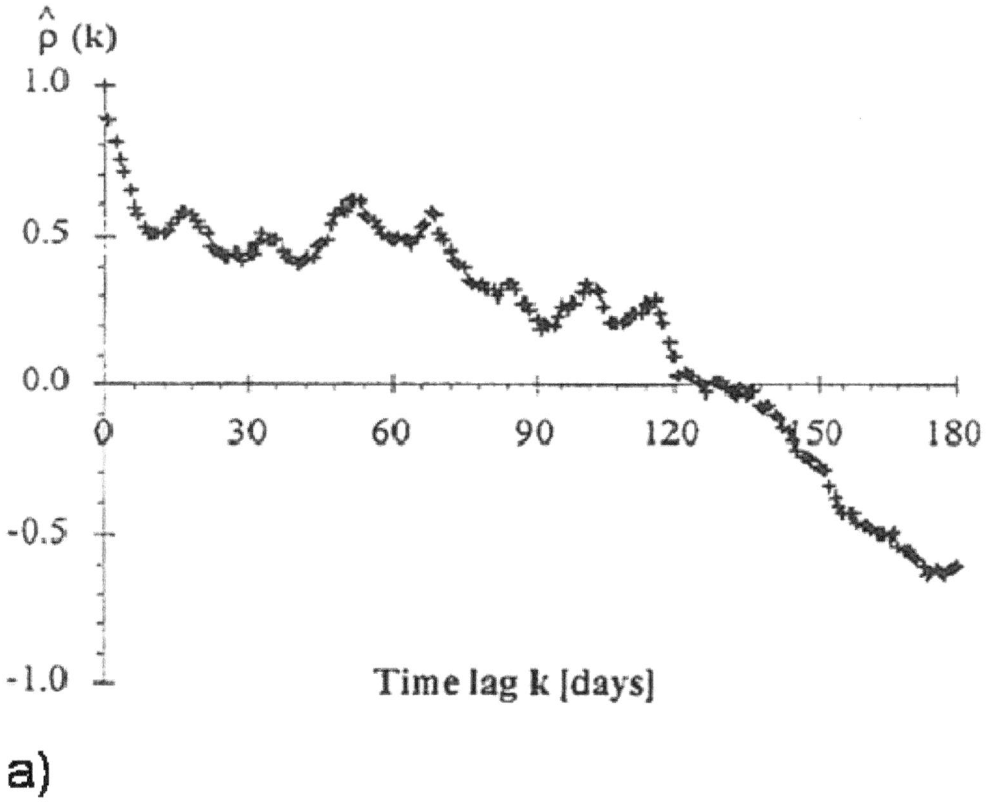

Figure 2a. Autocorrelogram of daily maximal ambient temperatures in 2005.

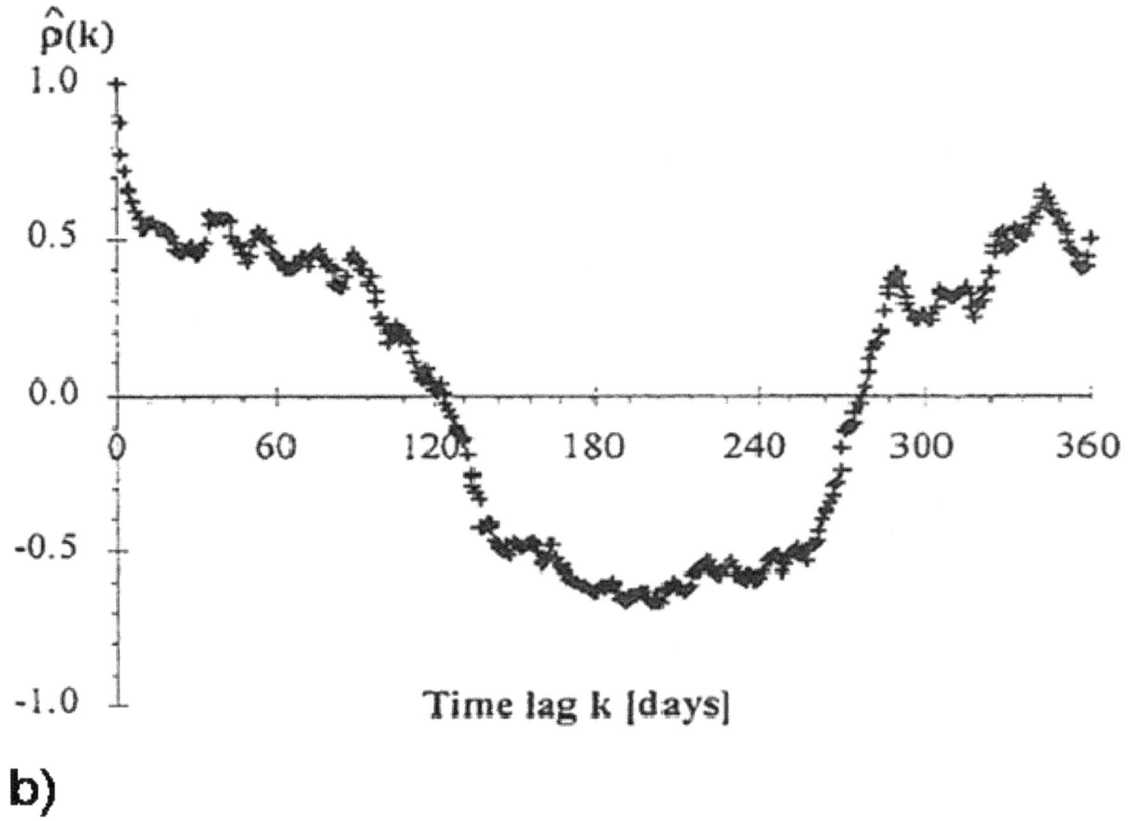

Figure 2b. Autocorrelogram of daily maximal ambient temperatures in 2004 and 2005.

described by a purely deterministic sinusoidal function of the form, then:

$$x_t = X\cos\omega t\,, \tag{5}$$

Where; X and ω are constants, its autocorrelation would evidence this cyclic variation, since for large sample lengths (N $\rightarrow\infty$) it would tend to:

$$\rho(k) = \cos\omega t\,. \tag{6}$$

Following the evidences of Figure 2 autocorrelogram and International Standards (IEC-354, 1991) suggestion, the deterministic component, $x_{det\ t}$, of model (1) was assumed to be given by a generic deterministic sinusoidal variation represented by:

$$x_{\det_t} = \bar{x}_d + \Delta x_d \cos(\omega t + \varphi_x)\,, \tag{7}$$

Where; $\bar{x}_d$, Δx_d and φ_x are constants.

From the analysis of time series residuals (random component x_{rant_t}) in Figure 3a and respective autocorrelogram in Figure 3b, one can extract clues towards its modelling. The interpretation of autocorrelogram does require considerably experience in time series analysis and, according to Chatfield (1975) and Popescu (2008), this is one of the hardest aspects of time series analysis.

Plot diagram of random component (Figure 3b) shows no evident cyclic or mean variations; as a first approximation, variable could be taken as random. However, a more accurate analysis shows that autocorrelogram represented on Figure 3b is not typical of a random variable since it should be $\rho(0) = 1$ and $\rho(k) = 0$ for $\forall k \neq 0$. Apart from small amplitude and of high frequency (probably) cyclicvariations, one can find considerably high $\hat{\rho}(k)$ values for initial time lags, k=1...6.

To determine the best model that fits a given sample au-tocorrelation function, the methodology proposed by Box and Jenkins (1970) consists of comparing sample autocorrelation function with the theoretical autocorrelation function of several models, and choosing the one which best agrees with the sample autocorrelation function. Most common models are the AutoRegressive models (AR), the moving Average Models (MA) and mixed mo-dels such as AutoRegressive Moving Average models (ARMA) and Autoregressive Integrated Moving Average models (ARIMA). As far as the objective of this work is concerned, one will limit oneself to present the AR model. The process $\{X_t\}$ is said to be AR of the order,

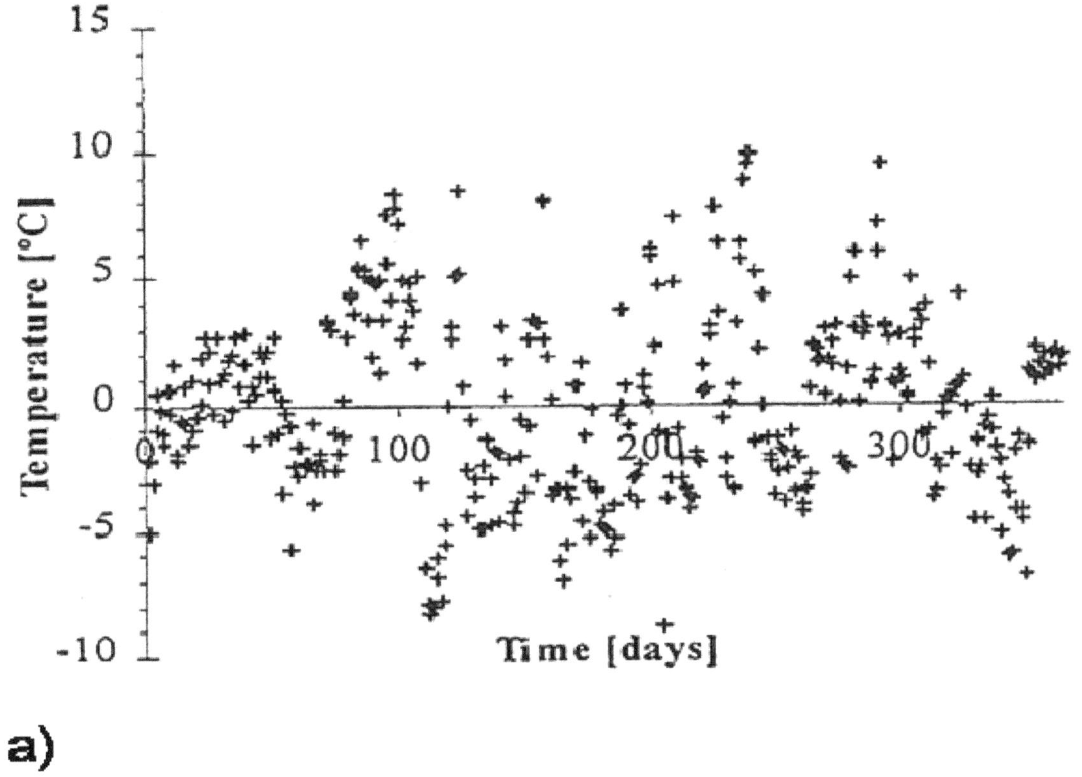

Figure 3a. Random component; and respective autocorrelogram.

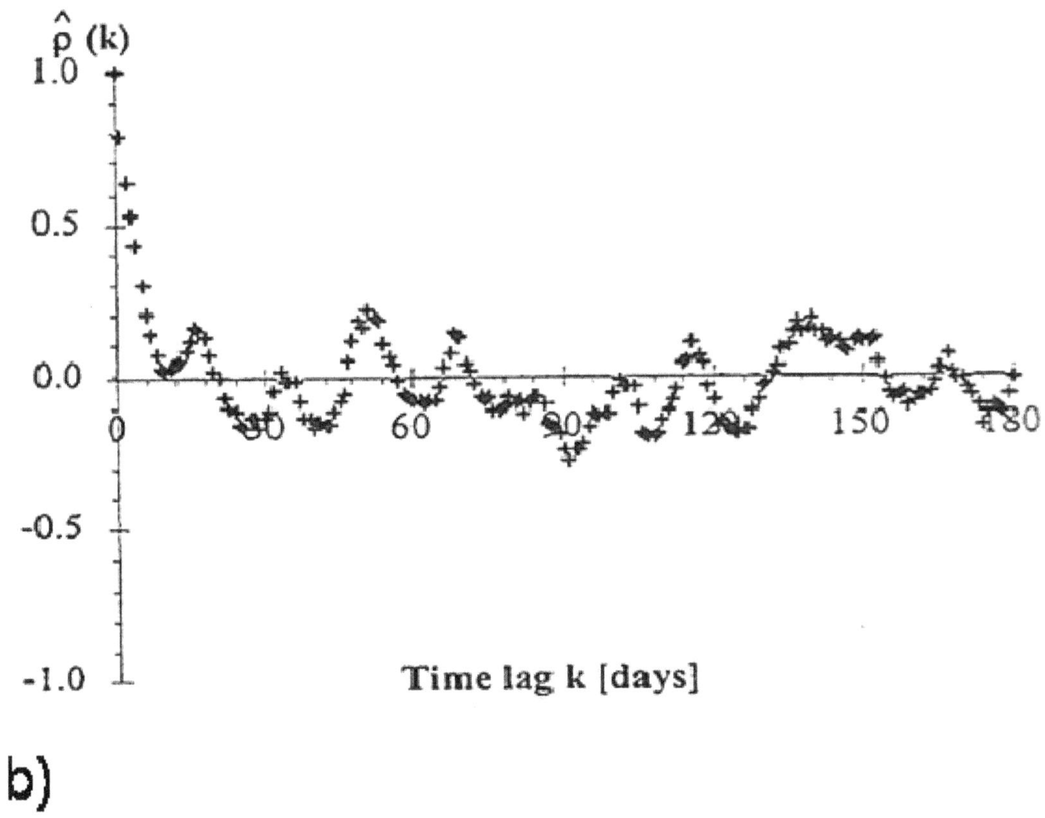

Figure 3b. Random component of maximal ambient temperature in 2005.

m if given a purely random process {Z_t} with null mean and variance σ_Z^2 (Chatfield, 1975; Popescu, 2009):

$$x_t = \alpha_1 x_{t-1} + ... + \alpha_m x_{t-m} + Z_t \,, \qquad (8)$$

Where; $\alpha_1 ... \alpha_m$ are constants.

For first order AR processes (also referred as Markov process [8]), the estimator of α_1 denoted by $\hat{\alpha}_1$, is (Chatfield, 1975):

$$\hat{\alpha}_1 = \hat{\rho}(0) \,. \qquad (9)$$

Autocorrelogram of Figure 3 is suspicious to correspond to a first order AR model since initial $\hat{\rho}(k)$ values appear to decrease geometrically (Chatfield, 1975; Popescu, 2009). If time series x_{rant} is a sample of a first order AR process { X_{rant} }, it must be:

$$X_{rand\,t} = \alpha_1 X_{rand\,t-1} + Z_t \,. \qquad (10)$$

Using estimator (9) and sample x_{rant}, the resulting z_t sample (being z_t a sample of the random variable, Z_t) and corresponding autocorrelogram are represented on Figure 4.

To determine whether Figure 4 corresponds to a sample autocorrelogram of a random variable ($\rho(k) = 0$ for $k > 0$) or not, confidence intervals must be determined. For large N values, being the sample autocorrelation, $\hat{\rho}(k)$, normally distributed with Mastorakis et al. (2009b) andPopescu et al. (2009b):

$$\mu_{\hat{\rho}(k)} = \rho(k) \text{ and } \sigma_{\hat{\rho}(k)} \approx \sqrt{\frac{1}{N}(1+2\hat{\rho}(0))} \text{ for } k>0 \qquad (11)$$

a $(1-\alpha_S)\%$ confidence interval for $\hat{\rho}(k)$, being α_S the significance level of the test, is given by:

$$P\left\{-\Phi^{-1}(1-\alpha_S)\% < \frac{\hat{\rho}(k)-\mu_{\hat{\rho}(k)}}{\sigma_{\hat{\rho}(k)}} < \Phi^{-1}(1-\alpha_S)\%\right\} \geq 1-\alpha_S \qquad (12)$$

For $k > 0$ being $\Phi^{-1}(1-\alpha_S)$ the inverse of the standardised normal distribution evaluated at $(1-\alpha_S)$. Attending to (11), and that for a random variable it is ($\rho(k)=0$ for $k>0$), probability expression (12) is traduced by the statement:

$$\hat{\rho}_k \in \left[-\sqrt{\frac{1}{N}(1+2\hat{\rho}(0))}\Phi^{-1}(1-\alpha_S);\sqrt{\frac{1}{N}(1+2\hat{\rho}(0))}\Phi^{-1}(1-\alpha_S)\right] \text{ for } k>0 \qquad (13)$$

On Figure 4, limits of (13) with α_S =5% are also represented. If $(1-\alpha_S)\%$ of the $\hat{\rho}(k)$ values, with $k > 0$, are within (13) limits, $\hat{\rho}(k)$ is accepted as representative of the autocorrelation of a random variable and therefore, z_t is accepted as a random variable. A second step for the complete modelling of time series x_t, is the determination of random variable z_t distribution function. This is achieved by testing the *p*robability *d*ensity *f*unctions (*pdf*) of theoretical (expected) random variables against the realistic (observed) *pdf* one obtains for z_t. These tests are referred to as goodness-of-fit tests (Bendat and Piersol, 1990; Gutmann and Wilks, 1982; Popescu, 2006; Ross, 1987). A key element associated to statistical tests is its p-value. According to Ross (1987) test formulation, the *p-value* represents the maximal significance level at which the hypothesis should be accepted. Its value measures the closeness of the observed *pdf* relatively to the theoretical *pdf;* the *p-value* will be as close to the unity as the observed *pdf* is close to the theoretical pdf. Justification to give relevance to AR models resides on their physical base. They represent memory systems in the sense that values at instant *t* are influenced by the memory of previous values at *t*-1,..., *t-m*. Due to earth thermal inertia, ambient temperature is expected to be a function of near past ambient temperatures; due to its correlation with ambient temperature similar behaviour can be expected on the load profiles of distribution transformers.

Case studies

Previously described techniques applied to four time series, representing maximal, Θ_M, minimal, Θ_m, average, Θ_{av}, and half-amplitude Θ_{am} values of daily ambient temperature in the Dolj (RO) region from 2002 to 2005 were:

$$\Theta_{av} \equiv (\Theta_M + \Theta_m)/2 \text{ and } \Theta_{av} \equiv (\Theta_M - \Theta_m)/2 \,. \qquad (14)$$

Samples length is, therefore, N = 365. In order to keep exposition as clear as possible, the previous generic notations *x* and *z* of §2.1 will be used, being $x, z \equiv \Theta_M, \Theta_m, \Theta_{av}, \Theta_{am}$. By means of discrete fourier transformer, parameters $\bar{x}_d, \Delta x_{d\,x}, \varphi_x$ and of deterministic model represented by (7) were determined (Mastorakis et al., 2009a; Popescu et al., 2009). Resulted random residuals, x_{rant}, were analysed. Although respective autocorrelograms revealed the presence of an AR model, the histogram of x_{rant} amplitudes passed a Chi-Square test, regarding a Gaussian distribution. If model:

$$x_t = \bar{x}_d + \Delta x_d \cos(\omega t + \varphi_x) + N(\hat{\mu}_x, \hat{\sigma}_x) \,, \qquad (15)$$

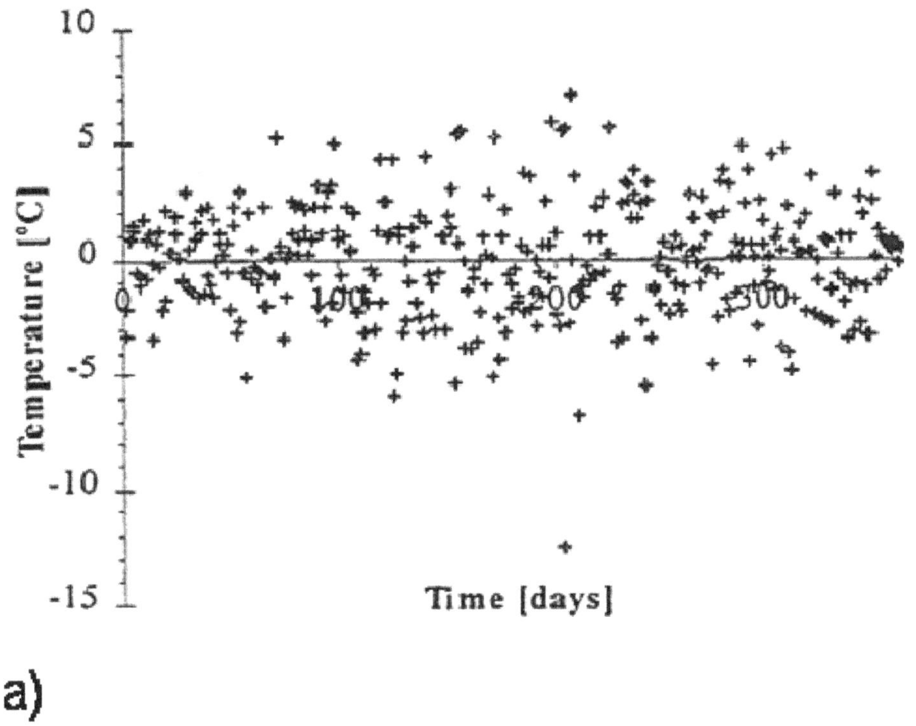

Figure 4a. Variable z_{ran_t} .

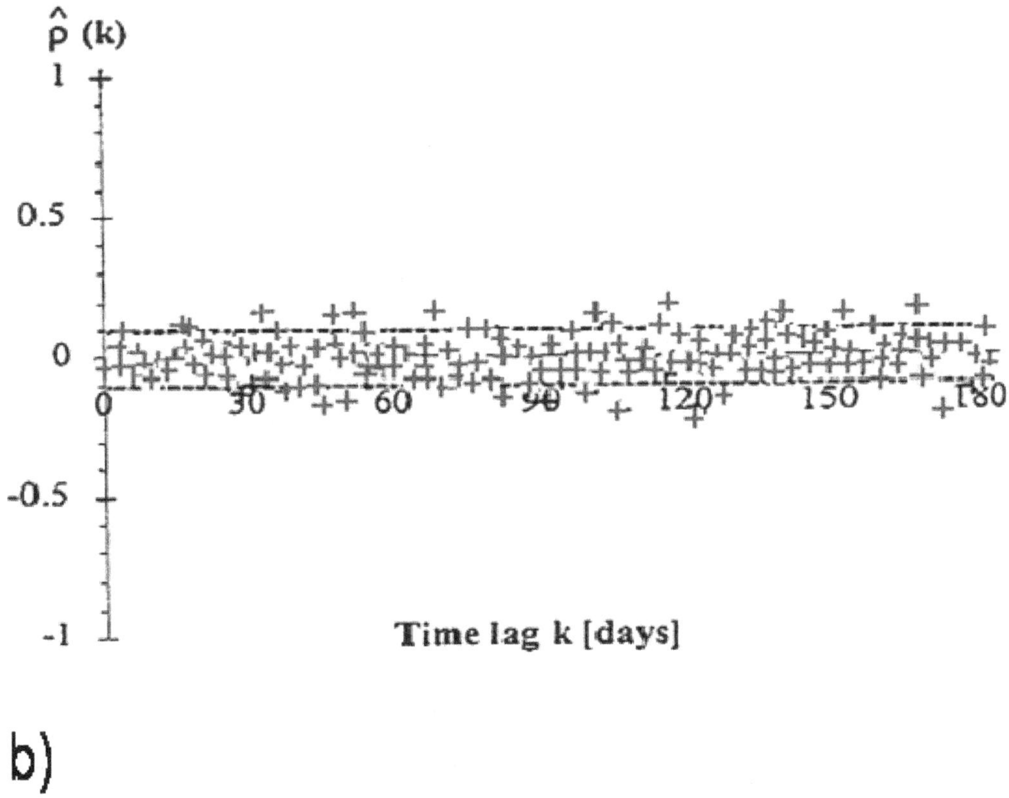

Figure 4b. Variable z_{ran_t} ; respective autocorrelogram.

Table 1. Deterministic model parameters, random component x_{ran_t} first moment estimators and *p-value*, for ambient temperature time series.

	$\bar{x}_d\left[{}^0C\right]$	$\Delta x_d\left[{}^0C\right]$	$\varphi_x[rad]$	$\hat{\mu}_x\left[{}^0C\right]$
2000: Θ_M	20.935	7.109	2.829	0.000
Θ_m	12.395	5.384	2.657	0.000
Θ_{av}	16.233	6.237	2.753	0.000
Θ_{am}	4.275	1.009	-2.969	0.000
2003: Θ_M	20.027	6.803	2.821	0.000
Θ_m	12.455	5.339	2.733	0.000
Θ_{av}	16.247	6.067	2.783	0.000
Θ_{am}	3.789	0.777	3.125	0.000
2004: Θ_M	21.011	6.283	2.733	0.000
Θ_m	12.975	4.699	2.531	0.000
Θ_{av}	16.993	5.465	2.647	0.000
Θ_{am}	4.021	0.963	3.243	0.000
2005: Θ_M	22.111	6.797	2.810	0.000
Θ_m	14.011	4.639	2.606	0.000
Θ_{av}	18.059	5.687	2.725	0.000
Θ_{am}	4.051	1.223	-3.082	0.000

	$\hat{\sigma}_x\left[{}^0C\right]$	$\overline{CV}_{x_t}(t)$ **[p.u.]**	**p-value[%]**
2002: Θ_M	3.529	0.169	53
Θ_m	2.084	0.167	40
Θ_{av}	2.435	0.145	77
Θ_{am}	1.567	0.366	52
2003: Θ_M	3.368	0.169	22
Θ_m	2.356	0.187	90
Θ_{av}	2.485	0.154	32
Θ_{am}	1.505	0.395	97
2004: Θ_M	3.162	0.152	9
Θ_m	2.183	0.169	11
Θ_{av}	2.331	0.136	75
Θ_{am}	1.395	0.345	83
2005: Θ_M	3.527	0.158	81
Θ_m	2.654	0.188	75
Θ_{av}	2.728	0.150	47
Θ_{am}	1.512	0.375	8

Where; $\hat{\mu}_x = 0$ is valid, x_t can be considered as a non-stationary random variable, which mean is time dependent, $\mu_{x_t}(t)$, according to:

$$\mu_{x_t}(t) = \bar{x}_d + \Delta x_d \cos(\omega t + \varphi_x) \tag{16}$$

and which standard deviation, generically denoted by $\sigma_{x_t}(t)$, results, in fact, in a time independent (stationary) function:

$$\sigma_{x_t}(t) = \hat{\sigma}_x . \tag{17}$$

From (16) and (17) one can obtain the variation coefficient, $CV_{x_t}(t)$:

$$CV_{x_t}(t) \equiv \frac{\sigma_{x_t}(t)}{\mu_{x_t}(t)} = \frac{\hat{\sigma}_x}{\bar{x}_d + \Delta x_d \cos(\omega t + \varphi_x)}, \tag{18}$$

which mean value, $\overline{CV}_{x_t}(t)$, is:

$$\overline{CV}_{x_t}(t) = \frac{\hat{\sigma}_x}{\bar{x}_d} . \tag{19}$$

From $\overline{CV}_{x_t}(t)$ values one can realise the degree of x_t concentration around its mean $\mu_{x_t}(t)$. Deterministic model parameters, $\bar{x}_d, \Delta x_d$ and φ_x, estimators of residuals first moment, $\hat{\mu}_x$ and $\hat{\sigma}_x$, mean value of variation coefficient, $\overline{CV}_{x_t}$, and *p-value* from the Chi-Square test are resumed in Table 1 for the 4 analysed years.

For these four analysed years, the model reproduces very well each year, although the number of considered years is insufficient to draw generalised conclusions or forecasts for the coming years. All samples passed with relatively high p-values Chi-square tests, regarding the hypotheses of being Gaussian distributed. From the low values of $\overline{CV}_{x_t}$ one can conclude that random component x_{rant} is relatively concentrated around the deterministic component. The histograms and respective theoretical probabilistic density functions (*pdf*) of a Gaussian distribution with parameters $\hat{\mu}_x$ and $\hat{\sigma}_x$ are represented on Figures 5 and 6.

A deeper analysis of x_{rant} residuals with respective autocorrelation functions revealed the presence of possible first order of AR models. The resulted z_t variables, once the first order AR model was removed, were studied. All passed a randomness test with a confidence level of 5%. Concerning the probabilistic distributions, in some cases z_t variable gets closer to the Gaussian distria)

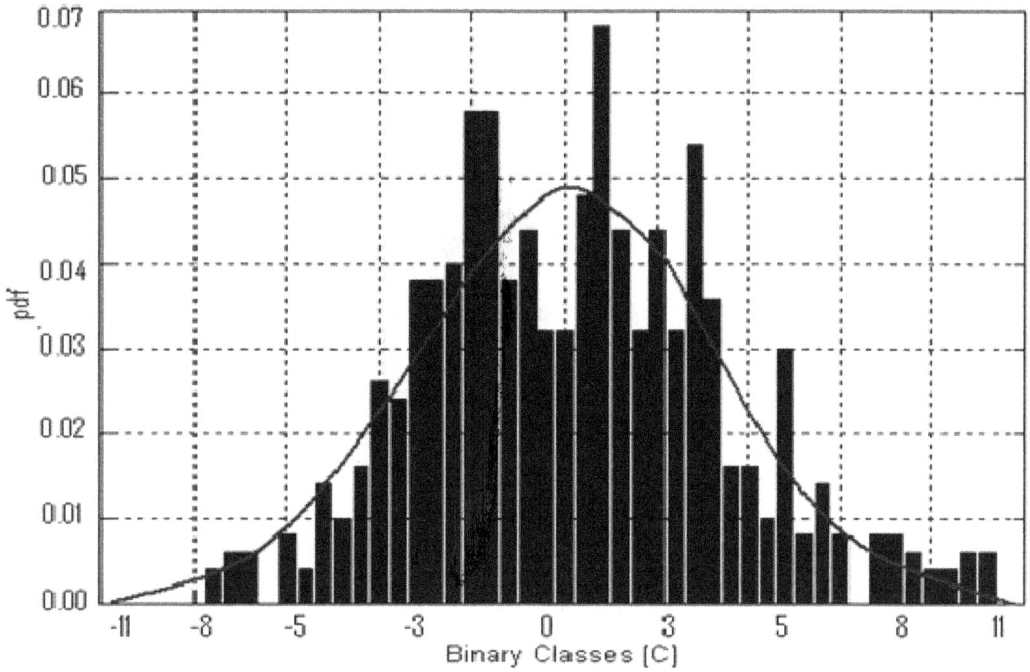

Figure 5a. Histogram and respective Gaussian *pdf* for random component of Θ_M and Θ_m.

b)

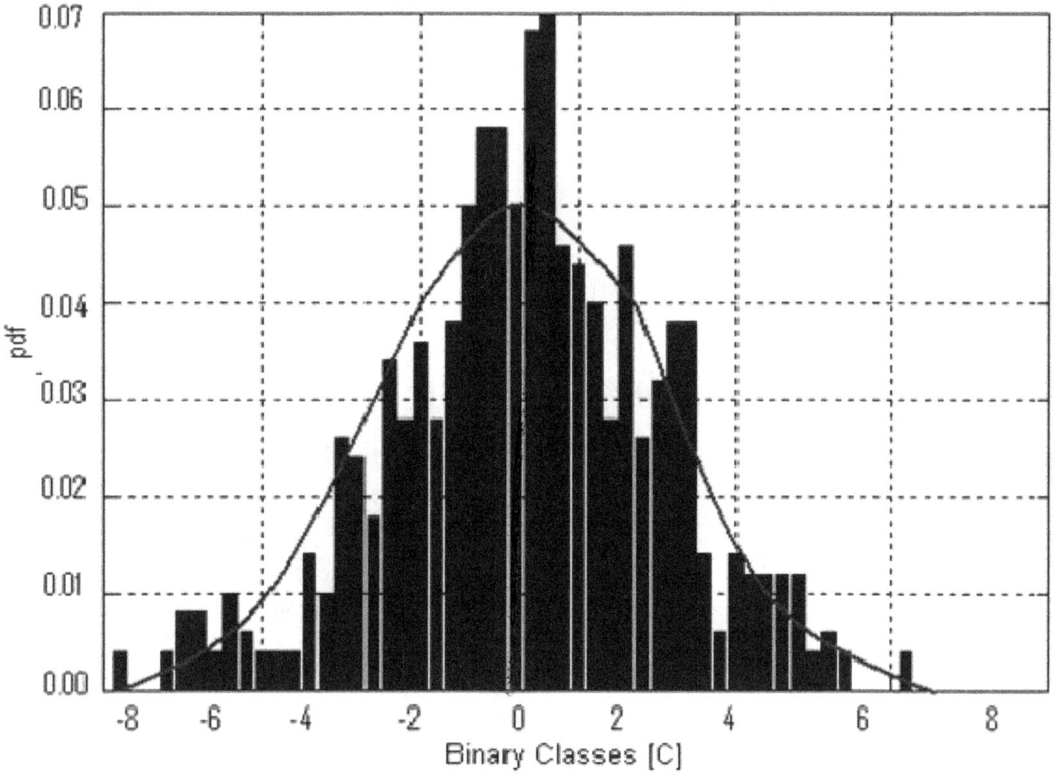

Figure 5b. Histogram and respective Gaussian *pdf* for random component of Θ_M ; Data from 2005.

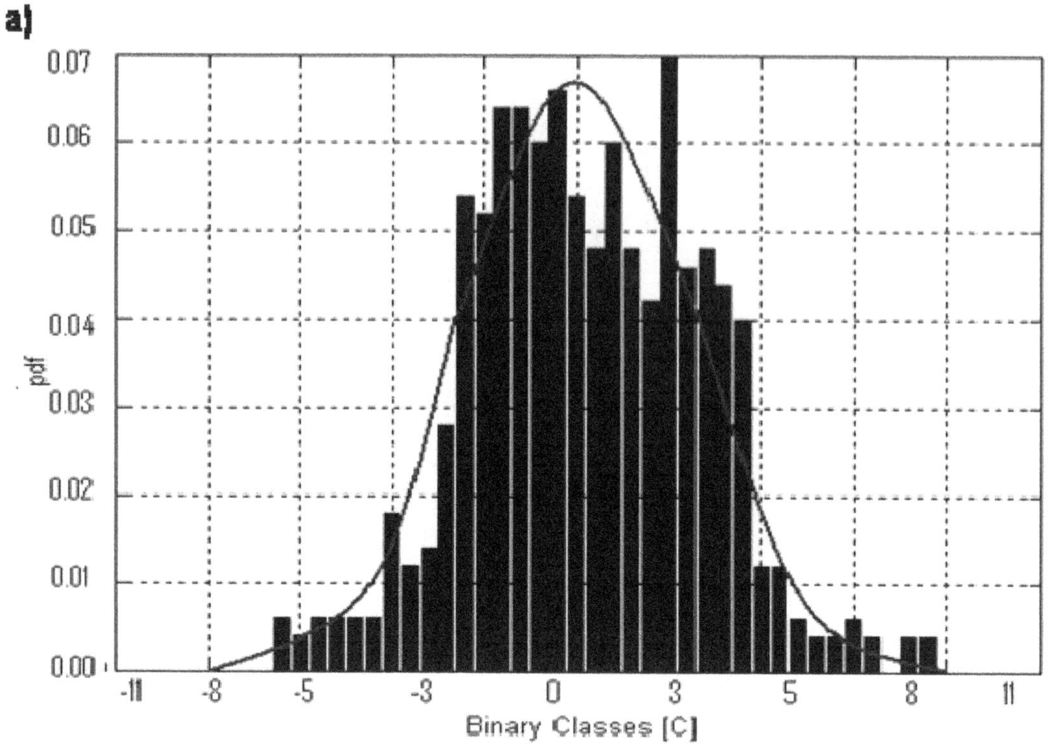

Figure 6a. Histogram and respective Gaussian *pdf* for random component of Θ_{av} and Θ_{am} .

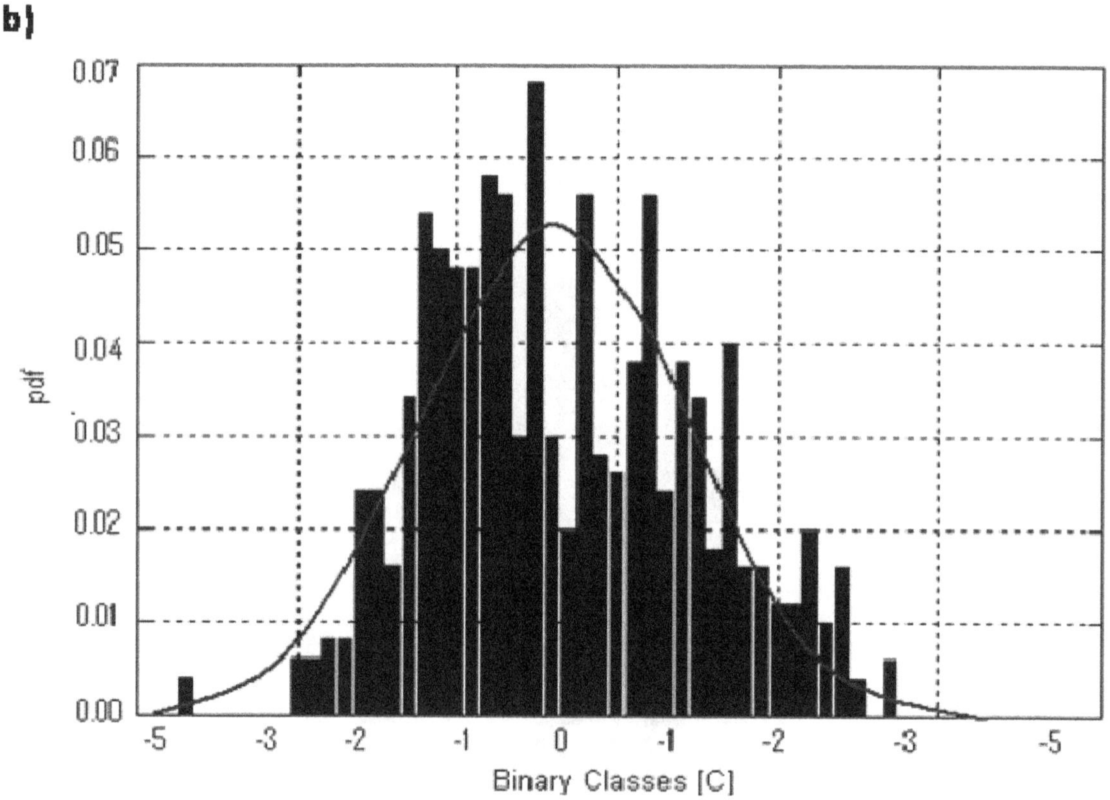

Figure 6b. Histogram and respective Gaussian *pdf* for random component of Θ_{av} ; Data from 2005.

Table 2. Random component z_{rant} first moment estimators and *p-value*, for ambient temperature time series.

	$\hat{\sigma}_x \left[{}^0C\right]$	$\hat{\sigma}_x \left[{}^0C\right]$	***p-value*[%]**	$\hat{\sigma}_x / \hat{\sigma}_z$ **[p.u.]**
2002 : Θ_M	0.000	2.556	43	1.381
Θ_m	0.000	1.613	78	1.289
Θ_{av}	0.000	1.612	7	1.505
Θ_{am}	0.000	1.361	4	1.153
2003: Θ_M	0.000	2.447	12	1.373
Θ_m	0.000	1.963	3	1.203
Θ_{av}	0.000	1.752	64	1.416
Θ_{am}	0.000	1.330	7	1.131
2004 : Θ_M	0.000	2.373	14	1.333
Θ_m	0.000	1.742	70	1.255
Θ_{av}	0.000	1.672	1	1.393
Θ_{am}	0.000	1.211	75	1.149
2005 : Θ_M	0.000	2.390	75	1.473
Θ_m	0.000	1.982	19	1.336
Θ_{av}	0.000	1.753	87	1.555
Θ_{am}	0.000	1.281	99	1.185

bution, while in other cases, gets far from the Gaussian distribution, even failing a Chi-square test at 5% level of confidence. These results indicate that more elaborate models are required to fully model these time series. First moment estimators, μ_z and $\hat{\sigma}_z$, of z_t variables, as well as the *p-value* of the Chi-square test concerning a Gaussian distribution, are represented on Table 2.

By comparing $\hat{\sigma}_x$ and $\hat{\sigma}_z$ values, one concludes that the taking into consideration the first order AR model reduces the variance level of random component ($\hat{\sigma}_z < \hat{\sigma}_x$). However, it is $\hat{\sigma}_x / \hat{\sigma}_z \approx 1$, meaning that supplementary information carried by the AR model is quite reduced.

Although the model is traduced by:

$$x_t = \bar{x} + \Delta x_d \cos(\omega t + \varphi_x) + \hat{\rho}_0 x_{t-1} + \mathrm{N}\ (\hat{\mu}_x, \hat{\sigma}_x) \qquad (20)$$

Where; $\hat{\mu}_z = 0$ results more precise for some of the analysed time series, the model traduced by:

$$\bar{x}_d + \Delta x_d \cos(\omega t + \varphi_x) + \mathrm{N}\ (\hat{\mu}_x, \hat{\sigma}_x), \qquad (21)$$

Where; $\hat{\mu}_x = 0$ can describe in a more generic way, although less accurate, the time series representative of maximal, minimal, average and half-amplitude values of daily ambient temperatures of analysed years.

A similar analysis was performed for the load profiles of two distribution transformers, denoted by ES1 and BI1. Available data include daily maximal, K_M, minimal, K_m, average, K_{av}, and half-amplitude, K_{am}, load factor for the 2003, 2004 and 2005 years, being, analogously to ambient temperature:

$$K_{av} \equiv (K_M + K_m)/2 \text{ and } K_{av} \equiv (K_M - K_m)/2. \qquad (22)$$

During this period, no structural network changes occurred in the network. As an example, maximal and minimal load factor values of ESI transformer, relatively to 2005, are represented on Figure 7. The loads served by this transformer are mainly of the residential type with a small component of industry.

Similar to ambient temperature modelling, the previous generic notations x and z will be used, being $x, z \equiv K_M, K_m, K_{av}, K_{am}$.

Deterministic cyclic component was assumed to follow also a sinusoidal variation as represented on (7) and resulted residuals, x_{rant}, were studied. Table 3 resumes obtained values for deterministic model parameters, $\bar{x}_d$, Δx_d and φ_x, estimators of residuals first moment, $\hat{\mu}_x$

Table 3. Deterministic model parameters, random component x_{ran_t} first moment estimators and *p-value*, for ESI distribution transformer.

	$\bar{x}_d$ [p.u.]	Δx_d [p.u.]	φ_x [rad]
2003: K_M	0.611	0.053	-0.029
K_m	0.277	0.016	0.292
K_{av}	0.443	0.034	0.025
K_{am}	0.165	0.023	-0.157
2004: K_M	0.582	0.055	0.115
K_m	0.273	0.016	0.571
K_{av}	0.426	0.037	0.209
K_{am}	0.156	0.023	-0.042
2005: K_M	0.602	0.035	-0.932
K_m	0.281	0.016	-0.773
K_{av}	0.440	0.025	-0.879
K_{am}	0.157	0.009	-1.089

	$\hat{\mu}_x$ **[p.u.]**	$\hat{\sigma}_x$ **[p.u.]**	$\overline{CV}_{x_t}$ **[p.u.]**	***p-value*** **[%]**
2003: K_M	0.000	0.045	0.074	47
K_m	0.000	0.025	0.086	0
K_{av}	0.000	0.033	0.074	35
K_{am}	0.000	0.018	0.104	15
2004: K_M	0.000	0.049	0.085	59
K_m	0.000	0.026	0.095	19
K_{av}	0.000	0.035	0.079	11
K_{am}	0.000	0.019	0.121	85
2005: K_M	0.000	0.040	0.065	25
K_m	0.000	0.018	0.061	12
K_{av}	0.000	0.024	0.053	0
K_{am}	0.000	0.019	0.118	67

and $\hat{\sigma}_x$, and *p-value* from the Chi-square test regarding a Gaussian distribution of the residuals.

On Figure 7, time series representative of K_M "looks" much more disperse than the K_m time series, which is a common fact in the three analysed years. Minimal values of distribution transformer load profiles are very well defined by (usually) night loads, corresponding to "base" equipment which is almost constant if no structural changes or accidents occur in the transformer network, while maximal values traduce temporary overloads due to residential/industrial activity. From $\overline{CV}_{x_t}$ values, represented on Table 3, one realises that load profiles are much more concentrated around respective deterministic components, than ambient temperature profiles are. Globally, results resumed on Table 3 can be considered as good although residuals from K_m in 2003 and K_{av} in 2005 can not be considered to follow a Gaussian distribution. In fact, from the study of the autocorrelation function, one realises the presence of higher frequencies than the fundamental frequency (annual) on the deterministic

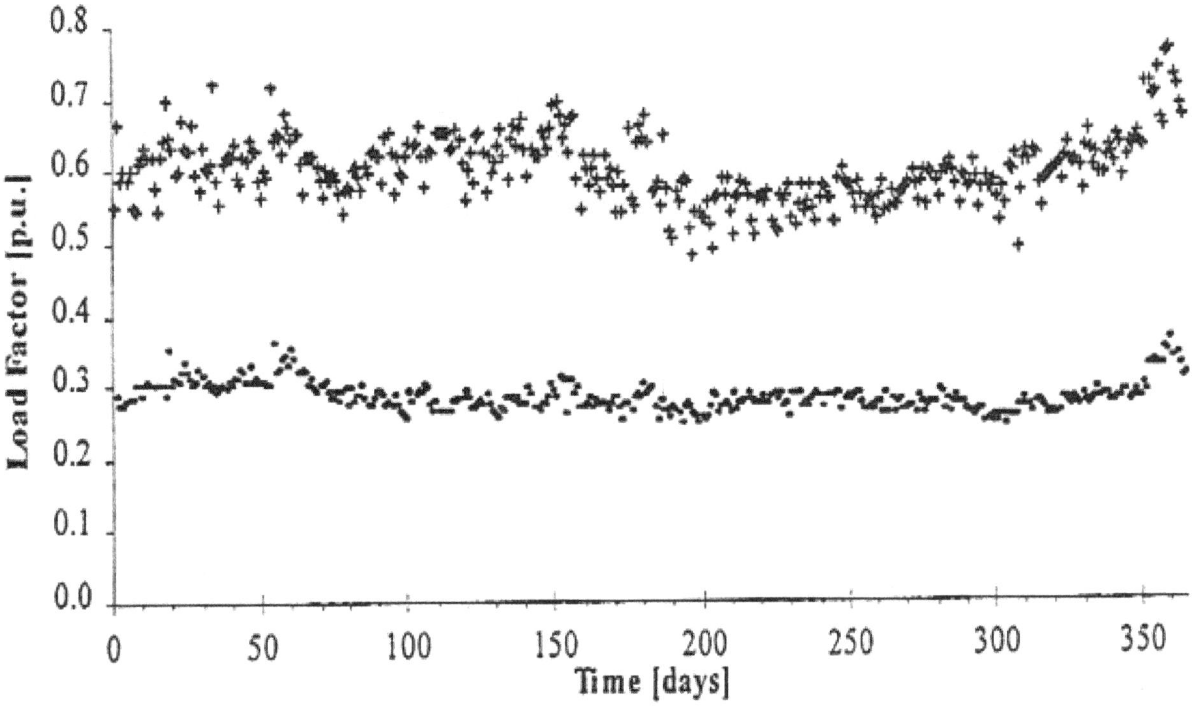

Figure 7. Annual time series representing daily maximal (denoted with +) and minimal (denoted with -) load factor of distribution transformer ESI. Data from 2005.

nistic sinusoidal model (7). This was an expected occurrence since load profiles are constrained to a much great diversity of factors than ambient temperature profiles are. One can detect, for example, the increasing appearance of a second harmonic (bi-annual) reflecting the increase in loads due to air-conditioning equipment during summer period. However, one should recall the purpose of this work: to give a physical justification for theoretical load and ambient temperature profiles used in following simulations and not fully modelling these profiles. The histograms and respective theoretical *pdf*s of Gaussian distributions, with parameters $\hat{\mu}_x$ and $\hat{\sigma}_x$, reproduced on Table 3 are represented on Figures 8 and 9, for the 2005 data set.

The great dispersion of K_M time series relatively to K_m time series can be visualised by the limits of histograms represented on Figure 8. The hypothesis that random component x_{ran_t} of time series representative of load profiles could be modelled by an AR model did not give as good results as with ambient temperature profiles. This fact is due, in part, to the already referred presence of other cyclic (bi-annual) variations in xran which were not taken into consideration on the deterministic model (7). Results are resumed on Table 4.

Results obtained with the second analysed distribution transformer, referred as BI1, are resumed on Table 5, where $\hat{\mu}_z$ values were omitted since it is $\hat{\mu}_z = \hat{\mu}_x = 0$ for all samples.

This transformer serves an area where loads are of residential and industrial types, in similar proportions. Residuals x_{ran_t} after removing the deterministic cyclic variation are not as normally distributed as residuals resulting from the ESI load profiles; in the 12 presented samples, 3 of them even fail the respective chi-square test. This fact does not invalidate the generic model represented by (20). Since 2004 is the year which data give the worst results, meaning lower *p-values* on the chi-square test for a Gaussian distribution of residuals, histograms of x_{ran_t} residuals and respective theoretical *pdf*s are represented on Figures 10 and 11. Although not passing the Chi-square test, the statistical distribution of random component x_{ran_t} relativeley to 2004 K_M, K_m, K_{av}, and K_{am}, values, is not far from a Gaussian distribution as can be visualised on Figures 10 and 11.

Although more elaborated models are required to fully model the load profiles of distribution transformers, it has been shown that (21) can be considered as a good generic model.

Global model

On this section a global model to represent the whole set of maximal, minimal and average temperatures (or load

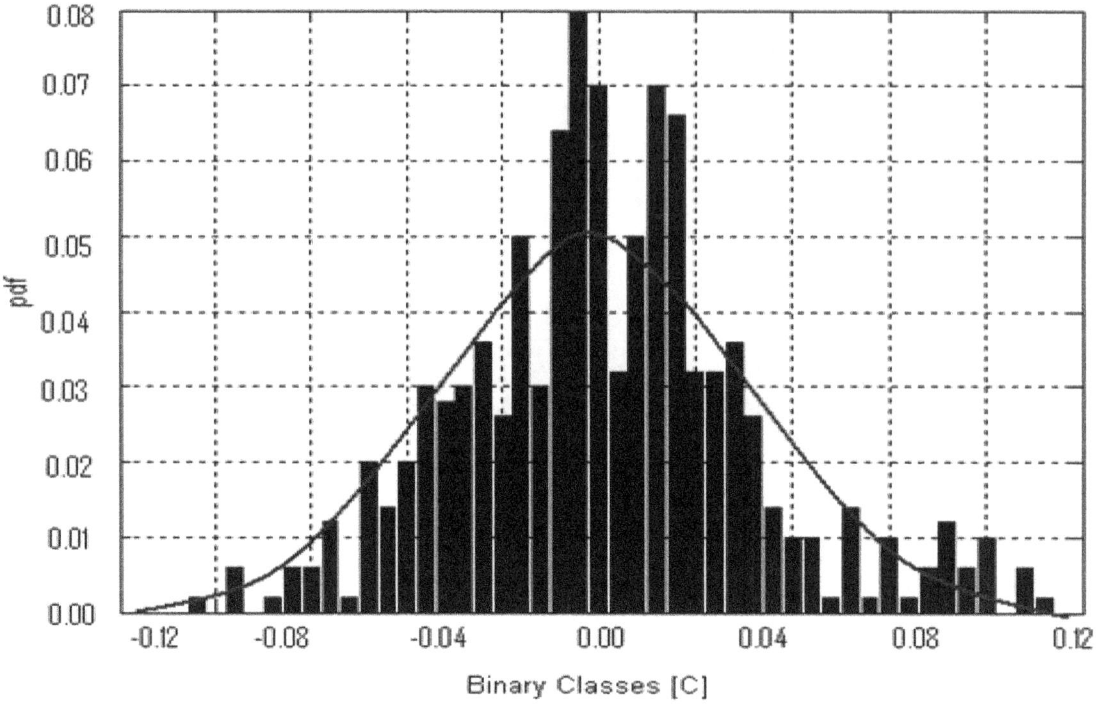

a)

Figure 8a. Histogram and respective Gaussian pdffor random component x_{ran_t} of K_M and K_m.

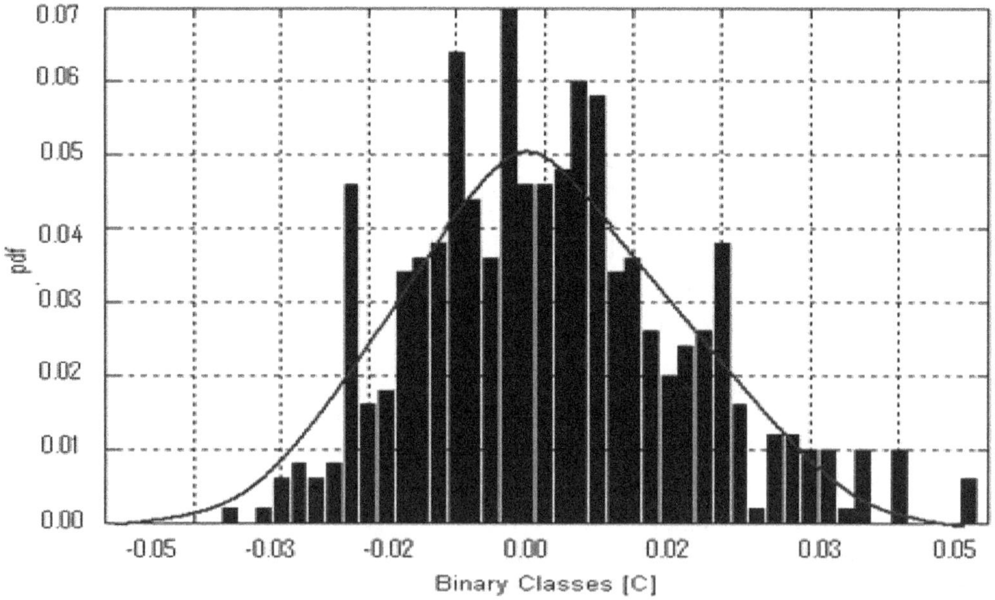

b)

Figure 8b. Histogram and respective Gaussian pdffor random component x_{ran_t} of K_M ; ESI transformer and data from 2005.

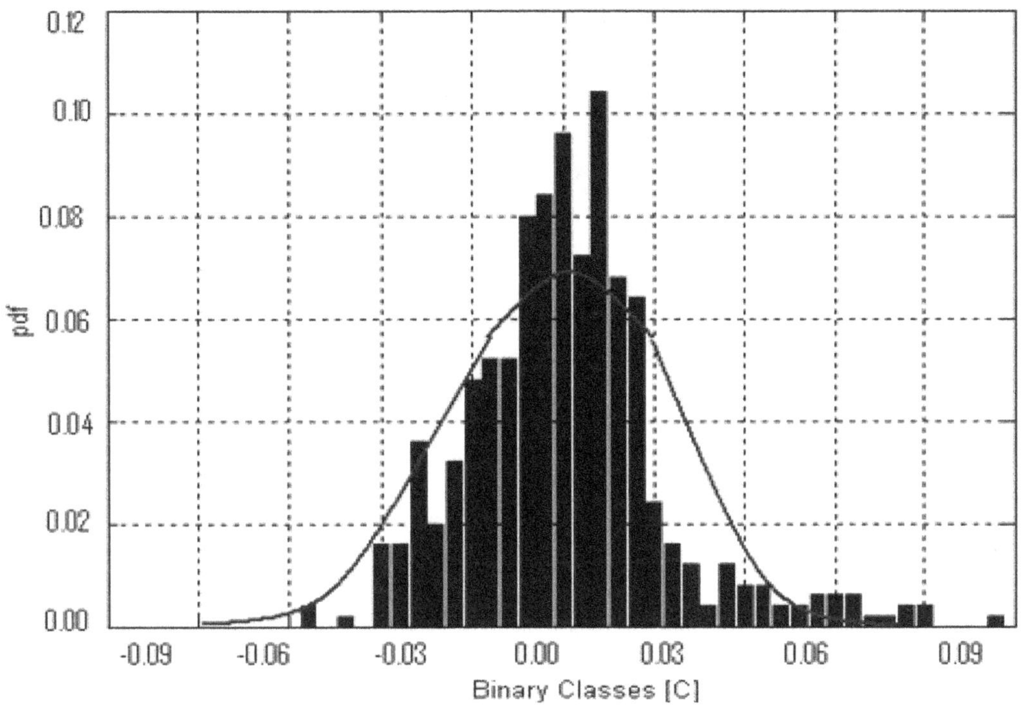

a)

Figure 9a. Histogram and respective Gaussian pdffor random component x_{ran_t} of K_{am} (a) and K_{av}.

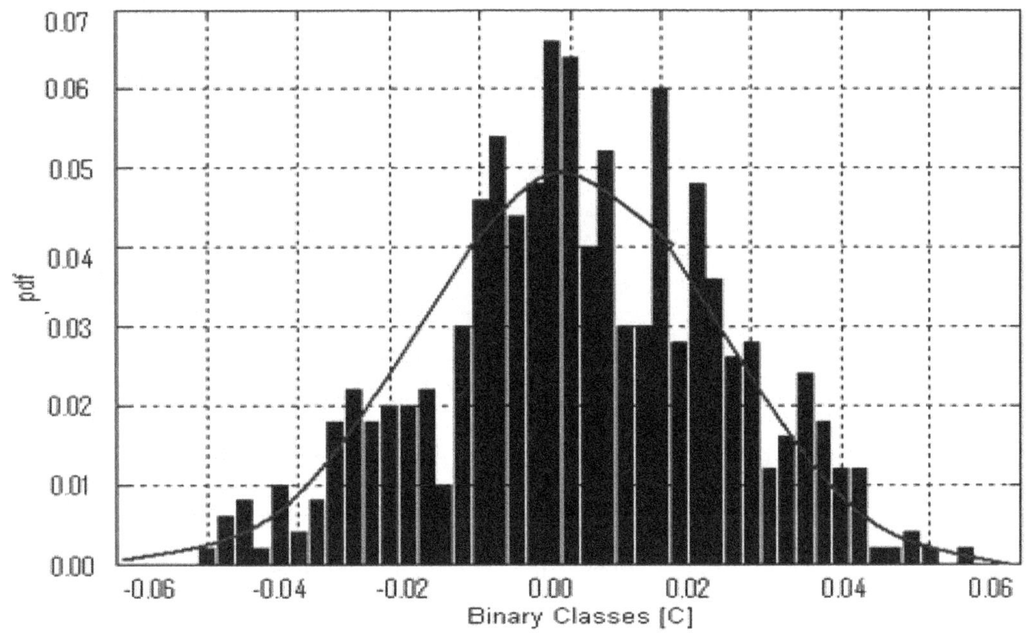

b)

Figure 9b. Histogram and respective Gaussian pdffor random component x_{ran_t} of K_{am} ; ESI transformer and data from 2005.

Table 4. Random component z_{ran_t} first momment estimators and *p-value*, for ESI load profiles.

	$\hat{\mu}_z \left[{}^0C\right]$	$\hat{\sigma}_x \left[{}^0C\right]$	*p-value*[%]	$\hat{\sigma}_x / \hat{\sigma}_z$ [p.u.]
2003: K_M	0.000	0.033	5	1.417
K_m	0.000	0.011	11	1.987
K_{av}	0.000	0.017	12	1.819
K_{am}	0.000	0.015	24	1.097
2004 : K_M	0.000	0.034	15	1.455
K_m	0.000	0.022	0	1.183
K_{av}	0.000	0.022	0	1.620
K_{am}	0.000	0.018	53	1.075
2005 : K_M	0.000	0.032	6	1.223
K_m	0.000	0.011	84	1.427
K_{av}	0.000	0.016	56	1.337
K_{am}	0.000	0.015	11	1.155

Table 5. Deterministic model parameters, random components x_{rant} and z_{rant} first moment estimators and p-values, for BI1 distribution transformer.

	$\overline{x}_d$ [p.u.]	Δx_d [p.u.]	φ_x [rad]	$\hat{\mu}_x$ [p.u.]	$\hat{\sigma}_x$ [p.u.]
2003: K_M	0.351	0.075	-0.196	0.000	0.035
K_m	0.166	0.014	-0.147	0.000	0.015
K_{av}	0.258	0.046	-0.189	0.000	0.023
K_{am}	0.093	0.033	-0.206	0.000	0.017
2004 : K_M	0.375	0.068	-0.088	0.000	0.037
K_m	0.188	0.011	0.545	0.000	0.015
K_{av}	0.282	0.039	-0.001	0.000	0.027
K_{am}	0.095	0.031	-0.238	0.000	0.019
2005 : K_M	0.391	0.071	-0.019	0.000	0.033
K_m	0.201	0.016	0.091	0.000	0.015
K_{av}	0.296	0.043	-0.001	0.000	0.022
K_{am}	0.097	0.029	-0.051	0.000	0.011

	$\overline{CV}_{x_t}$ [p.u.]	*p-value* [%]	$\hat{\sigma}_z \left[{}^0C\right]$	*p-value*[%]	$\hat{\sigma}_x / \hat{\sigma}_z$ [p.u.]
2003 K_M	0.102	63	0.033	51	1.118
K_m	0.083	31	0.012	1	1.307
K_{av}	0.089	61	0.017	21	1.236
K_{am}	0.171	5	0.016	14	1.028

Table 5 contd.

2004 K_M	0.095	6	0.027	0	1.387
K_m	0.085	5	0.015	0	1.195
K_{av}	0.101	0	0.023	0	1.264
K_{am}	0.191	0	0.018	0	1.068
2005 K_M	0.081	21	0.019	0	1.726
K_m	0.079	0	0.011	1	1.546
K_{av}	0.078	9	0.014	0	1.915
K_{am}	0.116	88	0.009	4	1.263

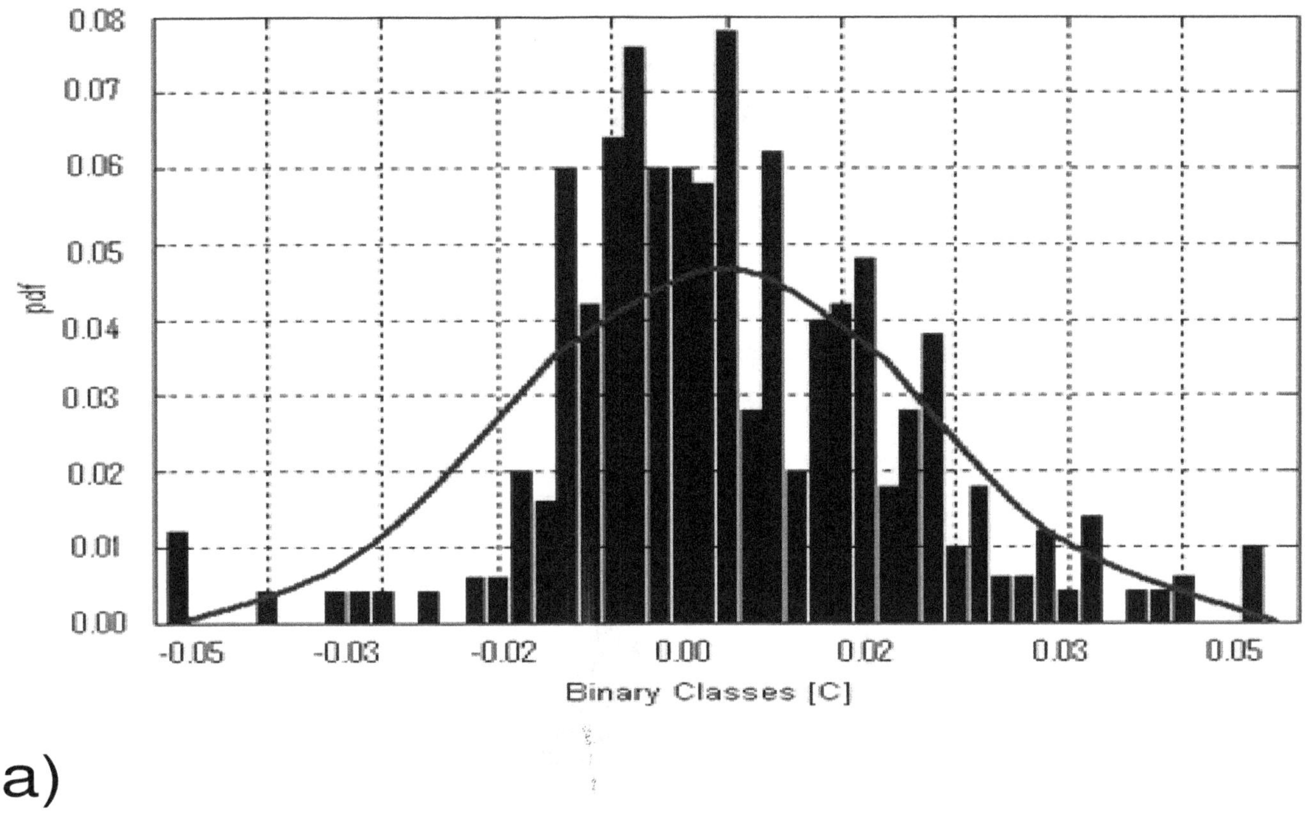

Figure 10a. Histogram and respective Gaussian *pdf* for random component xran of K_M (a) and K_m.

factors) along the year, will be described. The model is based on the previously studied x_{av} and x_{am} time series

Where; $x \equiv \Theta, k$ and is defined s a linear combination of these two:

$$x_t = x_{av} + \alpha_G x_{am}, \tag{23}$$

being α_G a real number and $\alpha_G \in [-1,1]$.

From x_{av} and x_{am} definition, (14) and (22), one can realise that the chosen α_G range, determines (23) to model variables from minimal to maximal values according to:

$$-1 \le \alpha_G \le 1 \Rightarrow x_m \le x_t \le x_M \tag{24}$$

If both x_{av} and x_{am} time series can be assumed to follow a deterministic and random components according to (21),

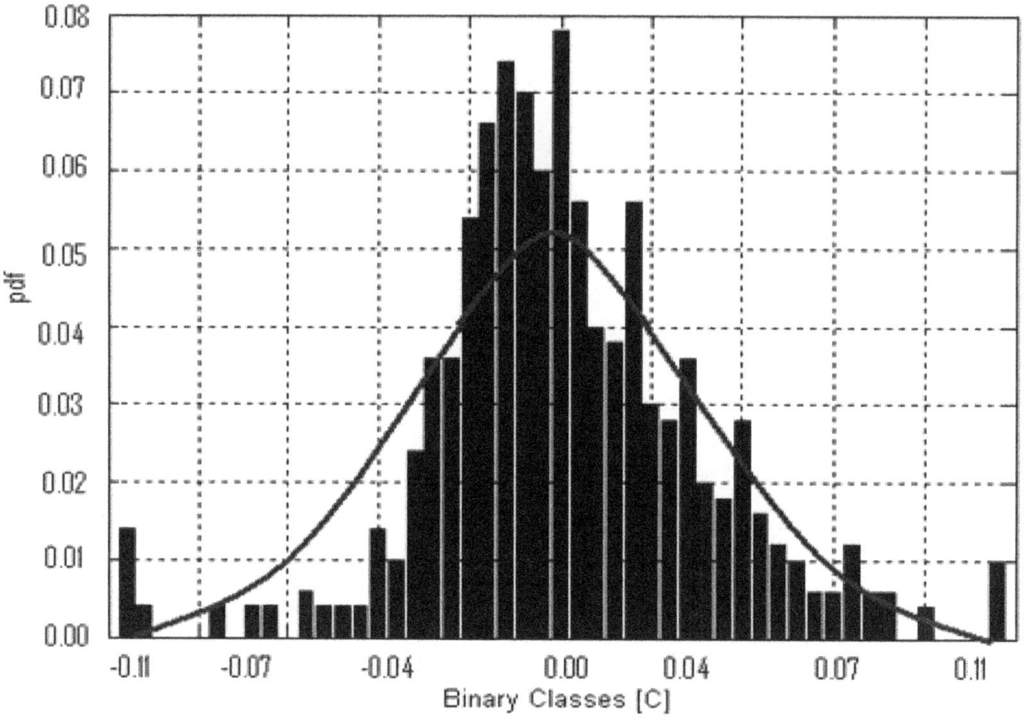

b)

Figure 10b. Histogram and respective Gaussian *pdf* for random component xran of K_M ; BI1 transformer and data from 2004

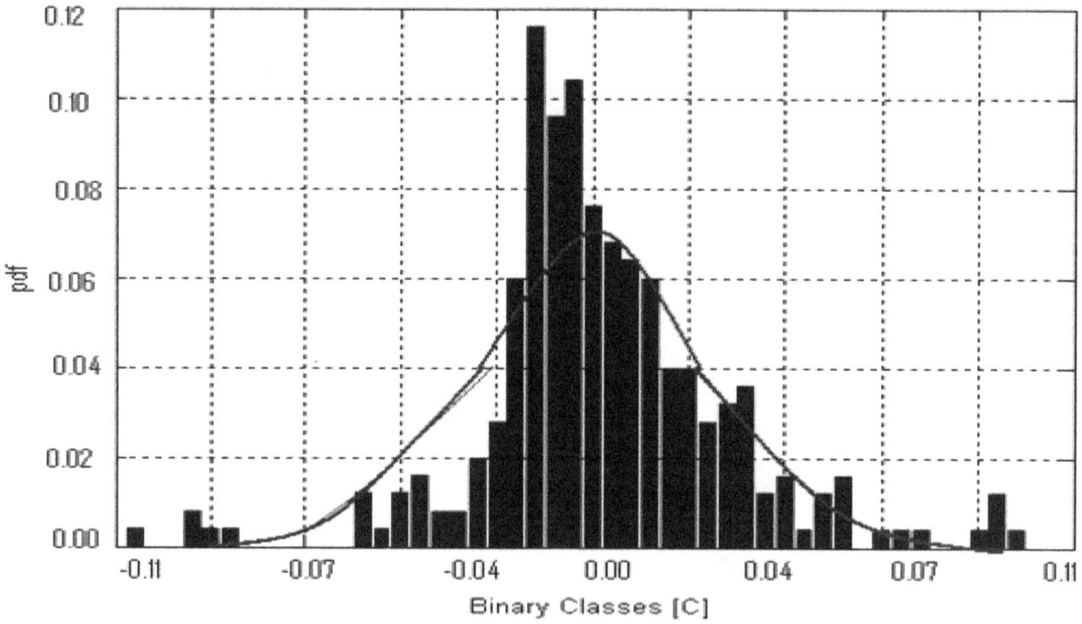

a)

Figure 11a. Histogram and respective Gaussian pdf for random component X_{ran} of $K(a)$ and $K_{am.}$

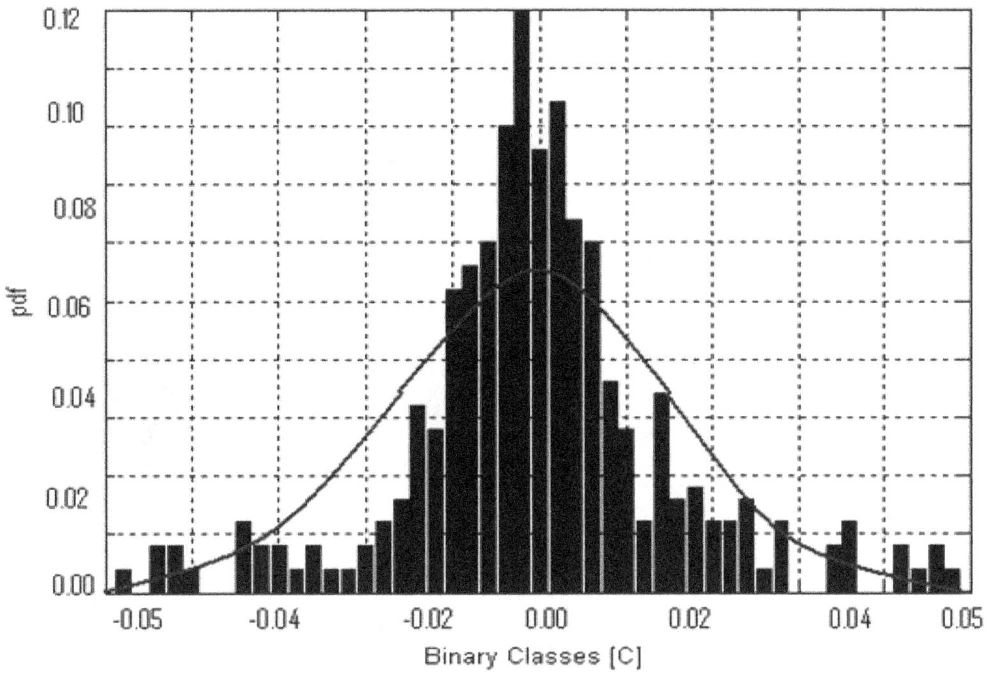

b)

Figure 11b. Histogram and respective Gaussian pdf for random component X_{ran} of *K*(a); BI1 transformer and data from 2004.

and attending to (23), x_t model will also result with deterministic and random components:

$$x_t = x_{\det_t} + x_{ran_t} \tag{25}$$

with:

$$x_{\det} = \bar{x}_d + \Delta x_d \cos(\omega t + \varphi_x) \quad 26)$$

$$x_{ran_t} = N(0, \sigma_x) \tag{27}$$

Each of the parameters $\bar{x}_d$, Δx_d, and φ_x can be analytically determined and result as:

$$\hat{x}_d = \hat{x}_{av} + \alpha_G \hat{x}_{amd} \tag{28}$$

$$\Delta x_d = \sqrt{(\Delta x_{av_d})^2 + (\alpha \Delta x_{am_d})^2 + 2\alpha_G \Delta x_{av_d} \Delta x_{am_d} \cos(\varphi_{xav} - \varphi_{xam})} \tag{29}$$

$$\varphi_x = arctg \frac{\Delta x_{av_d} \sin(\varphi_{xav}) + \alpha \Delta x_{am_d} \sin(\varphi_{xam})}{\Delta x_{av_d} \cos(\varphi_{xav}) + \alpha \Delta x_{am_d} \cos(\varphi_{xam})} \pm \pi \tag{30}$$

and

$$\sigma_x = \sqrt{(\sigma_{x_{av}})^2 + (\alpha \sigma_{x_{am}})^2 + 2CoV(x_{av_{rand}}, \alpha_G, x_{am_{rand}})} \tag{31}$$

Where; $COV(x_{av_{rand}}, \alpha_G, x_{am_{rand}})$ denotes the cova-riance (covariance function (3) with null time lag, *k*=0) between the random components $x_{av_{rand}}$ and $\alpha_G x_{am_{rand}}$. If profiles perfectly fitted model represented by (21), random components $x_{av_{rand}}$ and $x_{av_{rand}}$ would result as random variables and therefore uncorrelated from each other. Under this condition, (31) could be replaced by:

$$\sigma_x \approx \sqrt{(\sigma_{x_{av}})^2 + (\alpha \sigma_{x_{am}})^2}\,. \tag{32}$$

Since (21) is only an approximate model of profiles evolution, covariation between random components $x_{av_{rand}}$ and $x_{am_{rand}}$ is considerably. Since correlation is an image of covariation but normalised by variables respective variations, the strength of the link between $x_{av_{rand}}$ and $x_{am_{rand}}$ results clearer if correlation values are represented instead of covariation (Figure 12).

The usefulness of this global model resides on modelling compactness it traduces; by means of α_G parameter $-1 \le \alpha_G \le 1$, this single model is able to reproduce ambient temperature (or load factor profiles) models previously derived, from minimal to maximal

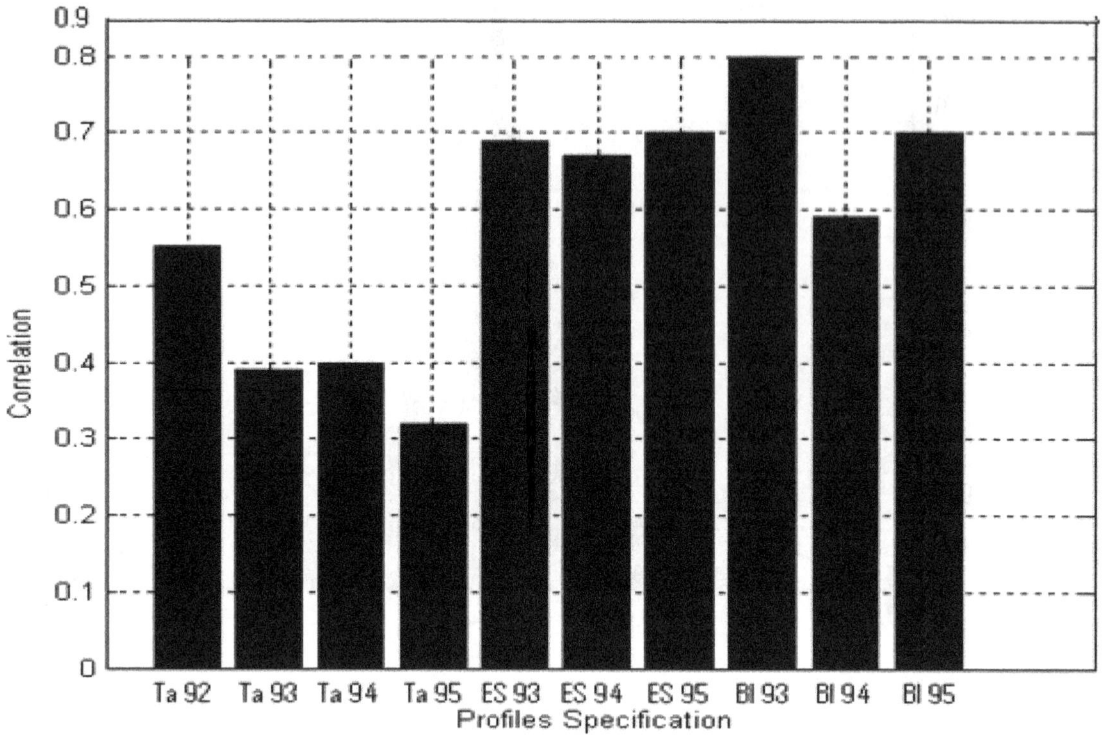

Figure 12. Correlation between random components $x_{av_{rand}}$ and $x_{am_{rand}}$.

Table 6. ESI deterministic and random correlation for 2004 data set.

Ambient temperature		ESI Load Profile					
		Deterministic			**Random**		
		Maximal	**Average**	**Minimal**	**Maximal**	**Average**	**Minimal**
Deterministic	Maximal	-0.471	-0.715	-0.771			
	Average	-0.569	-0.793	-0.841			
	Minimal	-0.636	-0.843	-0.885			
Random	Maximal				-0.074	-0.089	-0.087
	Average				-0.111	-0.135	-0.135
	Minimal				-0.112	-0.143	-0.143

values. Numerical validation of this model is not reproduced here, since obtained values are in agreement with those reproduced on Tables 1, 3 and 5.

Ambient temperature and load profiles correlation

From the time evolution of load and ambient temperature profiles the distribution transformers are subjected to, one can infer a relationship between them. For the analysed cases, when ambient temperature drops, loads increase, and when ambient temperature increases, transformer loads decrease. The strength of this relationship between loads and ambient temperature is measured by the correlation between them. Since models have a deterministic and a random part (21), correlation coefficient between each of these components, will be determined, to evidence that correlation between time series is mainly due to their deterministic components; random components are practically independent (uncorrelated) of each other. Correlation coefficients between transformer ESI load profile and 2004 ambient temperature are represented on Table 6.

Correlation between deterministic parts is clearly stronger than between random parts. The negative sign traduces the fact that, for the analysed data, models are inversely correlated; the ambient temperature increase implies loads decrease and vice-versa. Correlation strength increases as walking towards maximal values, which means that loads, and in particular, maximal ones, are much more "sensitive" to maximal ambient temperature than minimal ambient temperature. In fact, minimal loads along the year are almost constant and they traduce, in practice, a "base" load that is almost invariant with ambient temperature changes and depends most upon load characteristics of transformers distribution network. Previous considerations about correlation result clearer on Figure 13a, where Table 6 deterministic and random correlation values are graphically represented.

Although correlation coefficients are all negative, on Figure 13 correlation axis is in reverse order, so that graph visualisation results clearer. Similar relationship between deterministic and random correlation values can be obtained from BI1 transformer data (Figure 13b) and from 2003 and 2005 data sets (Figures 14 and 15). The con-stancy in the sign of correlation between random parts (all negatives in 2004 or all positives in ESI 2004) is an indication that random components x_{rand} of these profiles still carry deterministic behaviours that were not removed by the assumed deterministic model (21). If profiles were perfectly modelled by (21), correlation between any random component would result as null. Attending to the magnitude of correlation between deterministic parts and random parts and to results presented on the functional parameter modelling, it can be consider that,

$$x_t = \bar{x}_d + \Delta x_d \cos(\omega t + \varphi_x) + N\,(\hat{\mu}_x, \hat{\sigma}_x)\,, \qquad (33)$$

is a generic sufficiently accurate model to traduce the load annual evolution of distribution transformers as well as ambient temperature. Also, simulated load and ambient temperature profiles with random components will be used, to study the sensitivity of transformer thermal and loss of life models presented on Popescu (2006b), to such functional parameters.

FUNCTIONAL PARAMETERS SENSITIVITY

Probabilistic formulation

Input profiles

System inputs, K and Θ_a, are the transformer load and ambient temperature profiles which, by assumption, can be represented by an additive model of deterministic and random components, of the form:

$$K = K_{\det} + K_{ran} \text{ and } \Theta_a = \Theta_{a_{\det}} + \Theta_{a\,ran}\,. \qquad (34)$$

In fact, possible correlation can occur between K and T. In this general case, a non-stationary model must be considered. In this work this increase in model complexity will not be considered, since when the correlation exists, it derives, mainly, from a strong link between deterministic components (that is concomitant sinusoidal load and ambient temperature variations) and a weakest link between corresponding random components, as shown in Figure 13. The objective of this study is, under stationary conditions, to determine, on the output variable (*LOL*), its deterministic and random components, based on the previously referred additive model.

Methodology

The data acquisition frequency of a continuous type system must be carefully defined since it plays an important role on posterior analysis of data. Namely, the data acquisition set must represent faithfully the signal and, from this data set, one must be able to "rebuild" the original signal in a univocal way. The sampling theorem states that a continuous signal which Fourie Transform exists and is null out of the frequency interval [*-f, f*], should be sampled at a frequency f_s such that:

$$f_s > 2f \qquad (35)$$

Reciprocally, if the sampling frequency is f_s, no information can be inferred from the sampled data set, about signal occurrences with frequencies above the Nyquist, f_N, frequency, given by:

$$f_N = f_s / 2\,, \qquad (36)$$

Usually, in the case of long term forecasting, the acquisition period of data for analyses is long enough and therefore it is possible to neglect variables rapid fluctuations having a period of the same order of involved thermal time constants. Typically, it is τ_0=3 h and the windings constant $\tau_w \approx 5$ to 10 min (Asbury , 1975; IEC-354, 1991; Pierrat et al., 1996; Popescu, 2006a). Taking into account that input variables are approximately stationary, this simplification represents a second argument to consider a probabilistic stationary model, instead of a stochastic dynamic one. Both transformer thermal and ageing models, are strongly non-linear ones, which will determine the non-preservation of inputs statistical distribution structure (Bendat and Piersol, 1990; Bendat and Piersol, 1993; Popescu, 2006; Popescu, 2008). Nevertheless, provided each mathematical transfor-mation can be defined as a one-to-one function (with inverse) of an input random variable which *pdf* is known, output variable *pdf* can be analytically determined, either directly with recourse of characteristic functions. However, this methodology is not suitable for the system under study, since some transformations do not have an analytical exact expression for its inverse function:

$$y = \varphi(x)\,, \qquad (37)$$

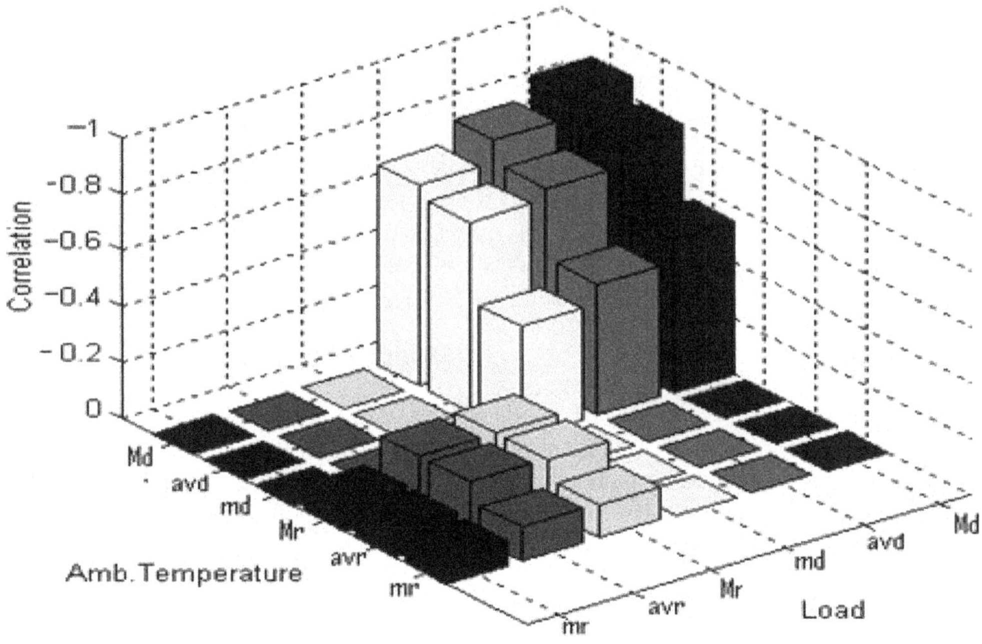

a)

Figure 13. Deterministic and random correlation between ambient temperature and ESI Data from 2004. (Table 6 for ESI transformer).

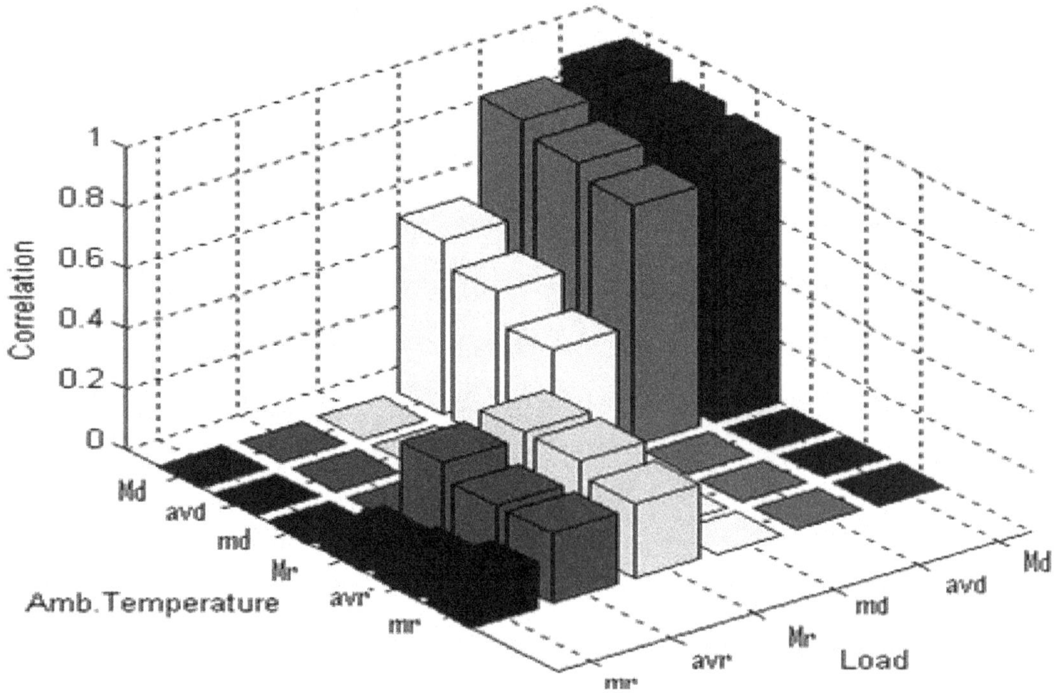

b)

Figure 13. Deterministic and random correlation between ambient temperature and BII. Data from 2004. (Table 6 for ESI transformer).

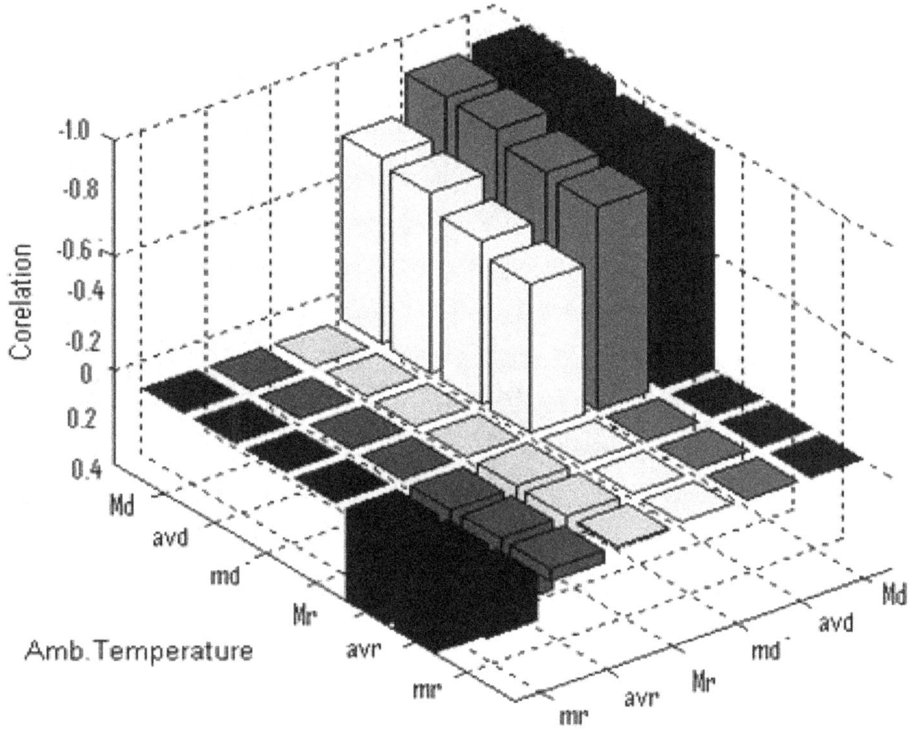

a)

Figure 14a. Deterministic and random correlation between ambient temperature and ESI profiles. Data from 2003.

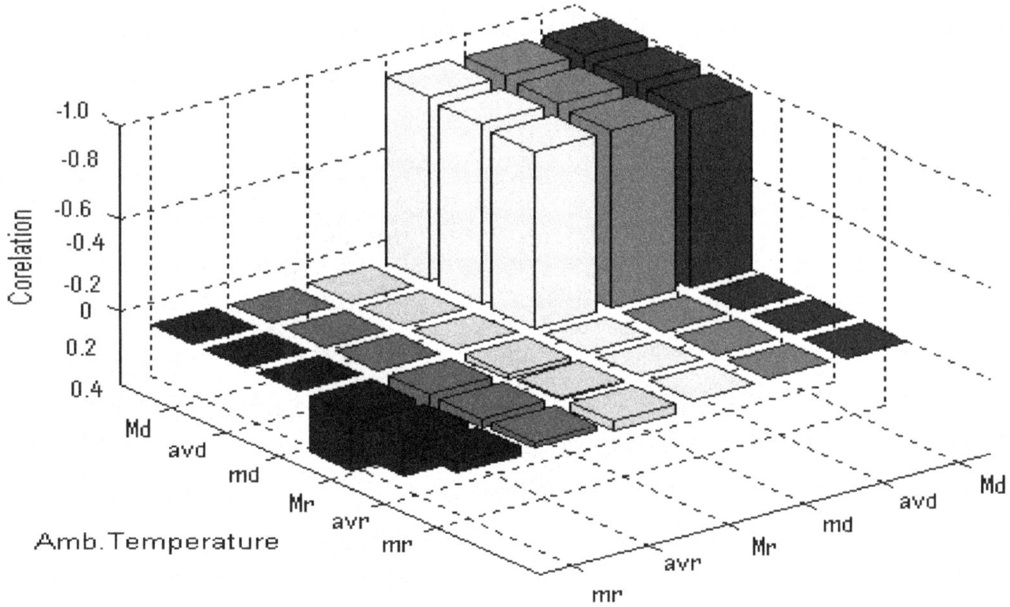

b)

Figure 14b. Deterministic and random correlation between ambient temperature and BII profiles. Data from 2003.

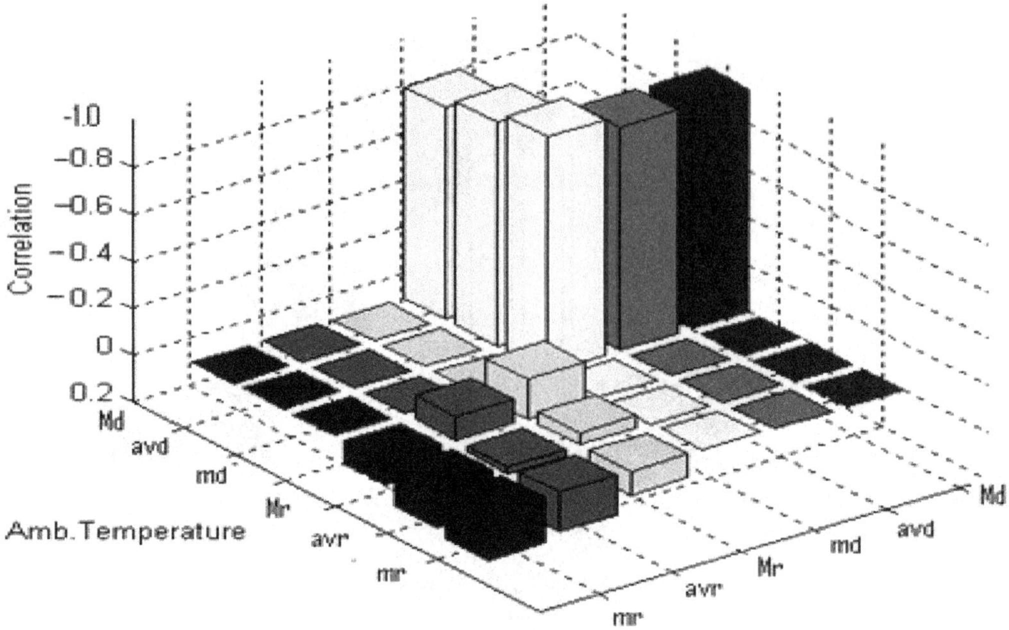

a)

Figure 15a. Deterministic and random correlation between ambient temperature and ESI profiles. Data from 2005.

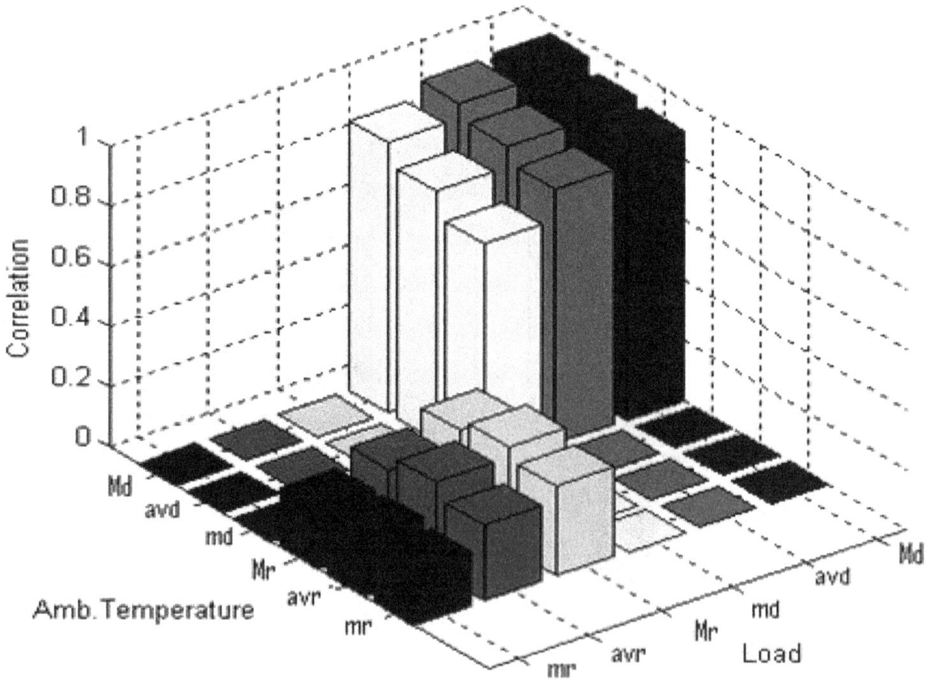

b)

Figure 15b. Deterministic and random correlation between ambient temperature and BII profiles. Data from 2005 which must be determined numerically.

The methodology used to estimate the stochastic output variable LOL, once the random inputs *K* and Θ_a are defined, is based on realistic characteristics of distribution transformers load profiles and ambient temperature ones. As already shown from the case studied, in a statistical sense, *K* and Θ_a can be considered as unimodal random variables concentrated around their modal values (mode) (Papoulis, 1984; Popescu, 2006) which means a reduced variation coefficient CV_x. Under this condition, it will be assumed as valid the linearisation of (37) in the vicinity of its input expected value μ_x, which first three terms are:

$$y \approx \varphi(\mu_x) + \frac{\partial\varphi(x)}{\partial x}\Big|_{x=\mu_x}[x-\mu_x] + \frac{\partial^2\varphi(x)}{\partial^2 x}\Big|_{x=\mu_x}[x-\mu_x]^2. \quad (38)$$

From (39) one can obtain estimators for y moments, denoted by $\hat{\mu}_y$ and $\hat{\sigma}_y$, as functions of x moments, denoted by μ_x and σ_x. Second order estimators will be given by:

$$\hat{\mu}_y = \varphi(\mu_x) + \frac{1}{2}\frac{\partial^2\varphi(x)}{\partial^2 x}\Big|_{x=\mu_x}\sigma_x^2 \quad (39)$$

$$\hat{\sigma}_y^2 = \left[\frac{\partial\varphi(x)}{\partial x}\Big|_{x=\mu_x}\right]^2\sigma_x^2 + \frac{1}{2}\left[\frac{\partial^2\varphi(x)}{\partial^2 x}\Big|_{x=\mu_x}\right]^2\sigma_x^4 \quad (40)$$

The errors one commits by considering the first order estimators, against the second order ones, can approximately be bounded by:

$$\varepsilon_\mu = \frac{1}{2}\sigma_x^2\frac{\partial^2\varphi(x)}{\partial^2 x}\Big|_{x=\mu_x}\frac{1}{\varphi(\mu_x)} \quad \text{and}$$

$$\varepsilon_{\sigma^2} = \frac{1}{2}\sigma_x^2\left[\frac{\partial^2\varphi(x)}{\partial^2 x}\Big|_{x=\mu_x}\frac{1}{\frac{\partial\varphi(x)}{\partial x}\Big|_{x=\mu_x}}\right]^2 \quad (41)$$

If the linearisation of (37) is assumed to be valid, it will also lead to the preservation of input variable statistical structure. Being the *x* variable pdf defined, the output variable y will present a similar structure and its *pdf* can be determined, approximately, with recourse of its first moments, which estimators are given by (39) and (40).

Approximate analytical model

Linearisation error

In order to evaluate the validity of the linearisation traduced by (38), the errors ε_μ and ε_{σ^2}, (39) and (40), for a range of μ_x and σ_x corresponding to realistic values of distribution transformer load profiles, were studied: $\mu_x \in [0.1,1.5]$ and $\sigma_x \in [0.01,0.8]$. To μ_x=0.1 p.u. corresponds a very low load, while μ_x=1 p.u. corresponds to an overload of limited duration. The resulting variation coefficient ranges, approximately: $CV_x \in [0.007,8]$. Numerical results, presented on Figure 16, are determinant in concluding for the importance of second order estimators, as CV_x increases, traducing the limits of linearisation procedure, based on first order estimators.

Stationary normal inputs

Considering that both system input variables are normally distributed, with parameters:

$$k \sim N(\mu_k,\sigma_k) \text{ and } \Theta_a \sim N(\mu_k,\sigma_k), \quad (42)$$

Resulting that $\hat{\mu}_{\Theta_{hs}}$ will present an approximately normal distribution, which estimated parameters are:

$$\hat{\mu}_{\Theta_{hs}} = \hat{\mu}_{\Delta\Theta_{hs}} + \mu_{\Theta_a}, \quad (43)$$

$$\hat{\sigma}_{\Theta_{hs}}^2 = \hat{\sigma}_{\Delta\Theta_{hs}}^2 + \hat{\sigma}_{\Theta_a}^2, \quad (44)$$

since mutual independence between random parts was admitted. Under a probabilistic fonnulation, where time dependence does not exist and for stationary statistical distributions Vag is identical to *LOL*, and therefore (Popescu, 2008), being Θ_{hs} approximately Normal, LOL will result strictly as a lognormal distributed random variable:

$$pdf(LOL) = \frac{1}{\hat{\sigma}_{LOL}2\pi}\exp\left[-\frac{(\ln(LOL)-\hat{\mu}_{LOL})^2}{2\hat{\sigma}_{LOL}^2}\right] \quad (45)$$

Where;

$$\hat{\mu}_{LOL} = \frac{\ln 2}{6}(\hat{\mu}_{\Theta_{hs}} - 98) \text{ and } \hat{\mu}_{LOL} = \frac{\ln 2}{6}\hat{\sigma}_{\Theta_{hs}} \quad (46)$$

Stationary uniform inputs

Input variables are considered to be uniformly distributed:

$$K \sim U[K_1,K_2] \text{ and } \Theta_a \sim U(\mu_{\Theta_a},\sigma_{\Theta_a}). \quad (47)$$

Their first moments are given by:

$$\mu_X = \frac{X_2+X_1}{2} \text{ and } \sigma_X = \frac{X_2-X_1}{2\sqrt{3}}, \quad (48)$$

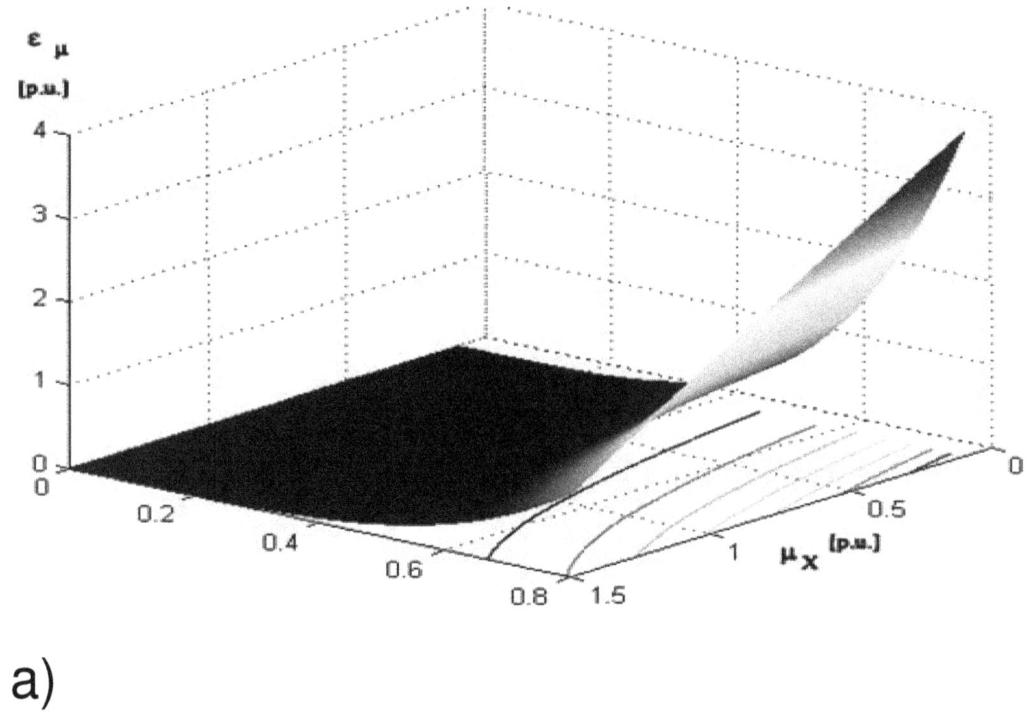

Figure 16a. First order linearisation error ε_{μ}.

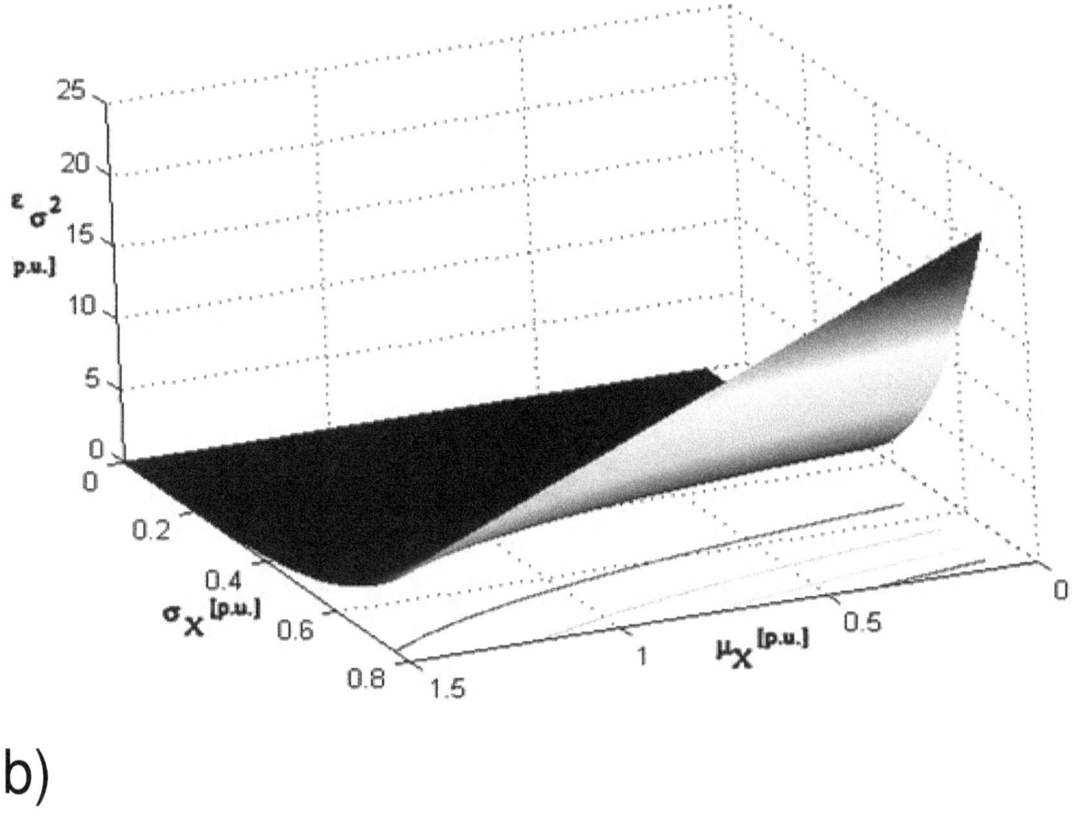

Figure 16b. First order linearisation error ε_{σ^2}.

Table 7. Limits of 95% confidence intervals of Normal inputs simulated by Monte Carlo.

	K		Θ_a	
	Lower	**Upper**	**Lower**	**Upper**
$\hat{\mu}$	0.802	0.809	19.762	20.117
$\hat{\sigma}$	0.098	0.105	4.923	5.175
$\hat{\gamma}$	0.123	0.128	0.246	0.263

Table 8. Input distribution parameters.

μ_k	σ_k	CV_K	μ_{Θ_a}	σ_{Θ_a}	CV_{Θ_a}
0.8	0.1	0.125	20	5	0.25

and the resulting variation coefficient by:

$$CV_X = \frac{1}{\sqrt{3}} \frac{X_2 - X_1}{X_2 + X_1}, \qquad (49)$$

with $X \equiv K, \Theta_a$. In this case, analytic *pdf* of output variable *LOL* is unknown because Θ_{hs} is a bounded random variable.

Simulation results and analysis

Simulation parameters and method

The results presented were obtained considering a distribution transformer rated 630 kVA, 10 kV/400V with copper windings. When needed parameters were omitted on transformer data sheet, the ones proposed on IEC-354 (1991) were assumed: $\Delta\Theta_{oR} = 55K$, $\Delta\Theta_{hsR} = 23K$, $R = 5$, $n = 0.8$, $m = 1.6$. Input variables sample length is N = 3000 and were simulated from a Monte Carlo Method (Popescu et al., 2009a; Popescu, 2007; Rubinstein, 1981). Histograms were drawn, considering 100 binary classes, for each variable. Table 7 represents the 95% confidence intervals of Normal inputs simulated by Monte Carlo.

In order to compare results from normal and uniform input distributions, random variables were simulated for similar expected and standard deviation values, on both sets of distributions (Table 8).

Their respective histograms are represented in Figures 17 and 18. Concerning uniformly distributed inputs, the bounds of their variation range are determined by (48), taking into account the same means and standard deviations of Table 8.

Results for normal and uniform inputs

For normally distributed load and ambient temperature profiles, which parameters are represented on Table 8, $\Delta\Theta_{hs}$, Θ_{hs} and *LOL pdf*'s results are represented in Figures 19 and 20(a), respectively; *LOL* cumulative distribution function (CDF) is represented on Figure 20b. For Uniform distributed input variables, corresponding variables are represented in Figures 20 and 21.

In the figures, dots represent simulated values and lines represent Normal *pdf*'s (and subsequent lognormal ones, for LOL), which parameters are the corresponding variables second order $\hat{\mu}$ and $\hat{\mu}$ estimators, obtained from the linearisation procedure (39) and (40).

Results analysis

Estimators determined from the linearisation method, corresponding to $\Delta\Theta_{hs}$, Θ_{hs} and LOL variables are represented in Table 9.

These estimators are, by definition, independent of input distributions (Normal and Uniform). Table 10 represents estimators obtained from the Monte Carlo method, for the variables $\Delta\Theta_{hs}$, Θ_{hs} and LOL, respectively for Uniform and Normal distributions.

Taking into account inherent errors of Monte Carlo methodology, these values can be considered as references. A general idea of linearisation precision can be drawn out from deviations of, for example, $\Delta\Theta_{hs}$ variable parameters; maximal deviations between Tables 9 and 10 values are 3%. Since these errors also include Monte Carlo inherent errors (Popescu et al., 2009a; Rubinstein, 1981) (Table 11), one can consider that second order estimators are of sufficient precision.

Comparison of Normal and Uniform parameters in Table 10 shows that LOL mean and standard deviation weakly depend upon input variables distribution. The strong non-linearity of *LOL* model leads to a great increase in *LOL* variation coefficient by reference to input variable ones. This tendency is more pronounced for normal distributions than for uniform ones; this is due to the fact that normal random variables domain do not respect positive values). From this point of view, since minimal information is available, the choice of uniformly distributed variables, leads to better results according to maximum entropy principle (Papoulis, 1984; Pierrat et al., 1997). On the other hand, Figures 20(a) and 20(b) show constrains associated to physical variables (bounded that linearisation and Monte Carlo methods sufficiently coincide for uniform and normal input variables, concerning CDF. The linearisation based on second order estimators allows the easily determination of *LOL* mean and standard mean and standard deviation, CDF and consequently, the probability of exceeding a certain level.

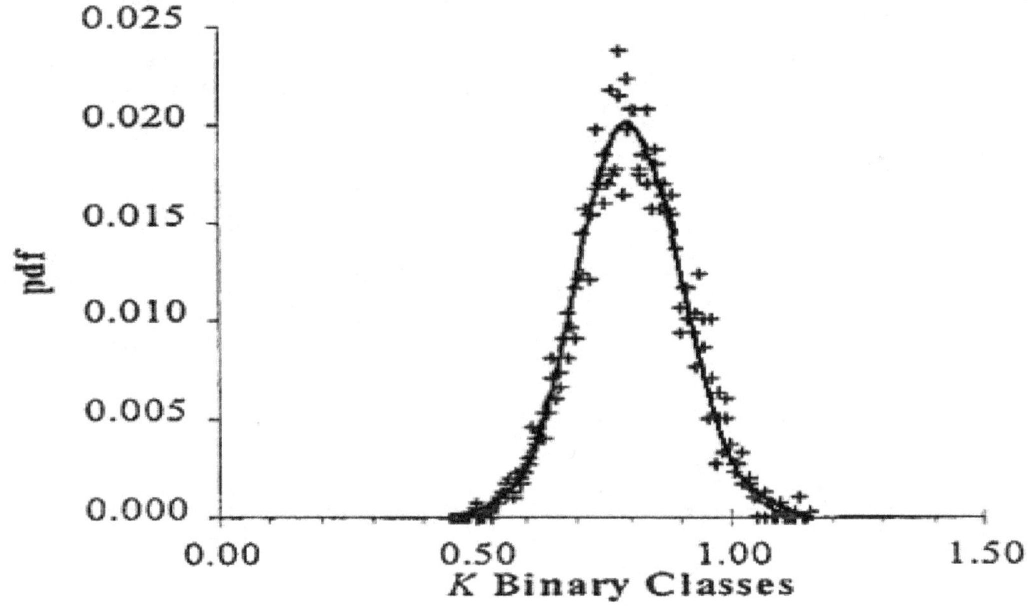

a)

Figure 17a. Histograms of Normal *K*.

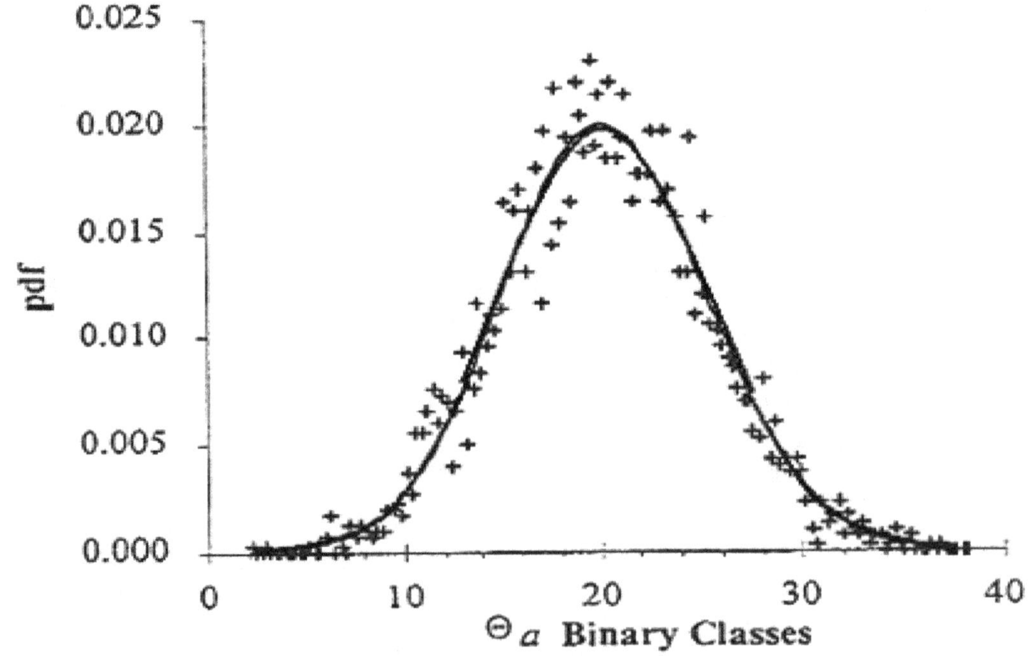

b)

Figure 17b. Histograms of Normal inputs Θ_a.

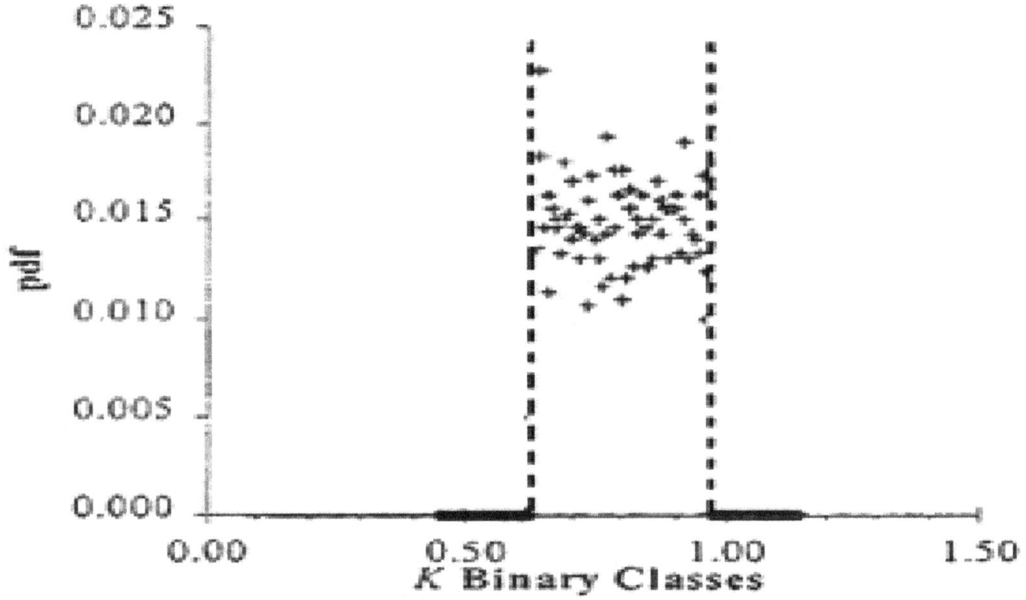

a)

Figure 18a. Histograms of Uniform *K* inputs.

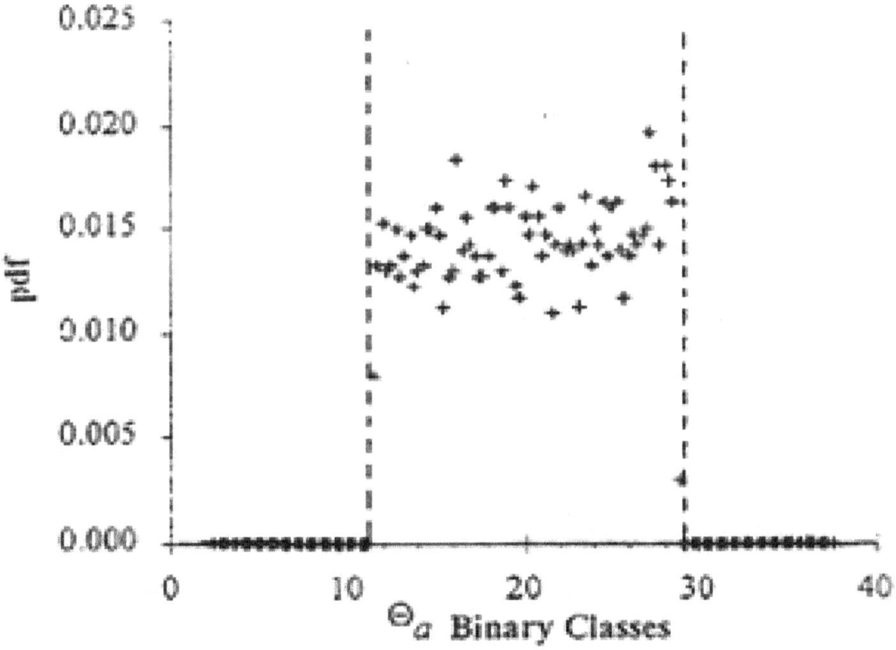

b)

Figure 18b. Histograms of Uniform Θ_a inputs.

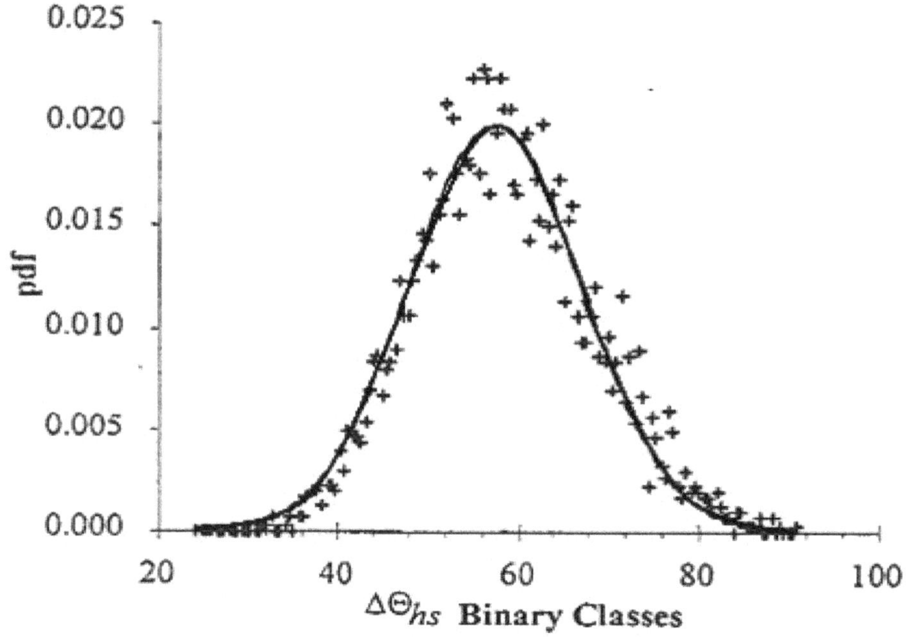

a)

Figure 19a. (a) $\Delta\Theta_{hs}$ pdf's for Normal inputs.

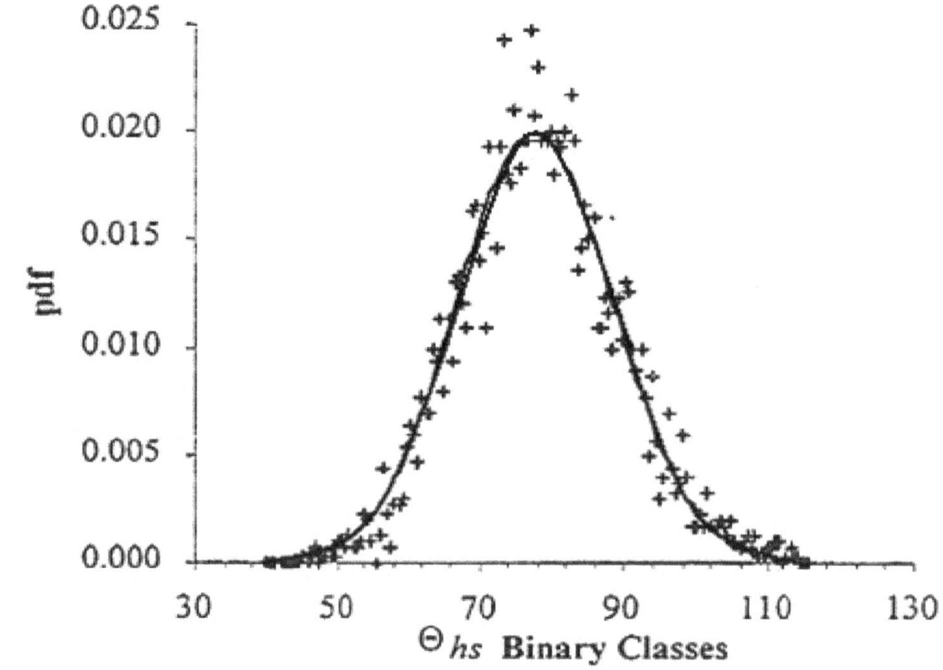

b)

Figure 19b. Θ_{hs} pdf's for Normal inputs.

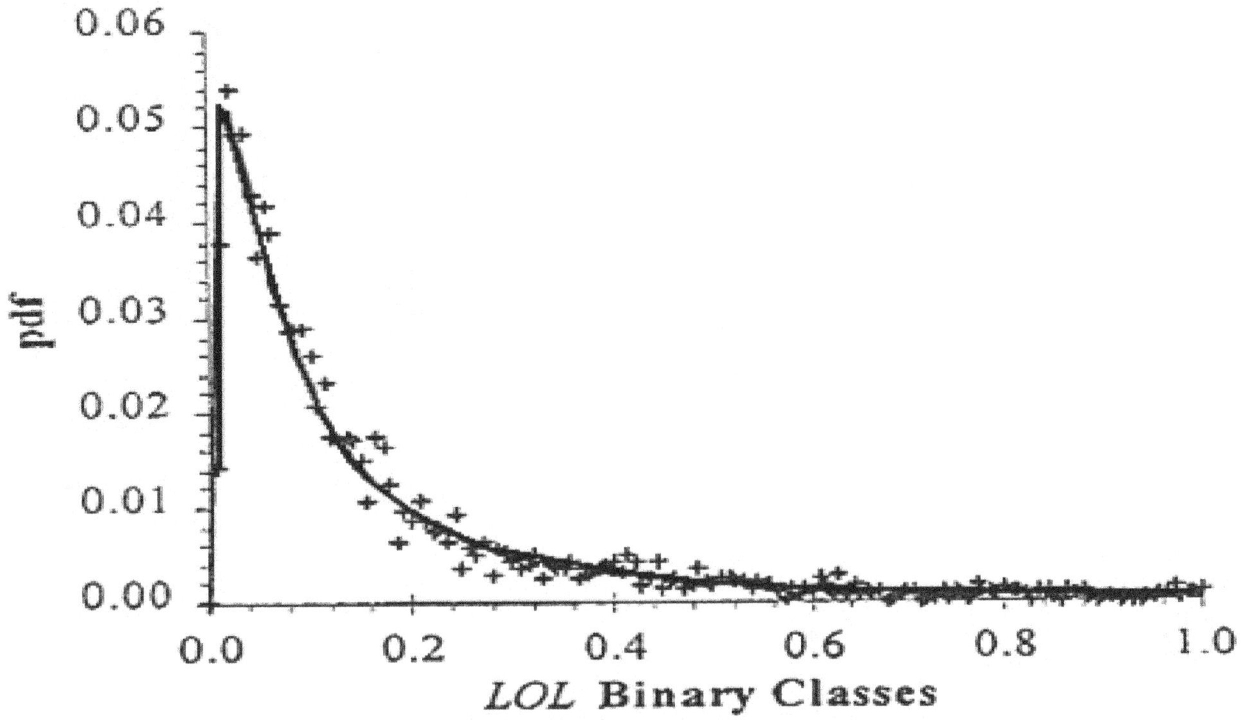

a)

Figure 20a. LOL and pdf for Normal inputs.

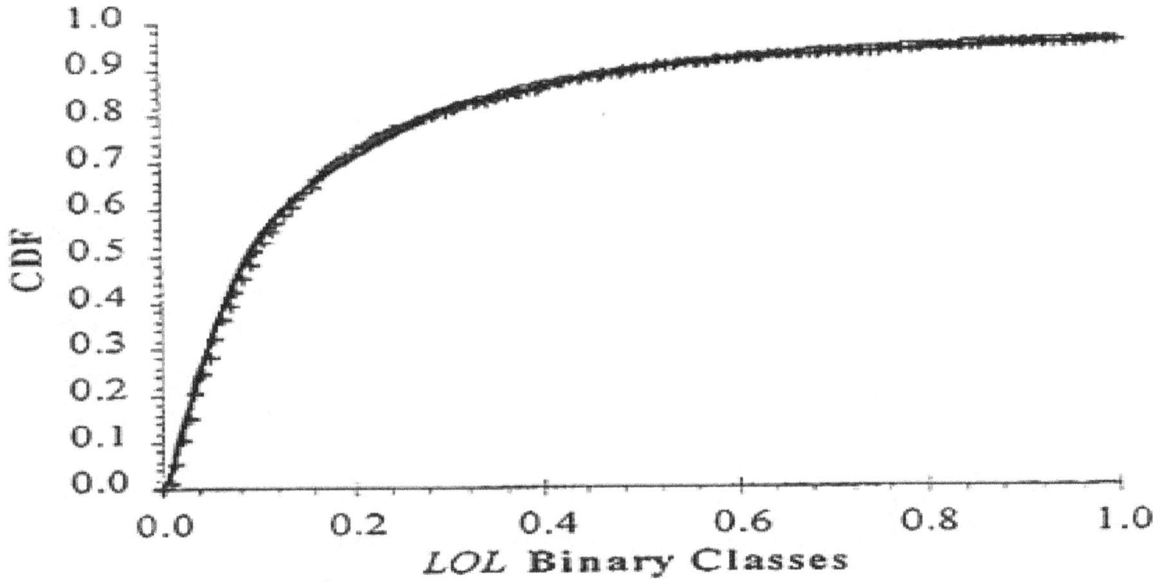

b)

Figure 20b. LOL and CDF for Normal inputs.

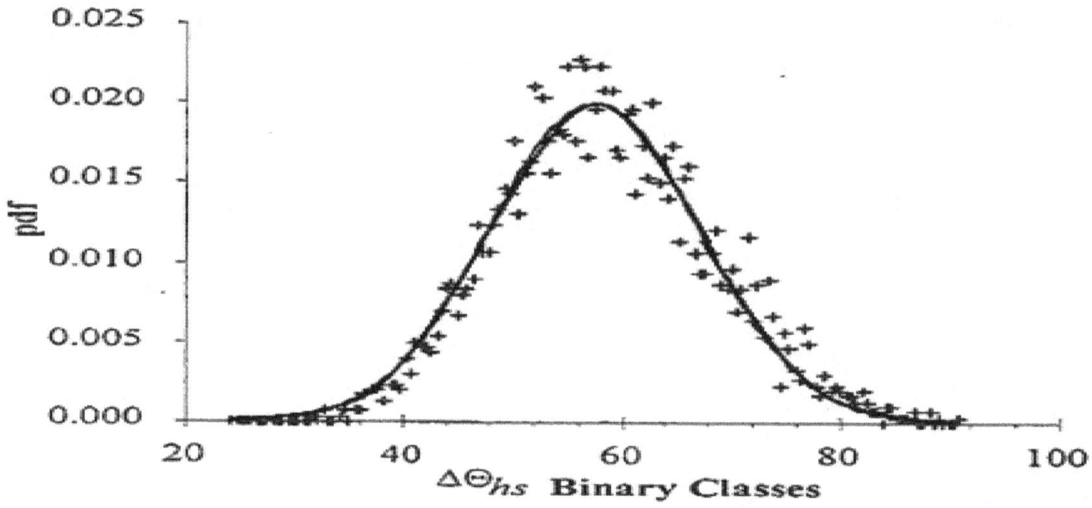

a)

Figure 21a. $\Delta\Theta_{hs}$ *pdf*s for Normal inputs.

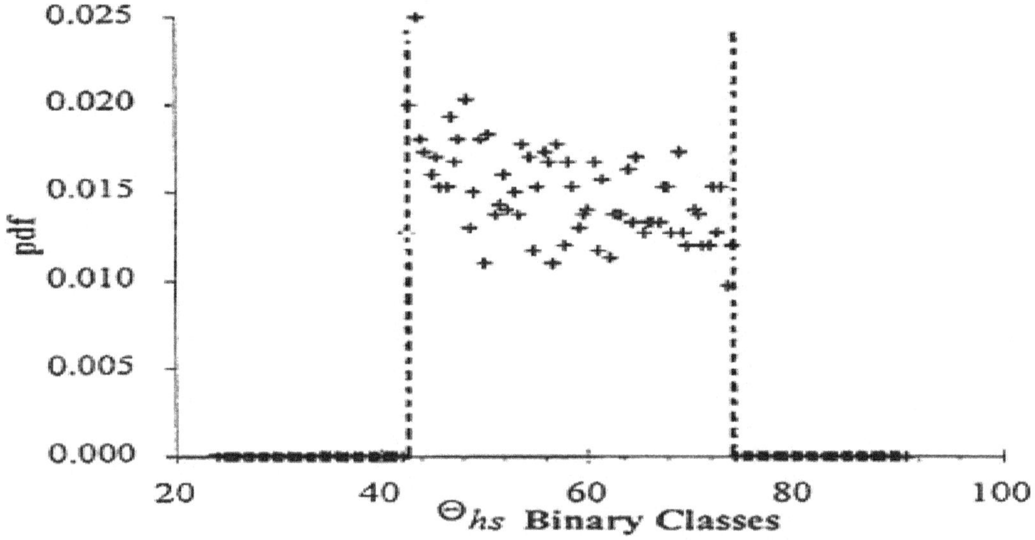

b)

Figure 21b. Θ_{hs} *pdf*s for Normal inputs.

Table 9. Second order LL, C and CV estimator values.

	$\Delta\Theta_{hs}$	Θ_{hs}	***LOL***
$\hat{\mu}$	57.950	78.131	0.222
$\hat{\sigma}$	9.579	10.869	0.433
$\widehat{CV}$	0.163	0.137	1.957

Table 10. Estimators from Monte Carlo simulations for Normal and Uniform inputs.

	Normal distribution			Uniform distribution		
	$\Delta\Theta_{hs}$	Θ_{hs}	*LOL*	$\Delta\Theta_{hs}$	Θ_{hs}	*LOL*
$\hat{\mu}$	58.268	78.209	0.236	57.835	77.835	0.211
$\hat{\sigma}$	9.656	10.870	0.477	9.536	10.767	0.406
$\widehat{CV}$	0.166	0.139	2.019	0.165	0.138	1.923

Table 11. Monte Carlo errors propagation for Normal inputs.

	$\Delta\Theta_{hs}$		Θ_{hs}		*LOL*	
	Lower	Upper	Lower	Upper	Lower	Upper
$\hat{\mu}$	57.524	58.208	77.286	78.327	0.281	0.419
$\hat{\sigma}$	9.391	9.935	10.602	11.203	0.000	1.681
$\hat{\gamma}$	0.161	0.173	0.135	0.145	0.000	5.971

STRUCTURAL PARAMETERS SENSITIVITY

This section objective is to study thermal and loss of life models sensitivity relatively to structural parameters, namely, $\Delta\Theta_{0R}$, $\Delta\Theta_{ohR}$, *R, n* and *m* (Resende et al., 1998).

Methodology

Functional inputs are the transformer load, *K*, and ambient temperature, Θ_a, profiles which, by assumption, will be represented by the additive model of (34). The deterministic component of functional inputs will be considered stationary. In fact, this basic representation on functional input profiles will have no influence on structural paramemeter sensitivity study, since these profileswill remain unchanged along the study. Both system functional input variables will be considered as normally distributed:

$$k \sim \mathrm{N}\,(\mu_k, \sigma_k) \text{ and } \Theta_a \sim \mathrm{N}\,(\mu_{\Theta_a}, \sigma_{\Theta_a}). \quad (50)$$

The values of two first variables moments are: μ_k =1 [p.u.], σ_k = 0.1 [p.u.], μ_{Θ_a} = 20°C and σ_{Θ_a} = 5°C. These values will lead to corresponding variation coefficient values of: CV_K =0.1 [p.u.] and CV_{Θ_a} =0.25 [p.u.]. For the reference scenario, structural parameters will be considered as deterministic variables, which values are those proposed by IEC-354 (1991):

$$\Delta\Theta_{0R} = 55\,, \Delta\Theta_{hsR} = 23K\,, R = 5\,, n = 0.8\ m = 1.6\,. \quad (51)$$

For this referential scenario, structural parameters will be considered as random variables. Other possible distribution could be envisaged, depending upon the available knowledge of parameters; due to its generality, it will be considered that structural parameters are random variables normally distributed:

$$\Delta\Theta_{0R} \sim \mathrm{N}\,(\mu_{oil}, \sigma_{oil})\,, \Delta\Theta_{hsR} \sim \mathrm{N}\,(\mu_{hs}, \sigma_{hs})\,, n \sim \mathrm{N}\,(\mu_n, \sigma_n)\,,$$
$$R \sim \mathrm{N}\,(\mu_R, \sigma_R) \text{ and } m \sim \mathrm{N}\,(\mu_m, \sigma_m) \quad (52)$$

which first moment values are those represented on (51) and second moment values are imposed by limiting physical conditions:

$$\Delta\Theta_{oR}, \Delta\Theta_{ohR},\ R,\ n,\ m>0,\ n<1 \text{ and } m<2 \quad (53)$$

Procedure

Using a Monte Carlo simulation method (Popescu et al., 2008; Rubinstein, 1981), load and ambient temperature profiles are simulated and, under the referential scenario, output variable two first moments μ_{LOL} and σ_{LOL} are determined. The model output sensitivity will be studied separately for each structural parameter. Therefore, five more simulations are performed where, one at the time, each structural parameter is considered as a random variable defined on (52), while the remain four, stay as deterministic ones; with this procedure, one is able to study output sensitivity due to each parameter, separately. The variability of each parameter was incremented up to the limits imposed by physical conditions stated in (53). This variability can be measured through the variation coefficient, *CV*. The output variable sensitivity is measured through the *LOL* variation coefficient, in per unit values based on those obtained under the referential

Table 12. Structural parameter values.

Set	$\Delta\Theta_{oR}$ [^{0}C]	$\Delta\Theta_{hsR}$ [^{0}C]	*R* [p.u.]	*n*	*m*
A	$\mu=55,\ \sigma\in[0;15]$	23	5	0.8	1.6
B	55	$\mu=23, \sigma\in[0;6.5]$	5	0.8	1.6
C	55	23	$\mu=5,\ \sigma\in[0;14]$	0.8	1.6
D	55	23	5	$\mu=0.8,\ \sigma\in[0;0.2]$	1.6
E	55	23	5	0.8	$\mu=1.6, \sigma\in[0;0.4]$

scenario:

$$CV_{LOL}[p.u.] = \frac{CV_{LOL}}{(CV_{LOL})_{\text{Re}\,ferencial\ Scenario}}. \quad (54)$$

Simulation results and analysis

The results presented were obtained considering a standardised distribution transformer rated 630 kVA, 10 kV/400 V with copper windings and mean values of structural parameters given by (51). Input variables sample length is N = 3000 and were simulated from a Monte Carlo Method (Popescu et al., 2009a; Rubinstein, 1981). In order to compare, separately, models sensitivity to each parameters, five set of input data were considered (Table 12).

Simulation results are represented in Figures 22, 23 and 24. *LOL* sensitivity to $\Delta\Theta_{ohR}$ and $\Delta\Theta_{hsR}$ parameters variability is important (Figure 22).

In fact these are the thermal model parameters that represent the transformer cooling conditions which are fundamental on hot-spot temperature estimation and so loss of life. Results show the importance of standardising with variation coefficients below 5%, the values obtained from tests for these two parameters. Under this condition, *LOL* sensitivity to them becomes negligible. Output variable sensitivity to *R* parameter variability is negligible (Figure 23) since, even with an increase in transformer losses (which would increase *LOL*), this *R* ratio stays almost constant. Parameter *R* is fundamental to optimise transformer efficiency as a function of load variability and losses economical value but its importance is reduced on hot-spot temperature estimation, at least assuming (IEC-354, 1991) thermal model. Parameter *R* is fundamental on economical models but its importance is reduced on thermal model. Sensitivity to *n* and *m* parameter is an important and actual subject since many discussions can be found on literature about these two parameters, for example Boteanu and Popescu (2008) and Zodeh and Whearty (1997). These parameters are of difficult measurement and therefore, one can find in specialised literature a great dispersion of measuring methodologies and correspondent obtained values.

Conclusion

The modelling of the time series representative of annual evolution of ambient temperature and transformer load showed that a non-complex additive model of deterministic and random components could genetically model such time series. Good results were obtained considering the deterministic component as a time varying function represented by a constant value (mean annual value) to which a first order sinusoidal function is added (annual added (annual cyclic variation). The model can easily be extended to daily, weekly or seasonally sinusoidal variations. Resulted residuals still denoted the presence of deterministic cyclic behaviours of higher than the first order but, generally, they could be approximated to random variables closely following a Gaussian distribution.

Most detailed models, such as the autorregressive models were experienced. They proved to mostly precise model some of the analysed time series but they could not be generalised for the analysed sample of profiles. The correlation between ambient temperature and distribution transformer load was also analysed. For the studied cases, the results obtained by splitting this analysis into correlation between deterministic compo-nents and correlation between random components, showed that ambient temperature and distribution trans-former load were inversely correlated and that this correlation derives mainly from a strong link between deterministic components rather than from random com-ponents. Due to their relative values, corre-lation be-ween random components is practically negligible, compared to that between deterministic com-ponents. Due to the strongly non-linearity of transformer thermal and loss of life models the statistical structure of input variables (load and ambient temperature) is not preserved on the output variable (loss of life). Moreover, the analytical determination of output statistical pdf is not possible either directly either with recourse of characteristic functions, since some mathematical transformations do not have an analytical exact expression for its inverse. Since, in a statistical sense, load variable is of reduced variability, meaning concentrated around its mean, a second order linearisation of the model, valid in the vicinity of load mean, was developed. The linearised model was validated for two different pdf's of the input variables: the Gaussian

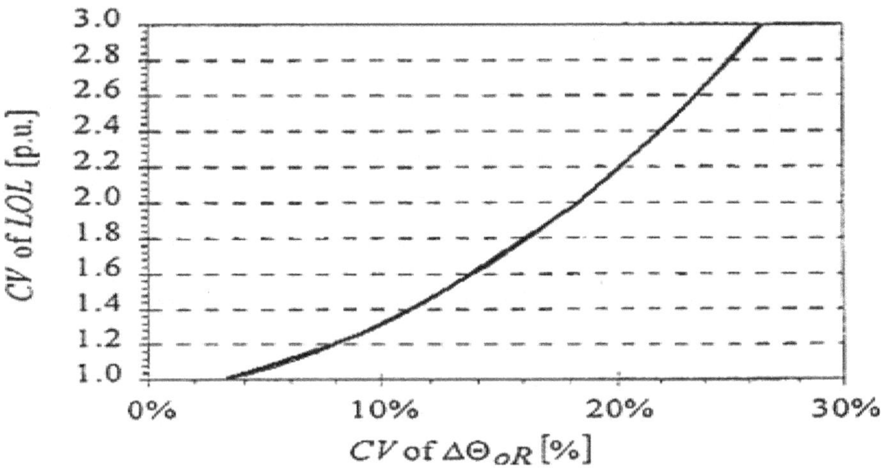

a)

Figure 22a. LOL sensitivity to $\Delta\Theta_{ohR}$ variability.

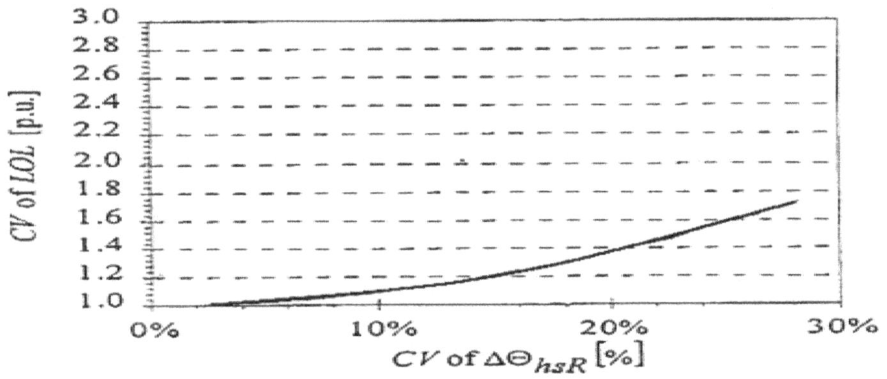

b)

Figure 22b. LOL sensitivity to $\Delta\Theta_{hsR}$ variability.

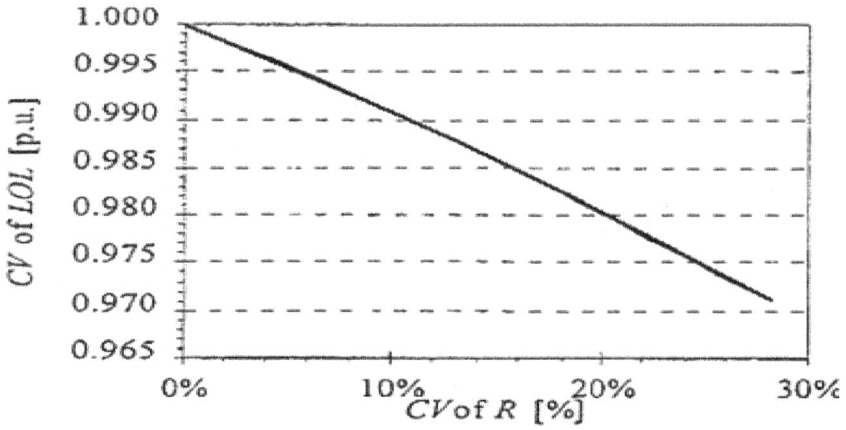

Figure 23. LOL sensitivity to R variability.

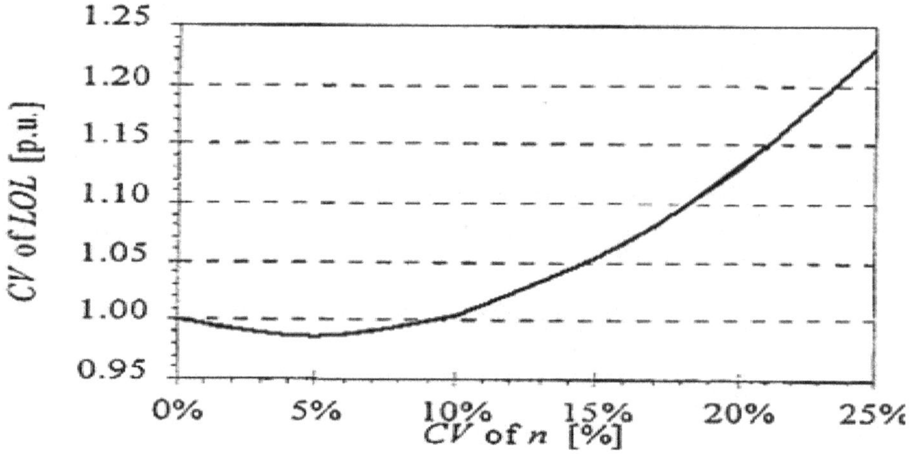

a)

Figure 24a. *LOL* sensitivity to *n*.

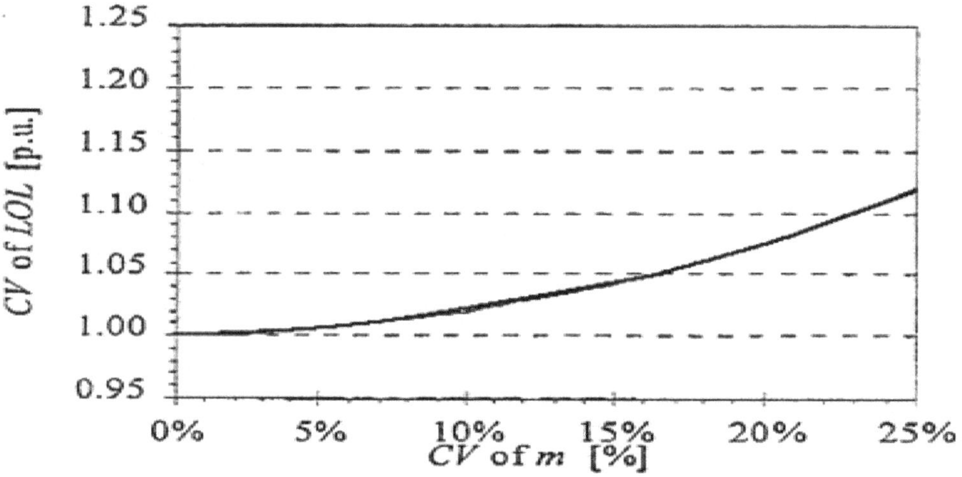

b)

Figure 24a. *LOL* sensitivity to *m*.

and the uniform pdf's. The input variables were simulated by a Monte Carlo method and results obtained from simulations are of good accuracy with those analy-tically estimated. Last study presented on this chapter refers to the sensitivity of transformer thermal and loss of life models, relatively to its structural parameters varia-bility. The existence of this variability has been shown on Borcosi et al. (2009:2008). The sensitivity was studied through the variability of output variable and was achieved by considering structural parameters as repre-sented by random variables normally distributed. This statistical structure was chosen attending to its generality. Variables were simulated by a Monte Carlo method and their mean values equalled those proposed by Inter-national Standards. Respective variation values were the maximum ones allowed by parameters' physical constrains. Results showed that the transformer thermal and loss of life assembly model is practically insensitive to the variability of the parameters *R, n* and *m*. On the other hand, its sensitivity to $\Delta\Theta_{hsr}$ and mainly to $\Delta\Theta_{or}$ var-iability is important. Justification for this sensitiveness resides on the fact that these two parameters are those which values directly reflect the cooling conditions of the transformer and therefoe are determinant on thermal loss of life estimation. For this reason, the study showed the importance of international standardisation of these parameters. If these parameters were standardised with variation coefficients below 5%, one could consider that loss of life sensitivity to them would be negligible.

LIST OF MOST IMPORTANT SYMBOLS

CV_x, Variation coefficient of x; **$C\hat{O}V_x$**, Estimator of X_t variable autocovariance (x_t sample autocovariance); $C\hat{O}V_{xy}$, estimator of covariance between X_t and Y_t X_t variables (covariance between x_t and y_t sample); **k**, time lag on chronological series [times]; **LOL**, relative loss of life over a period [p.u.]; *n*, oil exponent depending upon transformer refrigeration method [dimensionless]; N, generic integer number [dimensionless]; ***m***, hot-spot exponent depending upon transformer refrigeration method [dimensionless]; ***P*** (statement), probability of occurrence of statement between brackets [p.u]; ***R***, loss ratio (rated load loss on windings to no-load loss) [p.u]; ***x***, generic variable; $\bar{x}$, arithmetic averages of x_t; **Xt**, stochastic variable at instant t, with deterministic and random components; μ_x, first moment (mean or expected value) of variable x [same dimension as x]; $\hat{\rho}$, estimator of X_t variable autocorrelation (autocorrelation of x_t sample); $\hat{\rho}_{xy}$, estimator of correlation between X_t and Y_t variables (correlation between x_t and y_t samples); σ_Z^2, second moment (variance) of variable x (the square of x dimension); Θ_a, ambient temperature; $\Delta\Theta_{0R}$, top-oil temperature rise referred to ambient temperature under rated load [K]; ω, angular frequency [rad s^{-1}].

REFERENCES

Asbury CE (1975). Weather Load Model for Electric Demand and Energy Forecasting, IEEE Trans. Power Apparatus Syst. 94(4): 1111-1116.

Bendat JS, Pierso (1990). A Non–Linear System Analysis and Identification from Random Data, Wiley-Interscience, New Yory.

Bendat JS, Piersol A (1993). Engineering applications of Correlation and Spectral Analysis, Wiley-Interscience, New Yory.

Borcosi I, Olaru O, Popescu MC, Antonie N, Dinca A, Ionescu M (2009). Method to Protect from no Pulse for a Three-Phase Rectifier Bridge, International J. Math. Models Methods Appl. Sci. pp.473-482.

Borcosi I, Onisifor O, Popescu MC, Dincă A (2008). A Method to Protect from no Pulse for a Three-Phase Rectifier Bridge Connected with the Resistive-Inductive Load, Proceedings of the 10th WSEAS International Conference on Mathematical and Computational Methods in Science and Engineering pp.146-152.

Boteanu N, Popescu MC (2008). Optimal Control by Energetic Criterion of Driving Systems Proceedings of the 10th WSEAS International Conference on Mathematical and Computational Methods in Science and Engineering pp.45-51.

Box GEP, Jenkins GM (1970). Time Series Analysis, Forecasting and Control, Holden-Day, USA.

Chatfield C (1975). The Analysis of Time Series: Theory and Practice, Chapman and Hall, London.

Chong CY, Malhame RP (1984). Statistical Synthesis of Physically Based Models with Applications to Cold Load Pickup, IEEE Trans. Power Apparatus and Syst. 103 (7): 1621-1627.

Friedlander B, Francos J (1996). On the Accuracy of Estimating the Parameters of a Regular Stationary Process, IEEE Trans. Info. Theor. 42(4): 1202-1211.

Gutmann L, Wilks S, Hunter JS (1982). Introductory Engineering Statistics, 3rd Edition, John Wiley&Sons,.

IEC-354, International Electrotechnical Commission (1991). Loading Guide for Oil-Immersed Power Transformers, Second Edition,.

Mastorakis N, Bulucea CA, Popescu MC, Manolea Gh, Perescu L (2009). Electromagnetic and Thermal Model Parameters of Oil-Filled Transformers, WSEAS Trans. Circuits Syst. 8(6): 475-486.

Mastorakis N, Bulucea CA, Manolea Gh, Popescu MC, Perescu-Popescu L (2009). Model for Predictive Control of Temperature in Oil-filled Transformers, Proceedings of the 11th WSEAS International Conference on Automatic Control, Modelling and Simulation pp.157-165.

Papoulis A (1984). Probability Random Variables and Stochastic Processes, 2nd Ed., McGraw-Hill, New York,

Pierrat L, Resende MJ, Santana J (1996). Power Transformers Life Expectancy Under Distorting Power Electronic Loads, IEEE International Symposium on Industrial Electronics pp. 578-583.

Pierrat L (1997). Méthode d'Identification et Analyse d'Incertitude pour les Paramètres d'une Courbe d'Echauffement Tronquée, 8th International Congress of Metrology, Besacon.

Popescu MC, Mastorakis N, Bulucea CA, Manolea Gh, Perescu L (2009). Non-Linear Thermal Model for Transformers Study, WSEAS Trans. Circuits Syst. 8 (6): 487-497.

Popescu MC, Manolea Gh, Bulucea CA, Boteanu N, Perescu-Popescu L, Muntean IO (2009).Transformer Model Extension for Variation of Additional Losses with Frequency, Proceedings of the 11th WSEAS International Conference on Automatic Control, Modelling and Simulation pp.166-171.

Popescu MC (2006). Estimarea şi identificarea proceselor, Editura Sitech, Craiova,

Popescu MC, Manolea Gh, Manolea Gh, Perescu L, (2009). Parameters Modelling of Transformer, WSEAS Trans. Circuits Syst. 8 (8): 661-675.

Popescu MC (2008) Modelarea şi simularea proceselor, Editura Universitaria, Craiova,

Popescu MC (2009). Aplicaţii in informatica, Editura Universitaria Craiova.

Popescu MC (2006). Thermocuples-Comparative Study, Electrotehnică, Energetică, Electronică 4th International Conferecnce on Electrical and Power Engineering, Buletinul Institutului Politenic din Iaşi, Vol LII(LVI), Publicat de Universitatea Tehnică „Gh. Asachi", Tomul LII(LVI), Fasc. 5A: 1069-1074.

Popescu MC, Petrişor A (2006). The experimental Investigation of heat transfer and friction losses interrupted and way fins for fin-and-tube heat exchangers, Electrotehnică, Energetică, Electronică 4th International Conferecnce on Electrical And Power Engineering, Buletinul Institutului Politenic din Iaşi, Vol LII(LVI), Publicat de Universitatea Tehnică „Gh. Asachi", Tomul LII(LVI), Fasc. 5C: 1295-1301.

Popescu MC (2007). Simulation of a temperature control system with distributed parameters, International Conference on Electromechanical and Power Systems, Chişinău–Republica Moldova, Annals of the University of Craiova, Electrical Engineering, Editura Universitaria pp. 26-32.

Popescu MC (2008). Device for protection to the lack of the pulse for tri-phase rectifiers in bridge with load inductive resitive, Proceedings of the 14th national conference on electrical drives pp.201-204.

Popescu MC, Petrişor A, Drighiciu MA (2008). Modelling and Simulation of a Variable Speed Air-Conditioning System, IEEE International Conference on Automation, Quality and Testing, Robotics Proceedings pp.175-181,

Popescu MC (2008). Proiectarea unui stand automatizat pentru incercarea ventilatoarelor care echipeaza transformatoarele de putere. Contract nr.28033/2008. Beneficiar S.C. Electroputare SA-Divizia componente, Craiova.

Resende MJ, Pierrat L, Santana J (1998).Sensitivity Analysis as Accuracy Measure of Simplified Strongly Non-Linear Thermal Model Transformers Reliability Studies, 2nd International Symposium on Sensitivity Analysis of Model Output pp.227-230.

Ross S (1987). Introduction to Probability and Statistic for Engineers and Scientists, John Wiley & Sons, USA,

Rubinstein RY (1981). Simulation and the Monte Carlo Method, Wiley, New York.

Sachedev MS, Billinton R, Peterson CA (1977). Representative Biblio-

graphy on Load Forecasting, IEEE Trans. Power Apparatus Syst. 96(2): 697-700.
Srinivasan K (1975). Pronovost R., Short Term Load Forecasting Using Multiple Correlation Models, IEEE Trans. Power Apparatus Syst. 94(4): 1854-1858.
Zodeh OM, Whearty RJ (1997). Thermal Characteristics of a Mete-Aramid and Cellulose Insulated Transformer at Loads Beyond Nameplate, IEEE Trans. Power Delivery 12(1): 234-248.

2

Non-axisymmetric dynamic response of imperfectly bonded buried fluid-filled orthotropic thin cylindrical shell due to incident compressional wave

Rakesh Singh Rajput[1]*, Sunil Kumar[2], Alok Chaube[2] and J. P. Dwivedi[3]

[1]Department of Mechanical Engineering, Directorate of Technical Education Bhopal (M.P.), India.
[2]Department of Mecanical Engineering, Rajeev Gandhi Technical University, Bhopal, M. P., India.
[3]Department of Mechanical Engineering, IT-BHU, Varanasi, India.

The main aim of this paper is to assess and compare the relative importance of the effects of considering the fluid presence and the bond imperfection while evaluating the non-axisymmetric dynamic response of an imperfectly bonded empty as well as fluid filled orthotropic thin cylindrical shell buried under soil and excited by compressional wave (P-wave). While applying thin shell theory, the effect of shear deformation and rotary inertia need not to be considered. The pipeline had been modeled as an infinite cylindrical shell imperfectly bonded to surrounding. A thin layer is assumed between the shell and the surrounding medium (soil) such that this layer possesses the properties of stiffness and damping both. The effects of the fluid presence on the shell displacement have been studied for different soil conditions and at various angles of incidence of the longitudinal wave. It is observed that magnitude of the dynamic response of fluid filled pipeline is more than that of an empty pipeline. Axial and radial deflection of thin pipe is considerable even under hard soil conditions under imperfect bonding of pipe with soil. Numerical results have been presented for the longitudinal compressional wave (P- wave) only.

Key words: Orthotropic, Imperfect bond, seismic wave, non-axisymmetric, dynamic response, buried pipelines, thin shell.

INTRODUCTION

Growing urbanization has created congestion and problem of space for providing above ground utility services. In recent years, the use of underground power cabling, lying down of optic fiber communications line and water supply lines have been finding increasing use of thin shell pipes made of different types of orthotropic materials. After arrival of reinforced plastic mortar (RPM) pipes and its increasing use in providing utility services to ever-growing urban population, need was felt to analyze the pipe of orthotropic materials under static and dynamic conditions. The behavior of buried pipeline is observed to be significantly different from above ground pipes. Response of these buried pipes under seismic or other dynamic conditions requires to be analyzed.

During past few years, number of papers like Cole et al. (1979), and Singh et al. (1987) has appeared on the axisymmetric dynamic response of buried orthotropic pipe/shells. Later Chonan (1981), Dwivedi and Upadhyay (1989, 1990, 1991) and Dwivedi et al. (1991) have analyzed the axisymmetric problems of imperfectly bonded shell for the pipes made of orthotropic materials. Upadhyay and Mishra (1988) have presented a good account of work on non-axisymmetric response of buried thick orthotropic pipelines under seismic excitation. Results show that that there is negligible axial and radial

*Corresponding author. E-mail: rak6raj@yahoo.co.in.

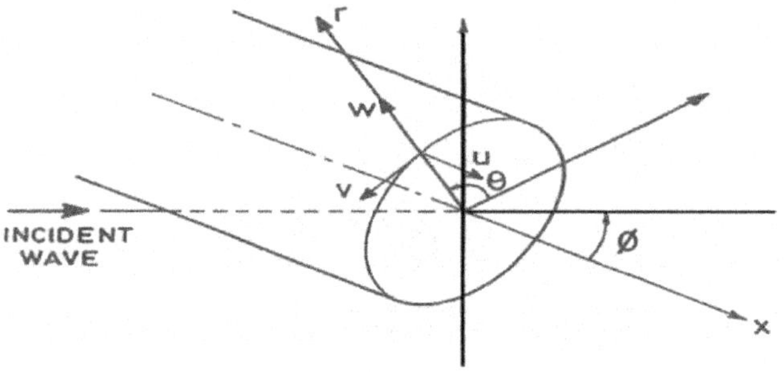

Figure 1. Geometry of the problem.

deflection of empty as well as fluid filled thick shell. Again Dwivedi et al. (1992a, 1992b), Dwivedi et al. (1993a, 1993b, 1996), and Dwivedi et al. (1998) have analyzed the non-axisymmetric problem of imperfectly bonded buried thick orthotropic cylindrical shells. Kouretzis et al. (2007) have presented analytical calculation of blast-induced strains on buried pipe lines. Hasheminajad and Kazemirad (2008) dynamic response of an eccentrically lined circular tunnel in poroelastic soil under seismic excitation. Lee et al. (2009) in their paper had done the risk analysis of buried pipelines using probabilistic method. But in all these analyses, pipeline had been modeled as thick shell. Rajput et al (2010) have reported comparison of non-axisymmetric dynamic response of imperfectly bonded buried orthotropic thick and thin fluid filled cylindrical shell due to incident shear wave (SH Wave) and have also presented non-axisymmetric dynamic response of imperfectly bonded buried orthotropic thin fluid empty cylindrical shell due to incident compressional wave.

As far as the non-axisymmetric dynamic response of thin shell is concerned, no work had been reported so far. Therefore, present paper attempts to analyze the effect of imperfect bond between pipe and surrounding medium on the non- axisymmetric dynamic response of buried orthotropic thin pipelines. A theoretical analysis of the non-axisymmetric steady state dynamic response of buried fluid-filled pipelines excited by seismic waves travelling in the surrounding infinite medium (soil) is presented. An infinite cylindrical shell model had been used for the thin pipeline. Comparisons of the numerical results for a fluid-filled shell with those for an empty shell have been presented and discussed.

BASIC EQUATIONS AND FORMULATIONS

The pipeline had been modeled as an infinitely long cylindrical shell of mean radius R and thickness h. It is considered to be buried in a linearly elastic, homogeneous and isotropic medium of infinite extent. Basic approach of the formulation is to obtain the mid plane displacements of the shell by solving the equations of motion of the orthotropic shell. Traction terms in the equations of motion are obtained by solving the three-dimensional wave equation in the surrounding medium. Appropriate boundary conditions are applied at the shell surfaces. Equations arising out of boundary conditions along with the equations of motion of the shell are simplified to yield a response equation in matrix form.

Equation governing the non axis-symmetric motion of an infinitely long orthotropic cylinder had been derived following the approach of Herrman and Mirsky (1957), Displacement at a particular point in the shell is taken in the form:

$$u_x(z, \theta, x, t) = u(\theta, x, t) + z_{\psi x}(\theta, x, t)$$

$$u_\theta(z, \theta, x, t) = v(\theta, x, t) + z_{\psi\theta}(\theta, x, t)$$

$$u_z(z, \theta, x, t) = w(\theta, x, t),$$

where u_z, u_θ, u_x are displacement component of a point in the shell.

Considering an infinitely long cylindrical shell of mean radius R and thickness h buried in a linearly elastic, homogeneous and isotropic medium of infinite medium, a thin layer is assumed between the shell and the surrounding medium (soil). The degree of imperfection of the bond is varied by changing the stiffness and the damping parameters of this layer. The shell is excited by a longitudinal wave (p-wave). A wave of wavelength Λ (=2Π/ξ) is considered to strike the shell at an angle Φ with the axis of theshell (as shown in Figure 1). Let a cylindrical polar co-ordinate system (r, θ, x) is defined such that x coincides with the axis of the shell and, in addition, z is measured normal to the shell middle surface, which is given as:

$$z = r - R, \qquad -h/2 \le z \le h/2 \tag{1}$$

The basic equations which describe the dynamic behavior of cylindrical shells with bending resistance under arbitrary loads are derived from the system of equations which had been presented by Upadhyay and Mishra (1988). But in the thin shell theory, effect of shear deformation and rotary inertia is not considered. After equating all the inertial and moments term equal to zero, the equilibrium equations of thick shell in stress form (from above reference) reduces to:

$$\frac{1}{R}\frac{\partial Q_\theta}{\partial \theta} + \frac{\partial Q_x}{\partial x} - \frac{N_{\theta\theta}}{R} + P_1^* = \rho h \frac{\partial^2 w}{\partial t^2}; \tag{2a}$$

$$\frac{1}{R}\frac{\partial N_{\theta\theta}}{\partial \theta} + \frac{\partial N_{x\theta}}{\partial x} + \frac{Q_\theta}{R} + P_2^* = \rho h \frac{\partial^2 v}{\partial t^2}; \tag{2b}$$

$$\frac{1}{R}\frac{\partial M_{\theta\theta}}{\partial \theta} + \frac{\partial M_{x\theta}}{\partial x} - Q_\theta = 0; \tag{2c}$$

$$\frac{1}{R}\frac{\partial N_{xx}}{\partial x} + \frac{1}{R}\frac{N_{\theta x}}{\partial \theta} + P_4^* = \rho h[\frac{\partial^2 u}{\partial t^2}]; \tag{2d}$$

$$\frac{\partial M_{xx}}{\partial x}+\frac{1}{R}\frac{\partial M_{\theta x}}{\partial \theta}-Q_x=0 \tag{2e}$$

Where $N_{xx};N_{\theta\theta};N_{x\theta};N_{\theta x}$ and $M_{xx};M_{\theta\theta};M_{x\theta};M_{\theta x}$ are stress resultants and moments respectively.

In connection with the equation of equilibrium, it can be argued that transverse shearing force Q_θ makes a negligible contribution to equilibrium of forces in circumferential direction. So after making Q_θ equal to zero in Equation 2(b), the values of Q_θ and Qx are determined from Equation 2(c) and (e) and putting it into Equation 2(b) and (d), above equations reduces to:

$$\frac{\partial^2 M_{xx}}{\partial x^2}+\frac{\partial^2 M_{x\theta}}{R\partial\theta\partial x}+\frac{\partial^2 M_{\theta x}}{R\partial\theta\partial x}+\frac{1}{R^2}\frac{\partial^2 M_{\theta\theta}}{\partial\theta^2}-\frac{N_{\theta\theta}}{R}+P_1^*=\rho h\frac{\partial^2 w}{\partial t^2}; \tag{3a}$$

$$\frac{1}{R}\frac{\partial N_{\theta\theta}}{\partial\theta}+\frac{\partial N_{x\theta}}{\partial x}+P_2^*=\rho h\frac{\partial^2 v}{\partial t^2}; \tag{3b}$$

$$\frac{\partial N_{xx}}{\partial x}+\frac{1}{R}\frac{\partial N_{\theta x}}{\partial\theta}+P_4^*=\rho h\{\frac{\partial^2 u}{\partial t^2}\}; \tag{3c}$$

For thin shell theory, shear deformation is not considered due to negligible thickness. So the shear strain components according to Herrman and Mirsky (1857) about z-axis in r-θ and r-x plane γ_{xz} and $\gamma_{z\theta}$ will be zero (no coupling is there due to negligible thickness) but at the same time shear stress component would be there due to Kirchhoff's hypothesis. So according to Herrman and Mirsky (1957)

$$\gamma_{xz}=\frac{\partial w}{\partial x}+\psi_x=0;$$

$$\gamma_{z\theta}=\frac{1}{R+z}\frac{\partial w}{\partial\theta}+\psi_\theta-\frac{1}{R+z}(v+z\psi_\theta)=0$$

So from the above equations:

$$\psi_x=-\frac{\partial w}{\partial x};$$

$$\psi_\theta=\frac{1}{R}(v-\frac{\partial w}{\partial\theta});$$

Here ψ_x and ψ_θ are angle of rotation in r-x and r-θ plane but in the r-θ plane the tangential deflection is negligible compared to component of radial deflection in that direction. So:

$$\psi_x=-\frac{\partial w}{\partial x};$$

$$\psi_\theta=-\frac{1}{R}(\frac{\partial w}{\partial\theta}); \tag{4}$$

From the above, stress resultants come out to be:

$$N_{xx}=E_p\frac{\partial u}{\partial x}-\frac{D}{R}\frac{\partial^2 w}{\partial x^2}+\frac{\nu_{\theta x}E_p}{R}(w+\frac{\partial v}{\partial\theta});$$

$$N_{\theta x}=G_{\theta x}[h\frac{\partial v}{\partial x}+\frac{1}{R}(h+I/R^2)\frac{\partial u}{\partial\theta}+(I/R^2)\frac{\partial^2 w}{\partial\theta\partial x}];$$

$$M_{xx}=\frac{D}{R}[\frac{\partial u}{\partial x}-R\frac{\partial^2 w}{\partial x^2}-\frac{\nu_{\theta x}}{R}\frac{\partial^2 w}{\partial\theta^2});$$

$$M_{\theta x}=G_{x\theta}[-2(I/R)\frac{\partial^2 w}{\partial\theta\partial x}-(I/R^2)\frac{\partial u}{\partial\theta}];$$

$$N_{x\theta}=G_{x\theta}[h\frac{\partial v}{\partial x}-(I/R^2)\frac{\partial^2 w}{\partial\theta\partial x}+(h/R)\frac{\partial u}{\partial\theta}];$$

$$N_{\theta\theta}=(\frac{E_p^{'}}{R}+\frac{D^{'}}{R^3})(w+\frac{\partial v}{\partial\theta})+\frac{D^{'}}{R^3}\frac{\partial^2 w}{\partial\theta^2}+\nu_{\theta x}E_p\frac{\partial u}{\partial x};$$

$$M_{x\theta}=G_{x\theta}(I/R)[\frac{\partial v}{\partial x}-2\frac{\partial^2 w}{\partial\theta\partial x}];$$

$$M_{\theta\theta}=-\frac{D^{'}}{R^2}\frac{\partial^2 w}{\partial\theta^2}-\frac{D^{'}}{R^2}(w+\frac{\partial v}{\partial\theta})-\nu_{\theta x}D\frac{\partial^2 w}{\partial x^2}; \tag{5}$$

Here, $G_{x\theta}$, G_{xz}, $G_{z\theta}$ are shear moduli of the shell Material. When these values of stress resultants are placed into above equations of equilibrium, it results in the required equation of motion of shell in the matrix form as:

$$[\{L\}\{U\}]+\{P^*\}=0 \tag{6}$$

where [L] is a 3×3 matrix operator and terms {P*} is column matrix

$$L_{11}=D\frac{\partial^4}{\partial x^4}+\frac{D^{'}}{R^4}\frac{\partial^2}{\partial\theta^2}+\frac{D^{'}}{R^4}\frac{\partial^4}{\partial\theta^4}+\frac{2\nu_{\theta x}D}{R^2}\frac{\partial^2}{\partial\theta^2\partial x^2}+4G_{x\theta}\left(\frac{I}{R^2}\right)\frac{\partial^4}{\partial\theta^2\partial x^2}+\left(\frac{E_p^{'}}{R^2}+\frac{D^{'}}{R^4}\right)+\rho h\frac{\partial^2}{\partial t^2};$$

$$L_{12}=\frac{D^{'}}{R^4}\frac{\partial^3}{\partial\theta^3}-G_{x\theta}\left(\frac{I}{R^2}\right)\frac{\partial^4}{\partial\theta^2\partial x^2}+\left(\frac{E_p^{'}}{R^2}+\frac{D^{'}}{R^4}\right)\frac{\partial}{\partial\theta};$$

$$L_{13} = -\frac{D}{R}\frac{\partial^3}{\partial x^3} + \frac{\nu_{\theta x}E_p}{R}\frac{\partial}{\partial x} + G_{x\theta}\left(\frac{I}{R^3}\right)\frac{\partial^3}{\partial\theta^2\partial x};$$

$$L_{21} = L_{12};$$

$$L_{22} = G_{x\theta}h\frac{\partial^2}{\partial x^2} + \left(\frac{E_p'}{R^2} + \frac{D'}{R^4}\right)\frac{\partial^2}{\partial\theta^2} - \rho h\frac{\partial^2}{\partial t^2};$$

$$L_{23} = \frac{G_{x\theta}h}{R}\frac{\partial^2}{\partial\theta\partial x} + \frac{\nu_{\theta x}E_p}{R}\frac{\partial^2}{\partial\theta\partial x};$$

$$L_{31} = L_{13};$$

$$L_{32} = L_{23};$$

$$L_{33} = E_p\frac{\partial^2}{\partial x^2} + \left(\frac{G_{x\theta}}{R^2}\right)\left(h + \frac{I}{R^2}\right)\frac{\partial^2}{\partial\theta^2};$$

and

$\{U\} = [w \;\; v \;\; u]^T$

Where, w, v and u are the displacement components of the middle surface of the shell in the radial, tangential and axial directions respectively. The elements of column matrix $\{P^*\}$ are given by Herrman and Mirsky (1957) as:

$$P_1^* = \left(1 + \frac{z}{R}\right)\sigma_{zz}\Big|_{-h/2}^{h/2}, \quad P_2^* = \left(1 + \frac{z}{R}\right)\sigma_{z\theta}\Big|_{-h/2}^{h/2},$$

$$P_3^* = z\left(1 + \frac{z}{R}\right)\sigma_{z\theta}\Big|_{-h/2}^{h/2}, \quad P_4^* = \left(1 + \frac{z}{R}\right)\sigma_{zx}\Big|_{-h/2}^{h/2},$$

$$P_5^* = z\left(1 + \frac{z}{R}\right)\sigma_{zx}\Big|_{-h/2}^{h/2}$$

where, σ_{ij} denotes the stresses with their usual meaning, but for thin shell P_3^* and P_5^* are zero. Different constants appearing in the expressions for L_{ij} are defined as:

$$E_p = \frac{E_x h}{1 - \nu_{x\theta}\nu_{\theta x}}, \quad E_p' = \frac{E_\theta h}{1 - \nu_{x\theta}\nu_{\theta x}}, \quad D = E_p\frac{h^2}{12}, \quad D' = E_p'\frac{h^2}{12},$$

Where, moment of inertia, $I = h^3/12$ and E_x, E_θ are elastic moduli, $\nu_{x\theta}$, $\nu_{\theta x}$ the Poisson ratio and ρ is the density of the shell material. 'n' indicate the mode in circumferential direction; n = 0 represents the axisymmetric mode.

For the evaluation of $\{P^*\}$, stress σ_{ij} at $z = \pm(h/2)$ must be determined in the terms of incident and scattered field in the surrounding ground. The total displacement field in the ground is written as:

$$d = d^{(i)} + d^{(s)}$$

Where, superscripts i and s represents the incident and scattered parts of deflection respectively. By solving the wave equation in the surrounding infinite medium, the components of incident and scattered fields can be written as (Chonan, 1981):

$$d_r^{(i)} = \left[\left\{\gamma I_n'\frac{\gamma r}{R}\right\}B_1 + \left\{-i\beta_1\delta I_n'\frac{\delta r}{R}\right\}B_3 + \left\{n\frac{R}{r}I_n\frac{\delta r}{R}\right\}B_5\right]\cos n\theta\exp[i\xi(x-ct)]$$

$$d_\theta^{(i)} = \left[\left\{-n\frac{Rr}{r}I_n\frac{\gamma r}{R}\right\}B_1 + \left\{in\frac{R}{r}\beta_1 I_n\left(\frac{\delta r}{R}\right)\right\}B_3 + \left\{-\delta I_n'\frac{\delta r}{R}\right\}B_5\right]\sin n\theta\exp[i\xi(x-ct)]$$

and

$$d_x^{(i)} = \left[\left\{i\beta_1 I_n\frac{\gamma r}{R}\right\}B_1 + \left\{\delta^2 I_n\frac{\delta r}{R}\right\}B_3\right]x\cos n\theta\exp[i\xi(x-ct)] \tag{7}$$

where, $B_1 = B_1'/R$, $B_3 = B_3'/R^2$ $B_5 = B_5'/R$ and (') denotes differentiation with respect to the argument of the Bessel functions. The constants B_1, B_3 and B_5 depend on the parameters of the incident wave and may be expressed as:

$$B_1 = (-1)^{n+1}\left(i\chi\frac{A_1}{\varepsilon_1}\right), \quad B_3 = (-1)^n\left(i\chi\frac{A_2}{\delta\varepsilon_2}\right), \quad B_5 = (-1)^n\left(\chi\frac{A_3}{\delta}\right) \tag{8}$$

$$d_r^{(s)} = \left[\left\{\gamma K_n'\left(\frac{\gamma r}{R}\right)\right\}B_2 + \left\{-i\beta_1\delta K_n'\left(\frac{\delta r}{R}\right)\right\}B_4 + \left\{n\left(\frac{R}{r}\right)K_n\left(\frac{\delta r}{R}\right)\right\}B_6\right]\cos n\theta\exp[i\xi(x-ct)]$$

$$d_\theta^{(s)} = \left[\left\{-n\left(\frac{R}{r}\right)K_n\left(\frac{\gamma r}{R}\right)\right\}B_2 + \left\{in\left(\frac{R}{r}\right)\beta_1 K_n\left(\frac{\delta r}{R}\right)\right\}B_4 + \left\{-\delta K_n'\left(\frac{\delta r}{R}\right)\right\}B_6\right]\sin n\theta\exp[i\xi(x-ct)]$$

$$d_x^{(s)} = \left[\left\{i\beta_1 K_n\left(\frac{\gamma r}{R}\right)\right\}B_2 + \left\{\delta^2 K_n\left(\frac{\delta r}{R}\right)\right\}B_4\right]\cos n\theta\exp[i\xi(x-ct)] \tag{9}$$

Where, d_r, d_θ, d_x components of displacement vector, $A_1; A_2; A_3$ are amplitudes of P, SV, SH waves respectively and $B_2 = B_2'/R$, $B_4 = B_4'/R^2$ and $B_6 = B_6'/R$. $B_1' \ldots\ldots B_6'$ are arbitrary constants. Stress field due to the incident wave can be obtained by plugging above equations into the stress-displacement relations of the medium, and is given by:

$$\sigma_{rr}^{(i)} = \frac{\mu}{R}\begin{bmatrix} \left\{(2\varepsilon_1^2 - \varepsilon_2^2) I_n\left(\frac{\gamma r}{R}\right) + 2\gamma^2 I_n''\left(\frac{\gamma r}{R}\right)\right\} B_1 \\ + \left\{-2i\beta_1\delta^2 I_n''\left(\frac{\gamma r}{R}\right)\right\} B_3 \\ + 2n\left(\frac{R}{r}\right)\left\{\delta I_n'\left(\frac{\delta r}{R}\right) - \left(\frac{R}{r}\right) I_n\left(\frac{\delta r}{R}\right)\right\} B_5 \end{bmatrix} \cos n\theta \exp\left[i\xi(x-ct)\right]$$

$$\sigma_{r\theta}^{(i)} = \frac{\mu}{R}\begin{bmatrix} 2n\left(\frac{R}{r}\right)\left\{\left(\frac{R}{r}\right) I_n\left(\frac{\gamma r}{R}\right) - \gamma I_n'\left(\frac{\gamma r}{R}\right)\right\} B_1 \\ + 2in\frac{R}{r}\beta_1\left\{\delta I_n'\left(\frac{\delta r}{R}\right) - \frac{R}{r} I_n\left(\frac{\delta r}{R}\right)\right\} B_3 \\ + \left\{-\delta^2 I_n''\left(\frac{\delta r}{R}\right) + \delta\left(\frac{R}{r}\right) I_n'\left(\frac{\delta r}{R}\right) - \left(\frac{nR}{r}\right)^2 I_n\left(\frac{\delta r}{R}\right)\right\} B_5 \end{bmatrix} \sin n\theta \exp\left[i\xi(x-ct)\right]$$

$$\sigma_{rx}^{(i)} = \frac{\mu}{R}\begin{bmatrix} \left\{2i\beta_1\gamma I_n'\left(\frac{\gamma r}{R}\right)\right\} B_1 + \left\{\delta(2\beta_1^2 - \varepsilon_2^2) I_n'\left(\frac{\delta r}{R}\right)\right\} \\ + \left\{in\left(\frac{R}{r}\right)\beta_1 I_n\left(\frac{\delta r}{R}\right)\right\} B_5 \end{bmatrix} \cos n\theta \exp\left[i\xi(x-ct)\right]$$

$$\sigma_{r\theta}^{(s)} = \frac{\mu}{R}\begin{bmatrix} \left\{(2\varepsilon_1^2 - \varepsilon_2^2) K_n\left(\frac{\gamma r}{R}\right) + 2\gamma^2 K_n''\left(\frac{\gamma r}{R}\right)\right\} B_2 + \\ \left\{-2i\beta_1\delta^2 K_n''\left(\frac{\gamma r}{R}\right)\right\} B_4 + \\ 2n\left(\frac{R}{r}\right)\left\{\delta K_n'\left(\frac{\delta r}{R}\right) - \left(\frac{R}{r}\right) K_n\left(\frac{\delta r}{R}\right)\right\} B_6 \end{bmatrix} \cos n\theta \exp\left[i\xi(x-ct)\right]$$

$$\sigma_{r\theta}^{(s)} = \frac{\mu}{R}\begin{bmatrix} 2n\left(\frac{R}{r}\right)\left\{\left(\frac{R}{r}\right) K_n\left(\frac{\gamma r}{R}\right) - \gamma K_n'\left(\frac{\gamma r}{R}\right)\right\} B_2 + \\ 2in\frac{R}{r}\beta_1\left\{\delta K_n'\left(\frac{\delta r}{R}\right) - \frac{R}{r} K_n\left(\frac{\delta r}{R}\right)\right\} B_4 + \\ \left\{-\delta^2 K_n''\left(\frac{\delta r}{R}\right) + \delta\left(\frac{R}{r}\right) K_n'\left(\frac{\delta r}{R}\right) - \left(\frac{nR}{r}\right)^2 K_n\left(\frac{\delta r}{R}\right)\right\} B_6 \end{bmatrix} \sin n\theta \exp\left[i\xi(x-ct)\right]$$

$$\sigma_{rx}^{(s)} = \frac{\mu}{R}\begin{bmatrix} \left\{2i\beta_1\gamma K_n'\left(\frac{\gamma r}{R}\right)\right\} B_2 + \left\{\delta(2\beta_1^2 - \varepsilon_2^2) K_n'\left(\frac{\delta r}{R}\right)\right\} B_4 \\ + \left\{in\left(\frac{R}{r}\right)\beta_1 K_n\left(\frac{\delta r}{R}\right)\right\} B_6 \end{bmatrix} \cos n\theta \exp\left[i\xi(x-ct)\right] \tag{10}$$

where, I_n ( ) are modified Bessel functions of first kind, J_n ( ) are Bessel function of first kind and K_n ( ) are modified Bessel functions of second kind

With the help of above equations, the stresses at the outer surface of the shell (z = h/2 or r = R + h/2) can be obtained. Thus $\{P^*\}$ in Equation (2) can be determined. For any disturbance propagating in the fluid governing linear acoustic equations are the continuity equation and the Euler equation of motion. These are given as follows:

$$\frac{\partial \rho_f}{\partial t} + \overline{\nabla}.(\overline{V_f}.\nabla)\overline{V_f} = \frac{1}{\rho_f}\overline{\nabla} p$$

Displacement d(r,θ, x, t) at any point, satisfied the equation of motion:

$$c_1^2 \underline{\nabla}(\underline{\nabla}.\underline{d}) - c_2^2 \underline{\nabla}\Lambda\underline{\nabla}\Lambda\underline{d} = \frac{\partial^2}{\partial t^2}(\underline{d}) \tag{11}$$

where, $c_1 = \left\{\frac{(\lambda + 2\mu)}{\rho_m}\right\}^{1/2}$ and $c_2 = \left\{\frac{\mu}{\rho_m}\right\}^{1/2}$ are the speeds of dilatational and shear waves respectively in the infinite medium. Further, λ and μ are the Lame's constant, and ρ_m is the density of the medium.

Now the mid plane displacement and slopes are assumed to be of the form:

$$w = w_0 \cos n\theta \exp[i\xi(x-ct)]$$

$$v = v_0 \sin n\theta \exp[i\xi(x-ct)]$$

$$u = u_0 \cos n\theta \exp[i\xi(x-ct)] \tag{12}$$

Plugging Equation (12) in Equation (2) and (11) along with the expression for {P*}, a set of three simultaneous algebraic equations were obtained. Four more equations were obtained by imposing the boundary conditions at the inner and outer surfaces of the shell, that is:

$$w = (d_r^{(i)} + d_r^{(s)})_{r=R+h/2}$$

$$v + (h/2)\Psi_\theta = (d_\theta^{(i)} + d_\theta^{(s)})_{r=R+h/2}$$

$$u + (h/2)\psi_x = (d_x^{(i)} + d_x^{(s)})_{r=R+h/2} \tag{13}$$

Boundary conditions at the outer surface of the shell (r = R + h/2) are obtained by assuming that the shell and the continuum are joined together by a bond which is thin, elastic and inertia less. This implies that the stress at the shell-soil interface is continuous. To take the elasticity of the bond into account, the stresses in the bond are assumed proportional to relative displacements between the shell and continuum. μ shear modulus of medium and ρ density of shell material

The inner surface of the shell continuity of the radial displacement had been assumed, that is,

$$\frac{\partial w}{\partial t} = \left[\frac{\partial d_r^f}{\partial t}\right]_{r=R-h/2}$$

$$(\sigma_{rx})_{r=R+h/2} = [(S_x + Z_x\frac{\partial}{\partial t})(\mu_x^i + \mu_x^s - u - (r-R)\psi_x]_{r=R+h/2}$$

$$(\sigma_{rr})_{r=R+h/2} = [(S_r + Z_r \frac{\partial}{\partial t})(\mu_r^i + \mu_r^s - w)]_{r=R+h/2}$$

$$(\sigma_{r\theta})_{r=R+h/2} = [(S_\theta + Z_\theta \frac{\partial}{\partial t})(\mu_\theta^i + \mu_\theta^s - u(r-R)\psi_\theta]_{r=R+h/2}$$

(14)

Where, $\zeta_R = \frac{\mu}{S_r.R}$, $\zeta_\theta = \frac{\mu}{S_\theta.R}$, and $\zeta_x = \frac{\mu}{S_x.R}$, are the non stiffness coefficient of the bond in radial, and axial direction, respectively; $\Gamma_r = \frac{\mu}{Z_r c_1}$, $\Gamma_\theta = \frac{\mu}{Z_\theta c_1}$ and $\Gamma_x = \frac{\mu}{Z_x c_1}$ are the non damping coefficient of the bond in radial, tangential and axial direction, respectively.

Thus, in-all seven algebraic equations are obtained. These seven equations when simplified give the final dynamic response equation, which may be put into the form

$$\{Q\}\{U_0\} = B_1\{F^1\} + B_3\{F^2\} + B_5\{F^3\}$$

(15)

Where [Q] is a (7×7)) matrix and $\{F^1\}$, $\{F^2\}$ and $\{F^3\}$ are (7×1) matrices. But for the response of longitudinal wave, the amplitudes due to shear waves B_3 and B_5 would be zero so the effect of $\{F^2\}$ and $\{F^3\}$ matrices would be eliminated. After putting values of $B_3 = B_5 = 0$ and substituting values of B_1 from Eq Equation (8), Equation (15) becomes

$$\{Q\}\{U_0\} = (-1)^{n+1}\left(i\chi \frac{A_1}{\varepsilon_1}\right)\{F^1\}$$

(16)

Now if the unknown matrix $\{U_0\}$ is non-dimensionalized with respect to the amplitude of the incident wave (A_1), the elements of above Q and F matrix are as follows:

$$Q_{11} = \frac{\bar{h}^3}{12\eta_2}[n^4\eta_1 N + 4n^2\beta_1^2\eta_3 + N\beta_1^4 + \eta_1 N - 2n^2\eta_1 N + 2\nu_{\theta x} Nn^2\beta_1^2] + \frac{\bar{h}\eta_1 N}{\eta_2} - \Omega^2;$$

$$Q_{12} = \frac{\bar{h}^3}{12\eta_2}[-n^3\eta_1 N + n\beta_1^2\eta_3 + \eta_1 Nn] + \frac{\bar{h}\eta_1 Nn}{\eta_2};$$

$$Q_{13} = \frac{i\bar{h}^3}{12\eta_2}[-n^2\eta_3\beta_1 + N\beta_1^3] + \frac{i\bar{h}\nu_{\theta x} Nn\beta_1}{\eta_2};$$

$$Q_{14} = -\left(1 + \frac{\bar{h}}{2}\right)\bar{\mu}\left[\left(2\in_1^2 - \in_2^2\right)K_n(\alpha_1) + 2\gamma^2 K_n''(\alpha_1)\right];$$

$$Q_{15} = \left(1 + \frac{\bar{h}}{2}\right)\bar{\mu}\left[2i\beta_1\delta^2 K_n''(\alpha_2)\right];$$

$$Q_{16} = -\left(1 + \frac{\bar{h}}{2}\right)\bar{\mu}\left[\frac{2n\{\alpha_2 K_n'(\alpha_2) - K_n(\alpha_2)\}}{\left(\frac{1+\bar{h}}{2}\right)^2}\right];$$

$$Q_{21} = -Q_{12};$$

$$Q_{22} = \frac{\bar{h}^3}{12\eta_2}[-n^2\eta_1 N] + \frac{\bar{h}}{\eta_2}[-n^2\eta_1 N - \beta_1^2\eta_3] + \Omega^2$$

$$Q_{23} = -\frac{i\bar{h}}{\eta_2}[-n\eta_3\beta_1 - 2\nu_{\theta x} Nn\beta_1];$$

$$Q_{24} = \left(1 + \frac{\bar{h}}{2}\right)\bar{\mu}\left[2n\frac{\{K_n(\alpha_1) - \alpha_1 K_n'(\alpha_1)\}}{\left(1 + \frac{\bar{h}}{2}\right)^2}\right];$$

$$Q_{25} = \left(1 + \frac{\bar{h}}{2}\right)\bar{\mu}\left[2in\beta_1\frac{\{\alpha_2 K_n'(\alpha_2) - K_n(\alpha_2)\}}{\left(1 + \frac{\bar{h}}{2}\right)^2}\right];$$

$$Q_{26} = \left(1 + \frac{\bar{h}}{2}\right)\bar{\mu}\left[2n\left\{\frac{\delta}{\left(1 + \frac{\bar{h}}{2}\right)^2}\right\}K_n'(\alpha_2) - \delta^2 K_n''(\alpha_2)\left\{\frac{n^2}{\left(1 + \frac{\bar{h}}{2}\right)^2}\right\}K_n(\alpha_2)\right];$$

$$Q_{31} = Q_{13};$$

$$Q_{32} = -Q_{23};$$

$$Q_{33} = \frac{\bar{h}}{\eta_2}[-N\beta_1^3 - n_3 n^2(\frac{1+\bar{h}^2}{12})] + \Omega^2;$$

$$Q_{34} = \left(1 + \frac{\bar{h}}{2}\right)\bar{\mu}\left[2i\beta_1\gamma K_n'(\alpha_1)\right];$$

$$Q_{35} = \left(1 + \frac{\bar{h}}{2}\right)\bar{\mu}\left[\delta\left(2\beta_1^2 - \in_2^2\right)K_n'(\alpha_2)\right];$$

$$Q_{36} = \left(1+\frac{\bar{h}}{2}\right)\bar{\mu}\left[\frac{in\beta_1 K_n(\alpha_2)}{\left(1+\frac{\bar{h}}{2}\right)}\right];$$

$Q_{41} = 1, Q_{42} = Q_{43} = 0,$

$$Q_{44} = -\gamma K_n'(\alpha_1), Q_{45} = -i\beta_1 \delta K_n'(\alpha_2),$$

$$Q_{46} = \frac{-n K_n(\alpha_1)}{\left(1+\frac{\bar{h}}{2}\right)};\ Q1_{51} = 0,\ Q_{52} = Q_{53} = 1;$$

$$Q_{54} = \frac{n K_n(\alpha_1)}{\left(1+\frac{\bar{h}}{2}\right)};\quad Q_{55} = \frac{-in\beta_1 K_n(\alpha_2)}{\left(1+\frac{\bar{h}}{2}\right)};$$

$$Q_{56} = \delta K_n'(\alpha_2);\ Q_{57} = Q_{47},\ ; Q_{58} = Q_{48}\ ; Q_{59} = 0;$$

$Q_{61} = Q_{62} = Q_{63} = 0$;

$$Q_{64} = -i\beta_1 K_n(\alpha_1);\ Q_{65} = -\delta^2 K_n(\alpha_2);$$

$$Q_{66} = -\gamma K_n'(\alpha_1) + \frac{\zeta_r\Gamma_r}{\Gamma_r - i\varepsilon_1\zeta_r}[(2\varepsilon_1^2 - \varepsilon_2^2)K_n(\alpha_1) + 2\gamma K_n''(\alpha_1)]$$

$$Q_{67} = i\beta\delta K_n'(\alpha_2) - \frac{\zeta_r\Gamma_r}{\Gamma_r - i\varepsilon_1\zeta_r}[(2i\beta\delta^2 K_n''(\alpha_2))]$$

$$Q_{68} = \{-nK_n(\alpha_2)/(1+\bar{h}/2)\} + \frac{\zeta_r\Gamma_r}{\Gamma_r - i\varepsilon_1\zeta_r}[(2n\{\alpha_2 K_n'(\alpha_2) - K_n(\alpha_2)\}/(1+\bar{h}/2)^2]$$

$$Q_{69} = 0, Q_{71} = 0, Q_{72} = Q_{73} = 1, Q_{74} = Q_{75} = 0$$

$$Q_{76} = \{nK_n(\alpha_1)/(1+\bar{h}/2)\} + \frac{\zeta_\theta\Gamma_\theta}{\Gamma_\theta - i\varepsilon_1\zeta_\theta}[(2n\{K_n(\alpha_1) - \alpha_1 K_n'(\alpha_1)\}/(1+\bar{h}/2)^2]$$

$$Q_{77} = \{-in\beta K_n(\alpha_2)/(1+\bar{h}/2)\}$$

$$F_1^1 = \left(1+\frac{\bar{h}}{2}\right)\bar{\mu}\left[\left(2\epsilon_1^2 - \epsilon_2^2\right)I_n(\alpha_1) + 2\gamma^2 I_n''(\alpha_1)\right];$$

$$F_2^1 = -\left(1+\frac{\bar{h}}{2}\right)\bar{\mu}\left[\frac{2n\{I_n(\alpha_1) - \alpha_1 I_n'(\alpha_1)\}}{\left(1+\frac{\bar{h}}{2}\right)^2}\right];$$

$$F_3^1 = -\left(1+\frac{\bar{h}}{2}\right)\bar{\mu}\,[2i\beta_1\gamma I_n'(\alpha_1)];$$

$$F_4^1 = \gamma I_n'(\alpha_1),\quad F_5^1 = \frac{-n I_n(\alpha_1)}{\left(1+\frac{\bar{h}}{2}\right)};$$

$$F_6^1 = -\{\gamma I_n'(\alpha_1) - \frac{\zeta_r\Gamma_r}{\Gamma_r - i\varepsilon_1\zeta_r}[2\varepsilon_1^2 - \varepsilon_2^2)\{I_n(\alpha_1) + 2\gamma^2 I_n''(\alpha_1)\}]$$

$$F_7^1 = -\{nI_n(\alpha_1)/(1+\bar{h}/2] - \frac{\zeta_\theta\Gamma_\theta}{\Gamma_\theta - i\varepsilon_1\zeta_\theta}[2n\{I_n(\alpha_1) - \alpha_1 I_n'(\alpha_1)\}/(1+\bar{h}/2)^2]$$

$$\{U\} = \left\lfloor \frac{w_0}{A_1}\ \frac{v_0}{A_1}\ \frac{u_0}{A_1}\ \frac{B_2}{A_1}\ \frac{B_4}{A_1}\ \frac{B_6}{A_1}\ \frac{B_f}{c_1} \right\rfloor^T = \left\lfloor \bar{W}\ \bar{V}\ \bar{U}\ \bar{B_2}\ \bar{B_4}\ \bar{B_6}\bar{B_f} \right\rfloor^T;$$

where

$$\bar{h} = \frac{h}{R},\ \eta_1 = \frac{E_\theta}{E_x},\ \eta_2 = \frac{G_{xz}}{E_x},\ \eta_3 = \frac{G_{x\theta}}{E_x},$$

$$\eta_4 = \frac{G_{z\theta}}{E_x},\ N = \frac{1}{\left(1 - \nu_{x\theta}\nu_{\theta x}\right)},\ \bar{\mu} = \frac{\mu}{G_{xz}},$$

$$\Omega^2 = \bar{h}\,\bar{\mu}\frac{\epsilon_2^2}{\bar{\rho}},\ \bar{\rho} = \frac{\rho_m}{\rho},\ \alpha_1 = \left(1+\frac{\bar{h}}{2}\right)\gamma_a$$

nd

$$\alpha_2 = \left(1+\frac{\bar{h}}{2}\right)\delta.$$

$$I_n'(\alpha_i),\ I_n''(\alpha_i),\ K_n'(\alpha_i)\ and\ K_n''(\alpha_i)$$

can be expressed as

$$I_n'(\alpha_i) = \left(\frac{n}{\alpha_i}\right)I_n(\alpha_i) + I_{n+1}(\alpha_i),$$

$$I_n''(\alpha_i) = \left[1 + \left(\frac{n^2}{\alpha_i^2}\right) - \left(\frac{n}{\alpha_i^2}\right)\right]I_n(\alpha_i) - \left(\frac{1}{\alpha_i}\right)I_{n+1}(\alpha_i),$$

$$K_n'(\alpha_i) = \left(\frac{n}{\alpha_i}\right)K_n(\alpha_i) - K_{n+1}(\alpha_i),$$

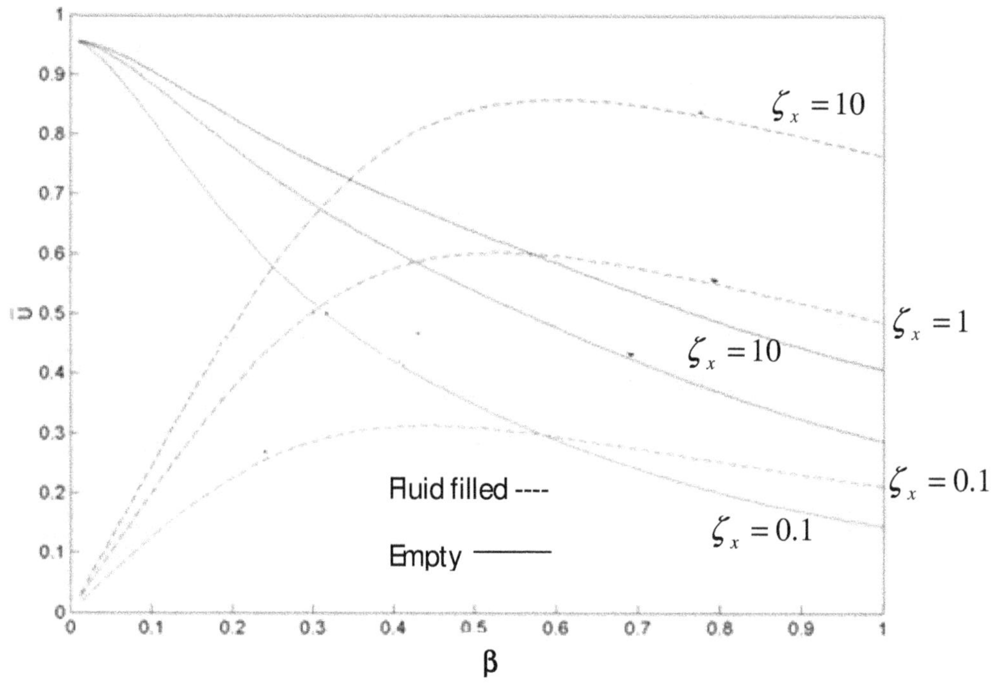

$\phi = 5°, \bar{\mu} = 0.01, \zeta_x = 0.1,1,10$

Figure 2. Axial displacement $(\bar{U})$ vs. wave number (β) with soil stiffness ζ_x as parameter.

$$K_n''(\alpha_i) = \left[1 + \left(\frac{n^2}{\alpha_i^2}\right) - \left(\frac{n}{\alpha_i^2}\right)\right] K_n(\alpha_i) + \left(\frac{1}{\alpha_i}\right) K_{n+1}(\alpha_i).$$

Here it must be pointed out that for an incident P-wave, strain $\in_1$ = β (a non-dimensional wave number of incident wave). Whereas, for an incident shear wave (SV-wave or SH-wave) $\in_2$ = β. In the present work, the non-dimensional wave number of the incident wave, that is, β (= 2π R/Λ) has been given as input, so either $\in_1$ or $\in_2$ is always known. The other $\in$ can be obtained by using the following relation:

$$\left(\frac{\in_2}{\in_1}\right)^2 = \frac{c_1^2}{c_2^2} = \frac{2(1 - \nu_m)}{(1 - 2\nu_m)} \qquad (17)$$

where ν_m is the Poisson ratio of the medium.

RESULTS AND DISCUSSION

Results are presented for a transversely isotropic shell with r-θ as the plane of isotropy. Consequently $E_\theta = E_z, G_{xz} = G_{x\theta}, \nu_{x\theta} = \nu_{xz}$, $\nu_{\theta z} = \nu_{z\theta}$, $G_{z\theta} = E_\theta / 2(1+\nu_{\theta z})$. Thus we have $\eta_3 = \eta_2$ and $\eta_4 = G_{z\theta}/E_x = \eta_1 / 2(1+\nu_{\theta z})$. In addition $\nu_{\theta Z} = \nu_{x\theta} = 0.3$ has been taken in the numerical calculations. Different values of shell orthotropy parameters η_1 and η_2 are used as 0.5, 0.01, 0.05 and 0.1, 0.05, 0.02, respectively. Soil parameter $\bar{\mu}$ had been varied from 0.1 to 10.0 to take into account different soil conditions around the pipe, representing soft to hard soil. For all the values of $\bar{\mu}$, ν_m = 0.25 had been assumed. Thickness to radius ratio of the shell ($\bar{h}$) had been taken as 0.01 and the density ratio of the surrounding medium to the shell ($\bar{\rho}$) had been taken as 0.75. Non-dimensional amplitude of the middle surface of the shell in the radial and axial directions ($\bar{W}$ and $\bar{U}$) have been plotted against the non-dimensional wave number of the incident P-wave (β=2πR/Λ). The shell response had been shown for empty and fluid filled shell for non-axisymmetric mode (flexural mode, n = 1) taking stiffness coefficients (ζ_x ζ_θ ζ_r) and damping coefficients (Γ_x Γ_θ Γ_r) as parameters. Figures 2 to 4 shows the effect of stiffness coefficient ζ_x on axial displacement $\bar{U}$ of the shell for soft, medium and hard types of soil respectively. At small angle of incident wave and for soft

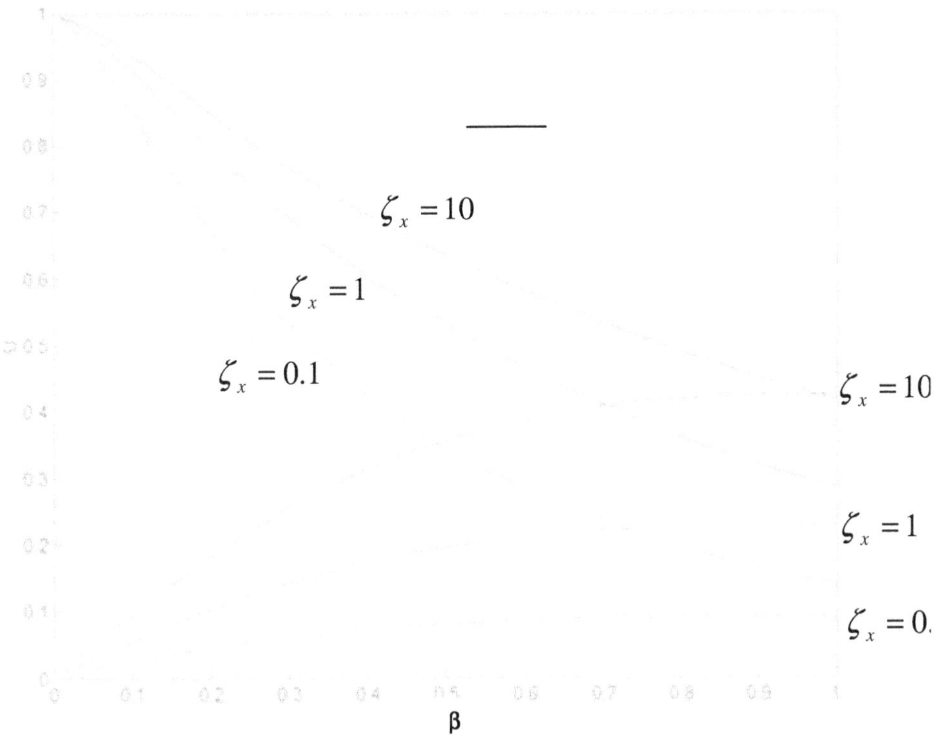

$\phi = 5^\circ, \overline{\mu} = 0.1, \zeta_x = 0.1,1,10$

Figure 3. Axial displacement $(\overline{U})$ vs. wave number (β) with ζ_x as parameter.

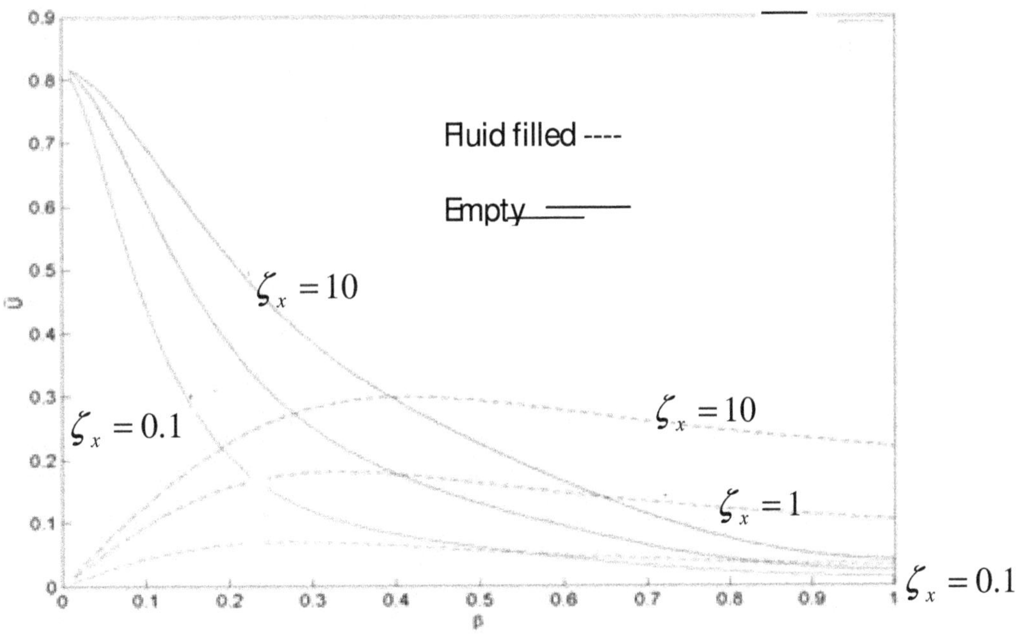

$\phi = 60^\circ, \overline{\mu} = 10, \zeta_x = 0.1,1,10$

Figure 4. Axial displacement $(\overline{\mathrm{U}})$ vs. wave number (β) with ζ_x as parameter.

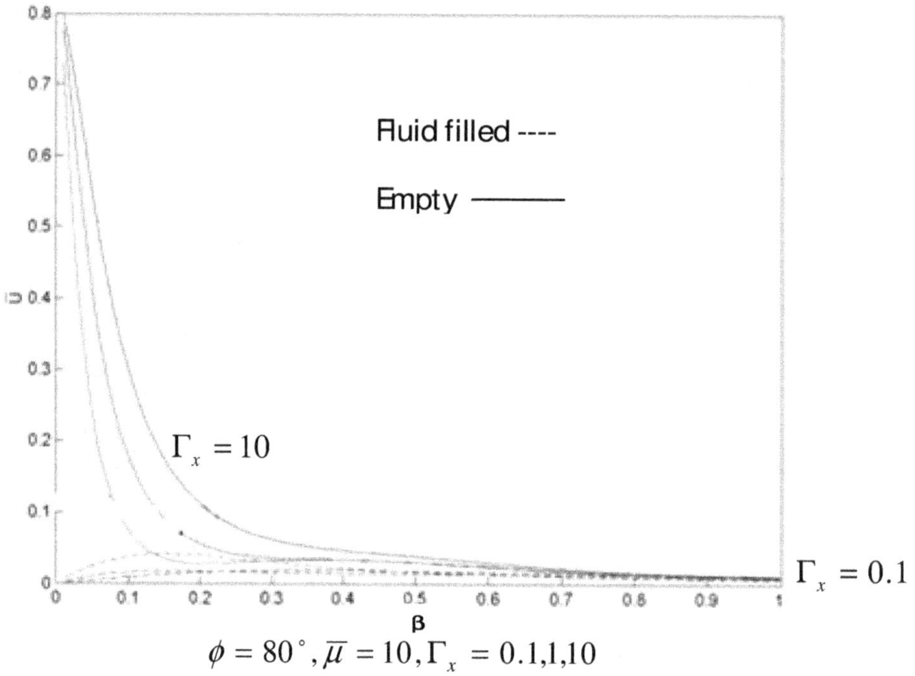

Figure 5. Axial displacement $(\overline{U})$ vs. wave number (β) with Γ_x as parameter.

soil, the effect of soil stiffness ζ_x is more in fluid filled shell as compare to empty shell, but at higher angle of incident wave and for hard soil, the effect of ζ_x on axial displacement is more in fluid filled shell as compare to empty shell as shown in Figure 4.

Figures 5 to 7 shows the effect of damping coefficient Γ_x on axial displacement $\overline{U}$ of the shell. At small angle of incident of the wave number and for soft soil the effect of Γ_x is more in fluid filled shell as compare to empty shell, but at higher angle of incident of the wave number and for hard soil the effect of Γ_x is more in fluid filled shell as compare to empty shell. Figure 5 shows that at higher wave number with higher angle of incidence under hard soil condition, the axial displacement is negligible both in the case of empty shell as well as fluid filled shell. The axial displacement is significant in fluid filled shell as compared to empty shell buried under soft soil.

Figures 8 to 9 shows the effect of stiffness coefficient ζ_x on radial displacement $\overline{W}$ of the shell with increasing wave number under different soil conditions. The radial displacement of fluid filled shell, first decreases then increases with increasing value of wave number. A reverse phenomenon can be seen in case of empty shell. Under imperfect bond conditions, radial displacement in empty shell is more predominant.

Figures 11 to 13 show the effect of damping coefficient Γ_x on radial displacement of the shell $\overline{W}$. As wave number increases radial displacement first decreases then increases with increasing value of Γ_r in medium soil in case of fluid filled shell but trend is reversed in empty shell at higher incidence angle.

Figure 14 shows the effect of orthotropy parameter η_2 on axial displacement of the shell with soil stiffness as another variable. Results show that orthotropy parameter η_2 has negligible effect on the response in the case of fluid filled and empty shell.

Figures 15 and 16 shows the effect of density of fluid on radial and axial displacement of the buried thin shell, respectively. Fluid density had been taken as variable and its value has been varied from 0.13 to 0.66. Results show that with increasing density of the fluid, radial displacement increases and axial displacement decreases.

Conclusions

To study the effects of the fluid presence on the thin shell displacement, under different soil condition at various angles of incidence of the longitudinal wave under imperfect bonding, parametric results in graphical

$$\phi = 5°, \overline{\mu} = 0.01, \Gamma_x = 0.1, 1, 10$$

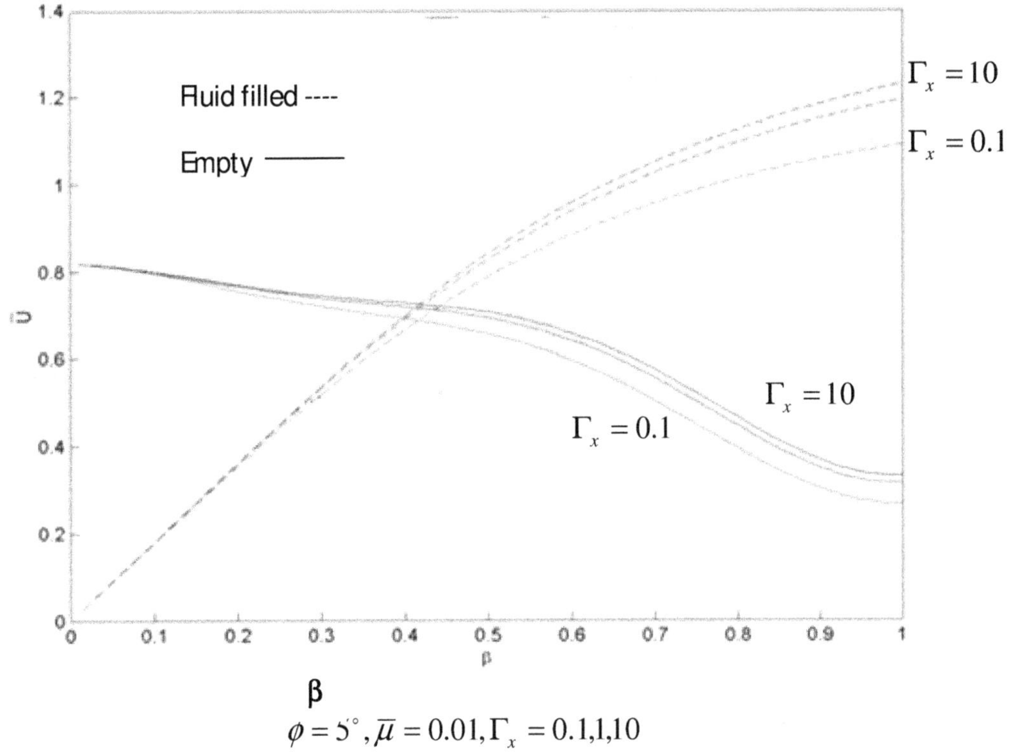

Figure 6. Axial displacement $(\overline{U})$ vs. wave number (β) with Γ_x as parameter.

$$\phi = 60°, \overline{\mu} = 0.1, \Gamma_x = 0.1, 1, 10$$

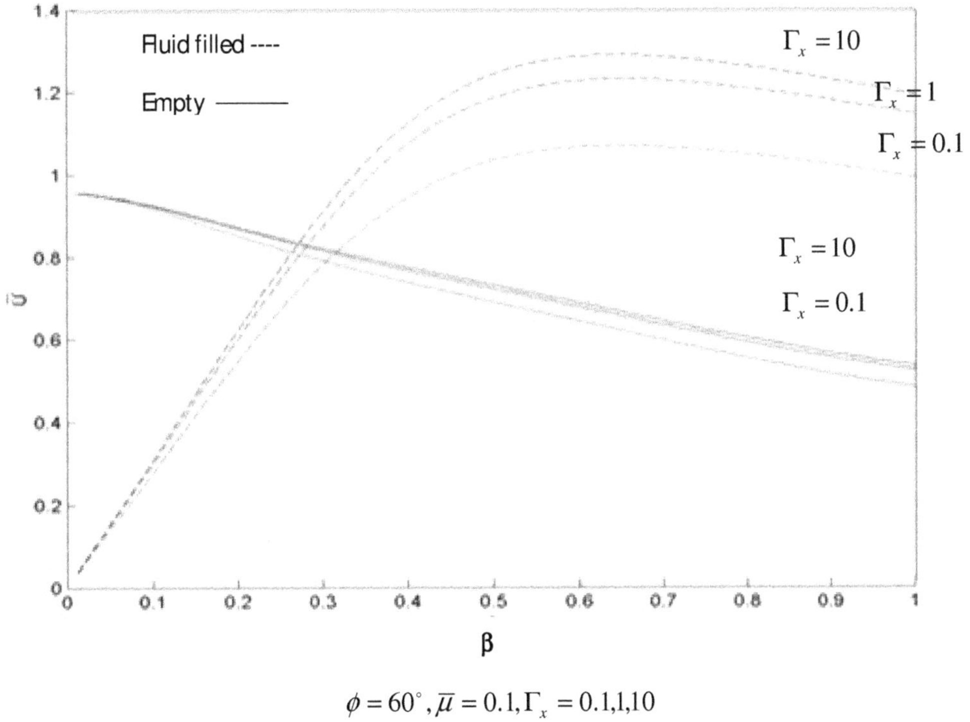

Figure 7. Axial displacement $(\overline{U})$ vs. wave number (β) with Γ_x as parameter.

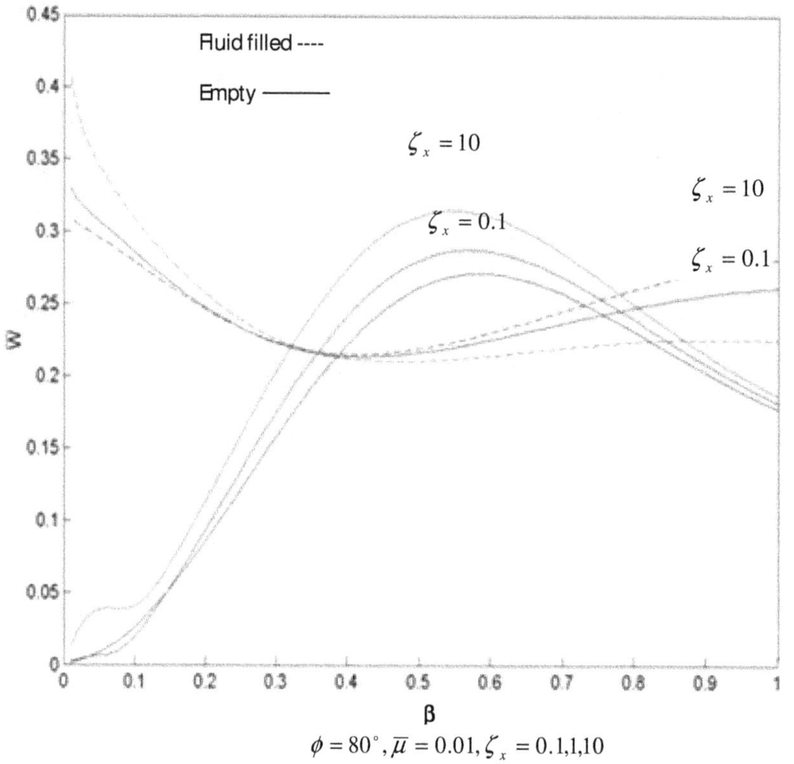

Figure 8. Radial displacement $(\overline{W})$ vs. wave number (β) with ζ_r as parameter.

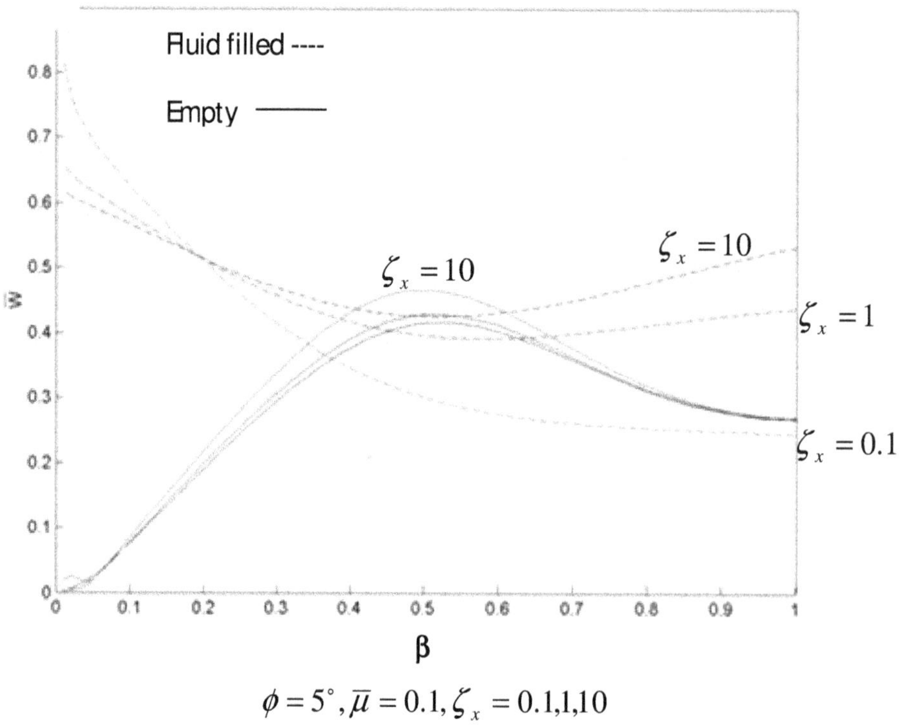

Figure 9. Radial displacement $(\overline{W})$ vs. wave number (β) with ζ_r as parameter.

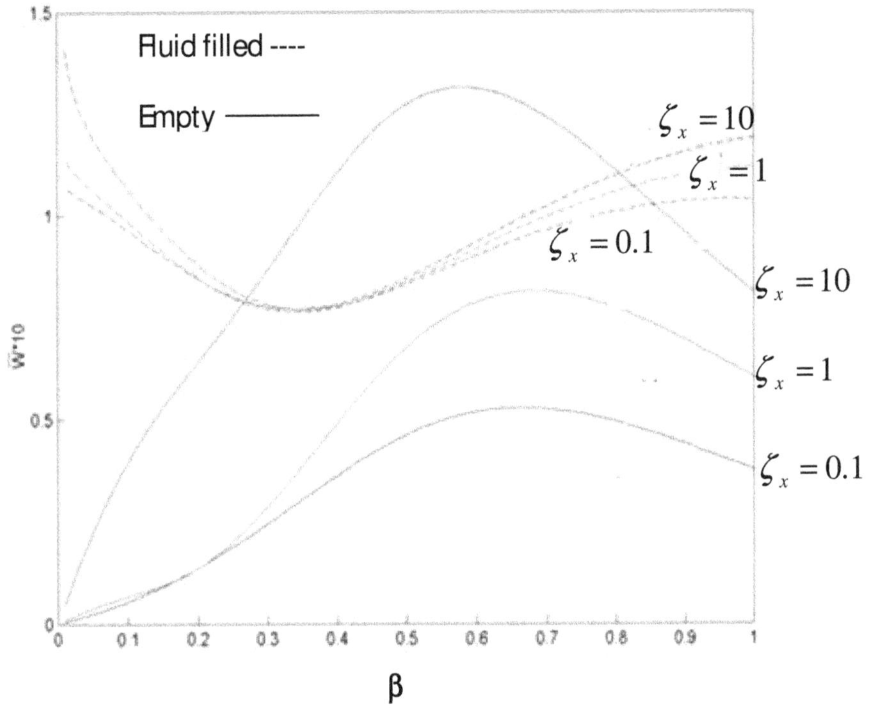

$\phi = 60^\circ, \overline{\mu} = 10, \zeta_x = 0.1,1,10$

Figure 10. Radial displacement $(\overline{W})$ vs. wave number (β) with ζ_r as parameter.

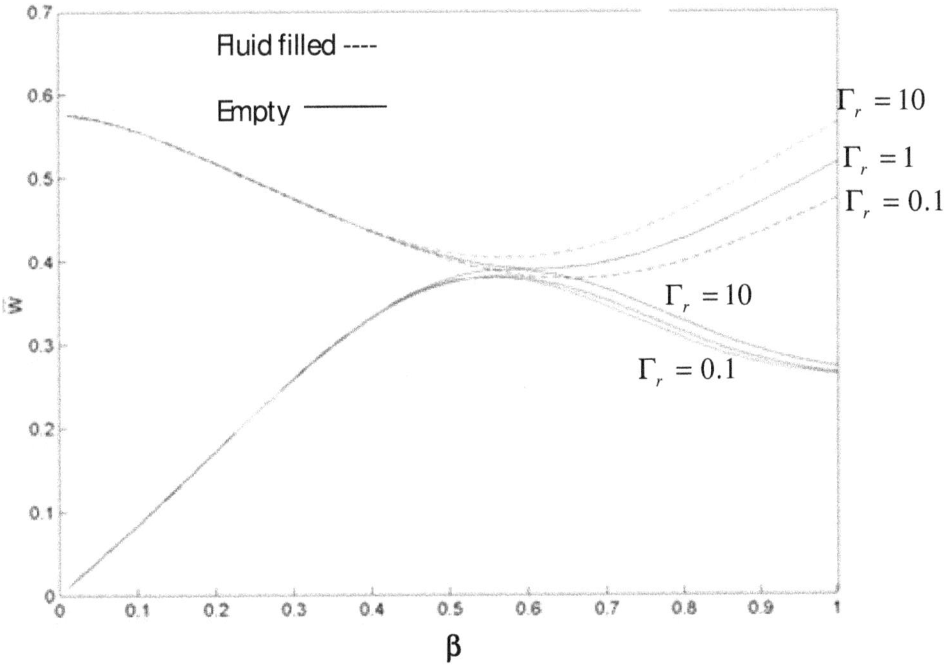

$\phi = 80^\circ, \overline{\mu} = 0.1, \Gamma_r = 0.1,1,10$

Figure 11. Radial displacement $(\overline{W})$ vs. wave number (β) with Γ_r as parameter.

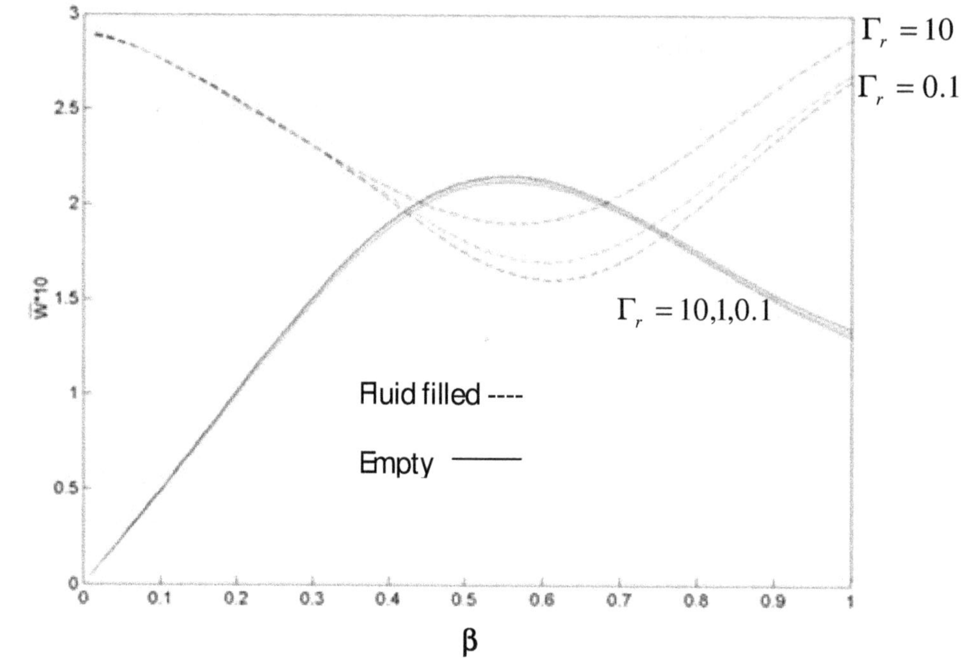

$\phi = 5^\circ, \bar{\mu} = 0.01, \Gamma_r = 0.1,1,10$

Figure 12. Radial displacement $(\overline{W})$ vs. wave number (β) with Γ_r as parameter.

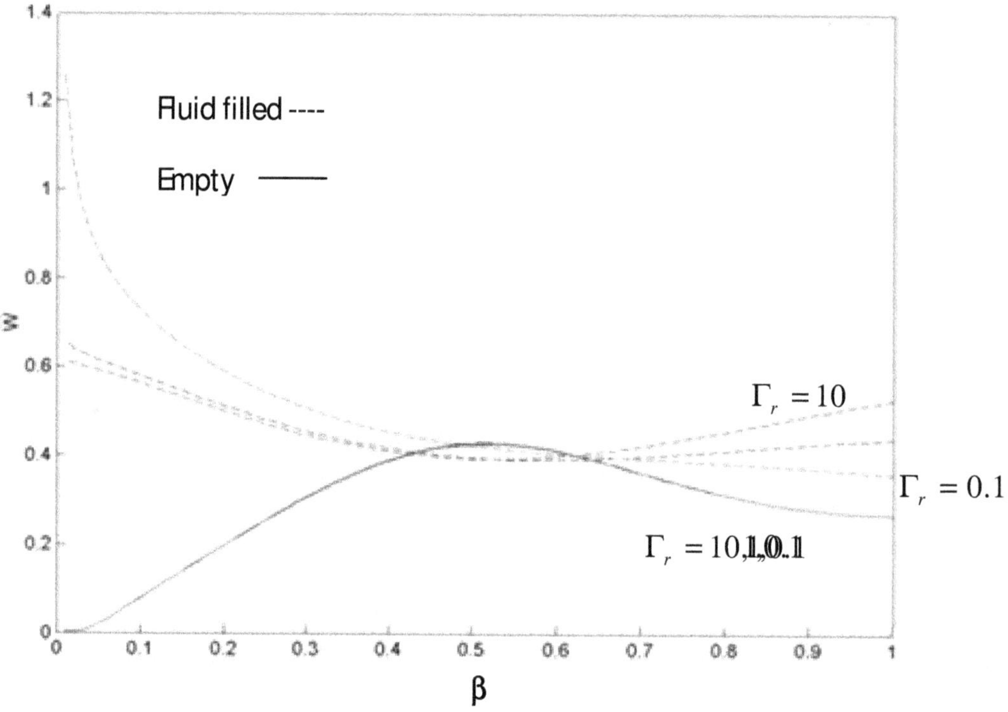

$\phi = 80^\circ, \bar{\mu} = 10, \Gamma_r = 0.1,1,10$

Figure 13. Radial displacement $(\overline{W})$ vs. wave number (β) with Γ_r as parameter.

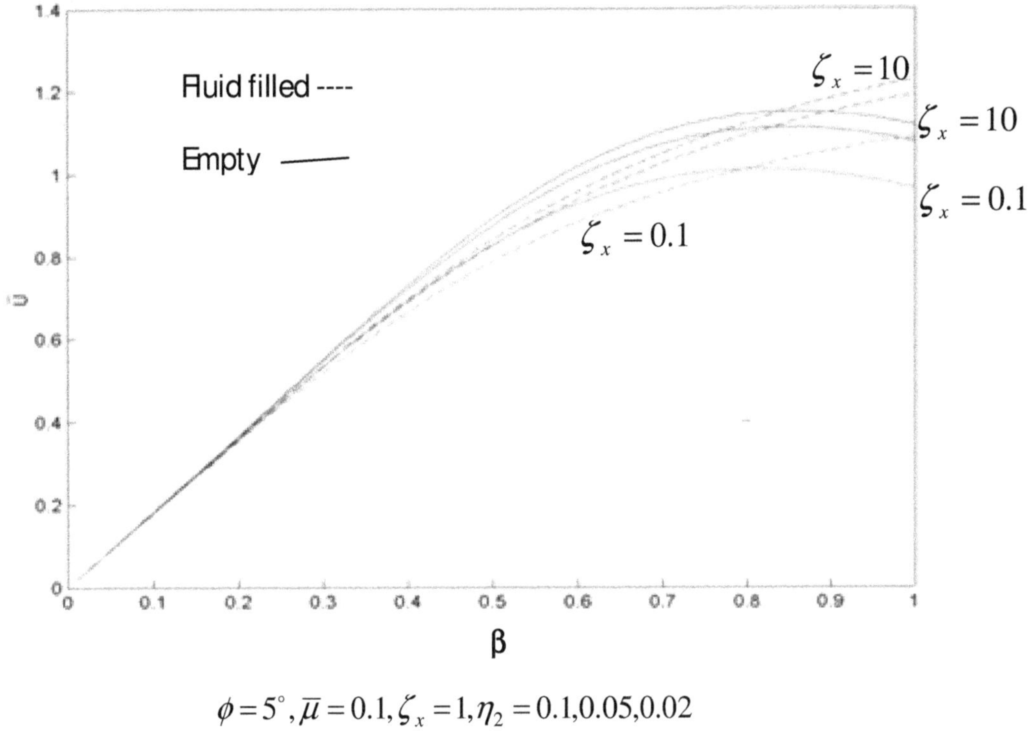

Figure14. Axial displacement $(\overline{U})$ vs. wave number (β) with η_2 as parameter.

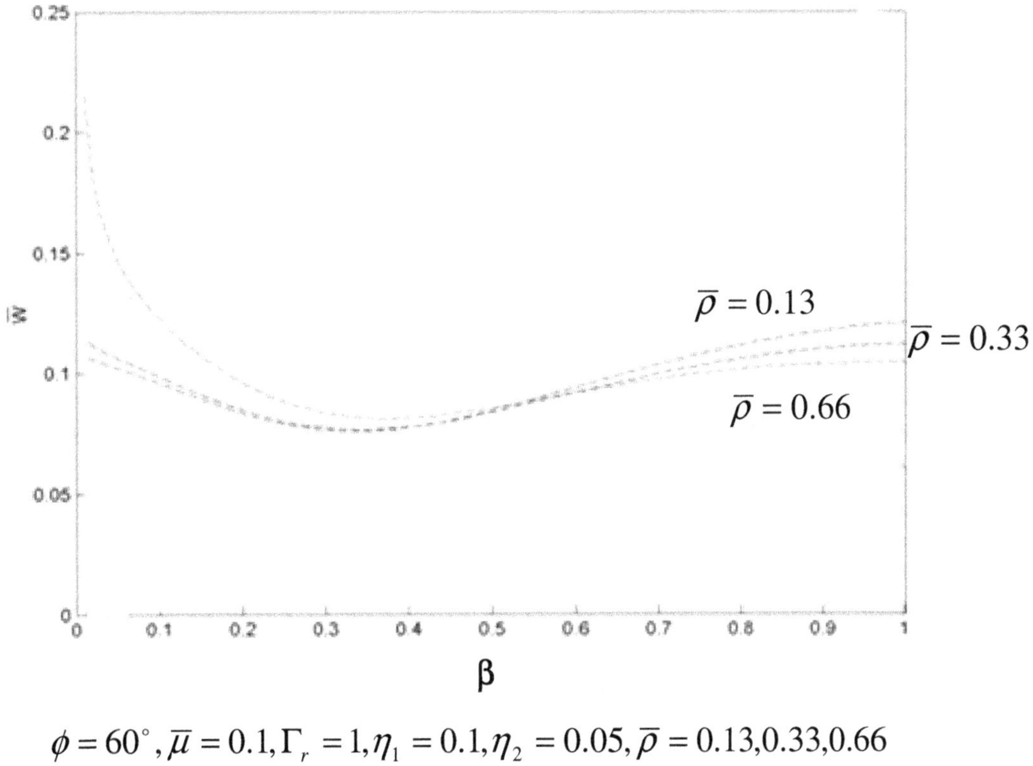

Figure 15. Radial displacement $(\overline{W})$ vs. wave number (β) with fluid density ($\bar{\rho}$) as parameter.

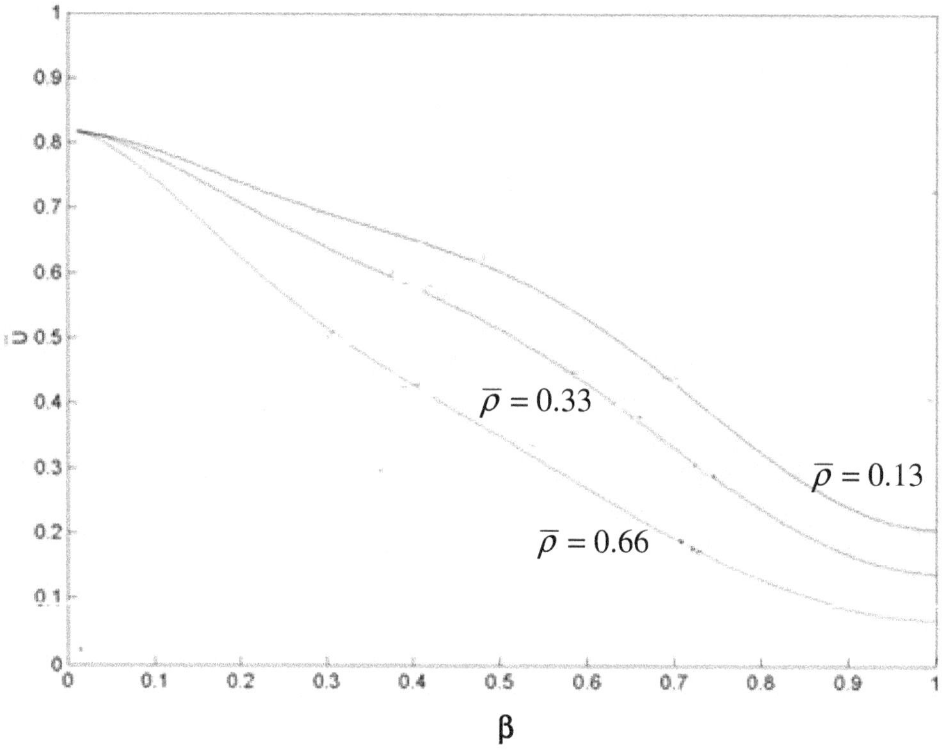

$$\phi = 5^\circ, \bar{\mu} = 0.1, \zeta_x = 1, \eta_1 = 0.1, \eta_2 = 0.05, \bar{\rho} = 0.13, 0.33, 0.66$$

Figure 16. Axial displacement $(\overline{U})$ vs. wave number (β) with fluid density ($\bar{\rho}$) as parameter.

form have been generated. Based on the results presented, following general conclusions could be drawn:

1. It is found that magnitude of the response of fluid filled pipeline can become even more than that of an empty pipeline, and hence, it cannot be assumed that a fluid filled pipeline will always furnish safe and conservative response.
2. Both the shell orthotropic parameters influence the radial displacement equally well but η_2 has a stronger influence on the axial displacement than η_1.
3. The density of the fluid becomes the important parameters in determining the shell response if incident longitudinal wave is of smaller wavelength.
4. The fluid filled pipeline response assumes considerable importance in soft soil condition and at higher apparent wave speed.
5. The fluid filled pipeline response due to incident longitudinal wave is significant only at large angle of incidence. Its response effect is small in hard shell.
6. For large angle of incidence, radial deflection is higher in fluid filled pipe as compared to empty shell. Thus for larger wavelength, empty pipe response is more important because the most common cause of pipeline failure is excessive axial deformation, while at smaller wavelength the fluid filled pipe has much importance for axial displacement.
7. Axial deflection and radial deflection both increase when the value of bonding parameter stiffness coefficient (ζ_x ζ_θ ζ_r) and damping coefficient ($\Gamma_x\,\Gamma_\theta\,\Gamma_r$) increase from zero to infinity (perfect to imperfect bonding) as variable.
8. The presence of fluid inside the shell, in general affects the radial displacement of the shell much more than the axial displacement, and in certain cases the change in radial displacement due to fluid presence is more prominent than that realized by variation of the bond parameter.

REFERENCES

Chonan S (1981). Response of a pre-stressed, orthotropic thick cylindrical shell subjected to pressure pulses. J. Sound Vib., 78: 257-267.

Cole BW, Ritter CJ, Jordon S (1979). Structural analysis of buried reinforced plastic mortar pipe. Lifeline earthquake engineering- buried pipelines, seismic risk and instrumentation. T. Ariman, S. C. Liu and R. E. Nickell, eds., ASME.

Dwivedi JP, Mishra BK, Upadhyay PC (1992). Non-axisymmetric dynamic response of imperfectly bonded buried orthotropic pipeline due to an incident shear wave. J. Sound Vib., 157(1): 81-92.

Dwivedi JP, Mishra BK, Upadhyay PC (1992b). Non-axisymmetric dynamic response of imperfectly bonded buried orthotropic pipelines due to incident shear wave (SH-wave). J. Sound Vib., 157(1): 177-182.

Dwivedi JP, Mishra BK, Upadhyay PC (1998). Non-axisymmetric dynamic response of imperfectly bonded buried orthotropic pipelines. Struct. Engg. Mechanics, 6(3): 291-304.

Dwivedi JP, Singh VP, Upadhyay PC (1991). Effect of fluid presence on the dynamic response of imperfectly bonded cylindrical shells due to incident shear-wave excitation. Comput. Struct., 40(4): 995-1001.

Dwivedi JP, Singh VP, Upadhyay PC (1993a). Non-axisymmetric dynamic response of imperfectly bonded buried fluid-filled orthotropic cylindrical shells due to incident shear wave. J. Sound Vib., 167(2): 277-287.

Dwivedi JP, Singh VP, Upadhyay PC (1993b). Effect of fluid presence on the non-axisymmetric dynamic response of imperfectly bonded buried orthotropic pipelines due to incident shear wave. Comput. Struct., 48(2): 219-226.

Dwivedi JP, Singh VP, Upadhyay PC (1996). Non-axisymmetric dynamic response of imperfectly bonded buried fluid-filled orthotropic cylindrical shells. ASME J. Press Vessel Technol., 118(1): 64-73.

Dwivedi JP, Upadhyay PC (1989). Effect of imperfect bonding on the axisymmetric dynamic response of buried orthotropic cylindrical shells. J. Sound Vib., 135: 477-486.

Dwivedi JP, Upadhyay PC (1990). Effect of fluid presence on the dynamic response of imperfectly bonded buried orthotropic cylindrical shells. J. Sound Vib., 139(2): 99-110.

Dwivedi JP, Upadhyay PC (1991). Effect of imperfect bond on the dynamic response of buried orthotropic cylindrical shells under shear-wave excitation. J. Sound Vib., 145(2): 333-337.

Hasheminajad SM, Kazemirad S (2008). Dynamic response of an eccentrically lined circular tunnel in poroelastic soil under seismic excitation. Soil Dyn. Earthquake Eng., 28: 277-292.

Herrman G, Mirsky J (1957). Non axially Symmetric Motion of Cylindrical Shell. J. Acoustical Soc. Am., 29(10): 1116-1123.

Kouretzis GP, Bouckovalas GD, Gantes CJ (2007). Analytical calculation of blast-induced strains on buried pipe lines. Int. J. Impact Eng., 34: 1683-1704.

Lee DH, Kim BH, Lee H and Kong JS (2009). Seismic behavior of buried gas pipelines under earthquake excitation. Eng. Struct., 31: 1011-1023.

Rajput RS, Sunil K, Chaubey A, Dwivedi JP (2010). Comparison of Non-Axisymmetric Dynamic Response of Imperfectly Bonded Buried Orthotropic Thick and Thin Fluid filled Cylindrical Shell due to Incident Shear Wave (SH Wave). IJEST, 2: 5845-5871.

Rajput RS, Sunil K, Chaubey A, Dwivedi JP (2010). Non-Axisymmetric Dynamic Response of Imperfectly Bonded Buried Orthotropic Thin Fluid Empty Cylindrical Shell due to Incident Compressional Wave. EJSR, 10: 443-464.

Singh VP, Upadhyay PC, Kishore B (1987). On the dynamic response of Buried Orthotropic Cylindrical Shells. J. Sound Vib., 113: 101-115.

Upadhyay PC, Mishra BK (1988). Non Axisymmetric Dynamic Response of buried Orthotropic shells. J. Sound Vib., 121: 149 -160.

3

An integral treatment for combined heat and mass transfer by mixed convection along vertical surface in a saturated porous medium

V. J. Bansod* and B. Ambedkar

Department of Mathematics, Technological University, Lonere India.

Combined free and forced convection heat and mass transfer over an impermeable vertical surface embedded in a saturated porous medium is considered. A similarity transformation has been used to study the mixed convection boundary-layer flow over a semi-infinite vertical flat plate. Integral solutions are derived for the coupled nonlinear similarity equations of coupled heat and mass transfer in porous media for the case where the free stream velocity and temperature and concentration near the wall are kept constant. The governing parameters for the problem under consideration are the Lewis number Le, the buoyancy ratio N and mixed convection parameter ($\frac{Ra}{Pe}$). The results for the heat and mass transfer coefficients in terms of Nusselt and Sherwood numbers are represented graphically for the various values of governing parameters of the problem. The heat and mass transfer in the boundary layer region has been analyzed for aiding and opposing flows.

Keywords: Convection in porous media, heat and mass transfer, integral method, boundary layer.

INTRODUCTION

The mechanisms of thermal and solutal transport by porous matrix are a phenomenon of great interest from the theory and application point of view. This is primarily because of the numerous applications of flow porous media, such as storage of radioactive nuclear waste materials, transpiration cooling, separation processes in chemical industries, filtration, transport processes in aquifers, ground water pollution, etc. The motivation of the present study is the fact that, both free and forced convection exist simultaneously in many of these applications. This is particularly relevant in situations where the Grashof number is large. Theories of mixed convection heat and mass transfer in porous media have been prepared for vertical surfaces (Lai, 1991; Chamkha and Khaled, 1999; Chamkha and Khaled, 2000; Postelnicu, 2007; Hassan, 2009; Oladapo, 2010; Bansod and Jadhav, 2010; Tak et al. , 2010), horizontal surfaces (Li and Lai, 1997; Bansod, 2003; Lakshmi and Murthy, 2008; Bansod and Jadhav, 2009) and inclined surfaces (Singh et al., 2002; Bansod, 2005; Shateyi, 2008; Beg et al., 2009). All these studies have been reported based on boundary layer analysis. The state of art concerning coupled heat and mass transfer by mixed convection in porous media has been summarized in the excellent monographs by Nield and Bejan (2006) and Ingham and Pop (2002).

In this chapter, an integral procedure is developed for combined heat and mass transfer by mixed convection from a vertical surface embedded in a saturated porous medium. The mixed convection flow is promoted by the uniform free stream and density variation due to the combination of temperature and concentration gradients. The Darcy model is considered and the porous medium porosity is assumed to be low. The heat and mass transfer in boundary layer region has been analyzed for aiding and opposing buoyancies for the aiding and opposing flows. The value of ($\frac{Ra}{Pe}$) is found to be the

*Corresponding author. E-mail: vjbansod@yahoo.co.in.

controlling parameter for the mixed convection. The results of heat and mass transfer in terms of Nusselt and Sherwood number, are presented for a wide range of governing parameters like the buoyancy ratio (N), -0.5 ≤ N ≤ 8, Lewis number (Le), 0.1 ≤ Le ≤ 100 and flow driving parameter ($\frac{Ra}{Pe}$), 0.1 ≤ $\frac{Ra}{Pe}$ ≤ 100.

Governing equations and transformation

Mixed convection heat and mass transfer from an impermeable vertical surface embedded in a fluid saturated porous medium was considered. The surface temperature and concentration are kept constant. At the same time, the temperature and concentration sufficiently far from the wall are T_∞ and C_∞, respectively. An external flow with a uniform velocity is introduced to the medium. For buoyancy induced by heat and mass transfer, the density is assumed constant everywhere, except in the body force term of Darcy's equation. Having invoked the Boussinesq and boundary layer approximation, the governing equations based on Darcy's law are given by:

$$\frac{\partial u}{\partial y} = \frac{Kg}{\upsilon}\left(\beta_T \frac{\partial T}{\partial y} + \beta_C \frac{\partial C}{\partial y}\right) \tag{1}$$

$$u\frac{\partial T}{\partial x} + v\frac{\partial T}{\partial y} = \alpha\frac{\partial^2 T}{\partial y^2} \tag{2}$$

$$u\frac{\partial C}{\partial x} + v\frac{\partial C}{\partial y} = D\frac{\partial^2 C}{\partial y^2} \tag{3}$$

In the aforestated equations, (u,v) are the Darcian velocities in the (x,y) directions. β_T and β_C are the thermal and concentration expansion coefficients respectively, υ is the kinematic viscosity, μ is the viscosity of the fluid, ρ is the density, g is acceleration due to gravity, K is the permeability of the porous medium, α and D are the equivalent thermal and mass diffusivities of the porous medium. T and C are the temperature and concentration. The boundary conditions at the wall are:

$$y = 0; v = 0, T = T_W, C = C_W \tag{4}$$

and at infinity are

$$y \to \infty; u = U_\infty, T \to T_\infty, C \to C_\infty \tag{5}$$

To solve the set of simultaneous equations defined previously, the following dimensionless variables are introduced:

$$\eta = \frac{y}{x}(Pe)^{1/2} \tag{6}$$

$$\Psi = \alpha (Pe)^{1/2} f(\eta) \tag{7}$$

$$\theta(\eta) = \frac{T - T_\infty}{T_W - T_\infty} \tag{8}$$

$$\phi(\eta) = \frac{C - C_\infty}{C_W - C_\infty} \tag{9}$$

In the aforestated equations, $Pe = \frac{U_\infty x}{\alpha}$ is Peclet number, ψ is stream function and f is its non-dimensional counterpart, θ is non-dimensional temperature distribution and ϕ is non-dimensional concentration distribution and η is similarity variable. After transformations, the resulting equations are

$$f'' = \frac{Ra}{Pe}(\theta' + N\phi') \tag{10}$$

$$\theta'' = -\frac{1}{2} f\theta' \tag{11}$$

$$\phi'' = -\frac{Le}{2} f\phi' \tag{12}$$

The parameter $N = \frac{\beta_C (C_W - C_\infty)}{\beta_T (T_W - T_\infty)}$ measures the relative strength of mass and thermal diffusion in the buoyancy-induced flow. It is clear that N = 0 for pure thermal buoyancy-induced flow, infinite for mass-driven flow, positive for aiding flow and negative for opposing flow. The diffusivity ratio, $Le = \frac{\alpha}{D}$ is nothing but the ratio of the Schmidt number $\left(\frac{\upsilon}{D}\right)$ and Prandtl number, $\left(\frac{\upsilon}{\alpha}\right)$ which denotes the relative rates of propagation of energy and mass within a system. The flow controlling parameters $\frac{Ra}{Pe} = \frac{Kg\beta_T (T_W - T_\infty)}{\mu U_\infty}$, measures the relative importance of the buoyancy effects and forced convection and is independent of x, the distance measured along the wall. $(\frac{Ra}{Pe}) \to 0$, represent the forced convection flow.

The transformed boundary conditions are

$$\eta = 0; \quad f = 0, \quad \theta = 1, \quad \phi = 1 \tag{13}$$

$$\eta \to \infty; \quad f' = 1, \quad \theta = 0, \quad \phi = 0 \tag{14}$$

In the equations, the primes indicate the derivative with respect to the similarity variable η. '∞' denotes the thickness of the boundary layer.

INTEGRAL SOLUTION AND DISCUSSION

The transformed energy Equation (10) to (12) together with the boundary conditions (13) and (14) can be integrated with respect to η from η = 0 to η = ∞, we get:

$$-\theta'(0) = \frac{1}{2}\int_0^\infty f'\theta \, d\eta \tag{15}$$

$$-\phi'(0) = \frac{Le}{2}\int_0^\infty f'\phi \, d\eta \tag{16}$$

The infinity is boundary layer thickness for temperature and concentration. We assume exponential temperature and concentration profiles as follows: -

$$\theta(\eta) = \exp(-\frac{\eta}{\delta_T}) \tag{17}$$

$$\phi(\eta) = \exp(-\frac{\xi\eta}{\delta_T}) \tag{18}$$

Here, δ_T is arbitrary scale for the thermal boundary layer thickness whereas ξ is its ratio to the concentration boundary layer thickness δ_C. With the help of the aforestated profiles and using Equation (10), Equations (15) and (16) can be obtained in two distinct expressions for δ_T^2 as:

$$\frac{1}{\delta_T^2} = \left(\frac{Ra}{Pe}\right)\left[\frac{\xi + 1 + 2N}{4(\xi+1)}\right] \tag{19}$$

$$\frac{1}{\delta_T^2} = \left(\frac{Ra}{Pe}\right)\left[\frac{2\xi + N(\xi+1)}{4\xi^2(\xi+1)}\right]Le \tag{20}$$

Equations (19) and (20) can be combined to give the following cubic algebraic equation for determining the boundary layer thickness ratio ξ as:

$$\xi^3 + (1+2N)\xi^2 - [(2+N)Le]\xi - N.Le = 0 \tag{21}$$

As ξ is determined using Newton-Raphson method from Equation (21), the results of practical interest in many applications are the heat and mass transfer coefficients, in terms of Nusselt and Sherwood numbers are given by equations

$$\frac{Nu}{(Pe)^{1/2}} = -\theta'(0)$$

$$= \frac{1}{\delta_T}$$

$$= 0.5\left[\frac{\xi+1+2N}{(\xi+1)}\right]^{1/2}\left(\frac{Ra}{Pe}\right)^{1/2} = \text{Nu}$$

$$\frac{Sh}{(Pe)^{1/2}} = -\phi'(0)$$

$$= \frac{\xi}{\delta_T}$$

$$= 0.5\,\xi\left[\frac{2\xi+N(\xi+1)}{2\xi(\xi+1)}\right]^{1/2} Le^{1/2}\left(\frac{Ra}{Pe}\right)^{1/2} = \text{Sh}$$

The notations Nu and Sh are used for the Nusselt number and Sherwood number respectively. As an indication of proper formulation and accurate calculation, the results thus obtained have been compared with the data published earlier with the case of thermally induced

$$\text{Nu} = 0.4448\left[\frac{\xi+1+2N}{(\xi+1)}\right]^{1/2}\left(\frac{Ra}{Pe}\right)^{1/2} \tag{22}$$

and

$$\text{Sh} = 0.4448\,\xi\left[\frac{2\xi+N(\xi+1)}{2\xi(\xi+1)}\right]^{1/2} Le^{1/2}\left(\frac{Ra}{Pe}\right)^{1/2} \tag{23}$$

The Nusselt and Sherwood numbers are plotted in Figures 1 (a, b and c) and 2 (a, b and c) respectively as a function of Lewis number. Approaching the forced convection dominated region (that is $\frac{Ra}{Pe}$ = 0.1), the heat transfer coefficient shows little dependence on the Lewis number, while the mass transfer coefficient shows the

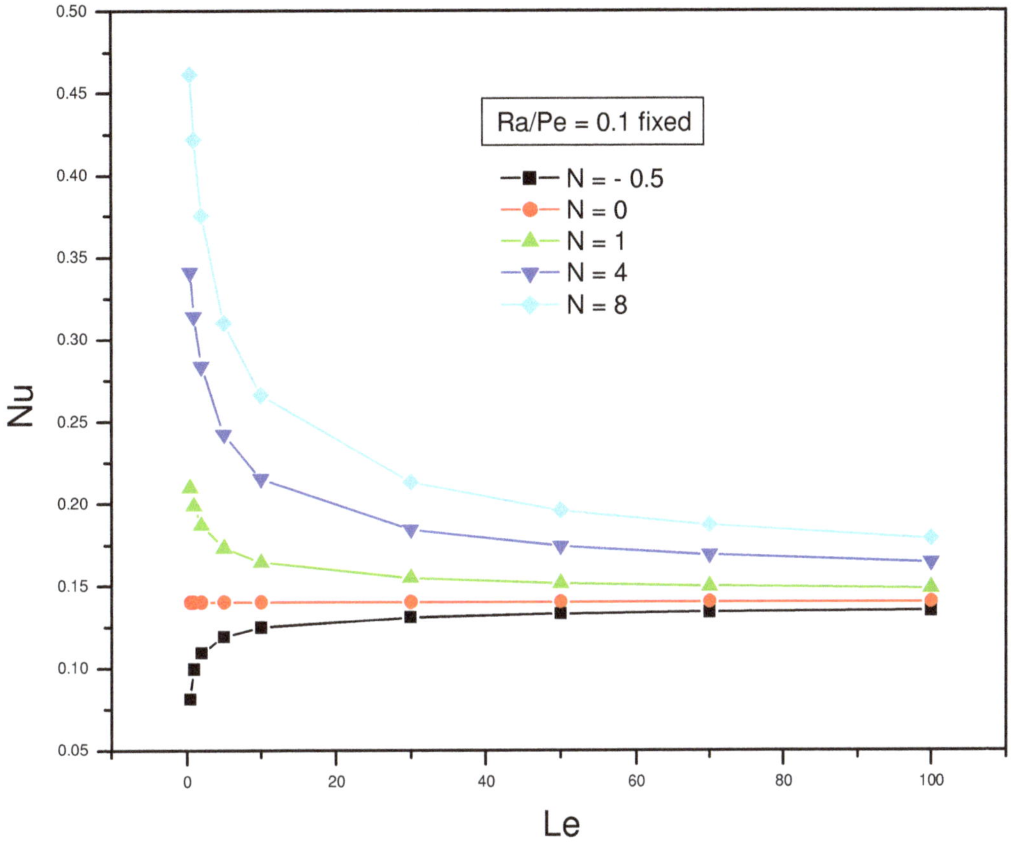
Ra/Pe = 0.1 fixed
N = - 0.5
N = 0
N = 1
N = 4
N = 8
Nu
Le
0.50
0.45
0.40
0.35
0.30
0.25
0.20
0.15
0.10
0.05
0
20
40
60
80
100

a

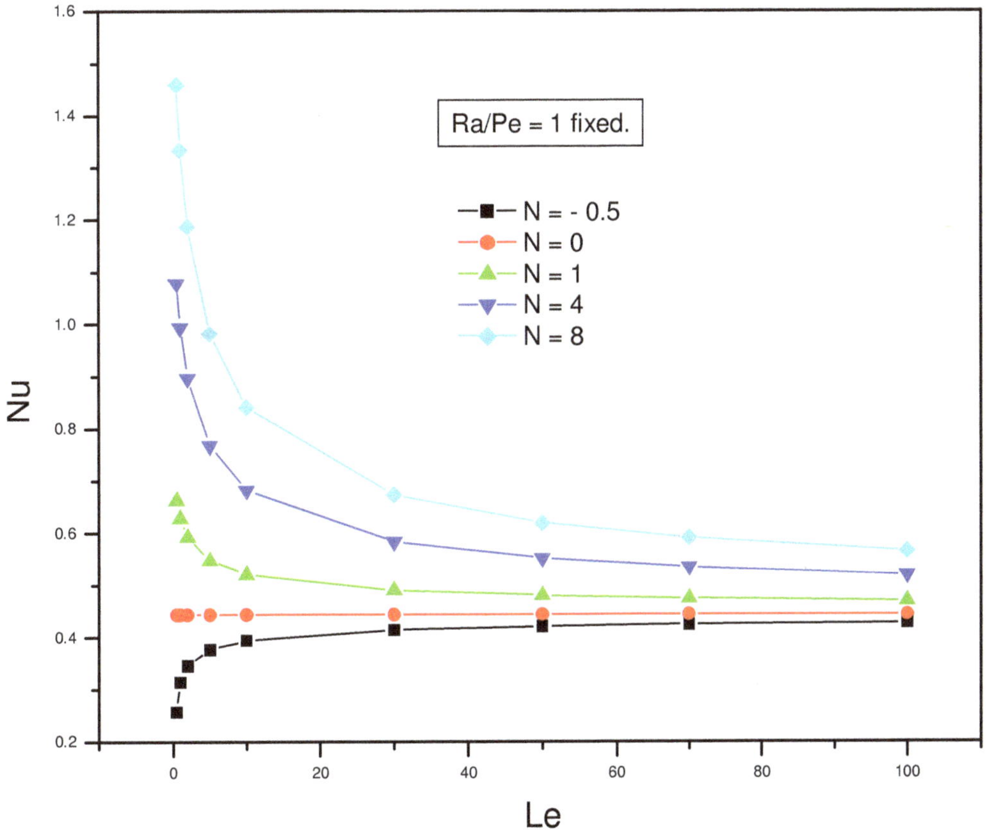
Ra/Pe = 1 fixed.
N = - 0.5
N = 0
N = 1
N = 4
N = 8
Nu
Le
1.6
1.4
1.2
1.0
0.8
0.6
0.4
0.2
0
20
40
60
80
100

b

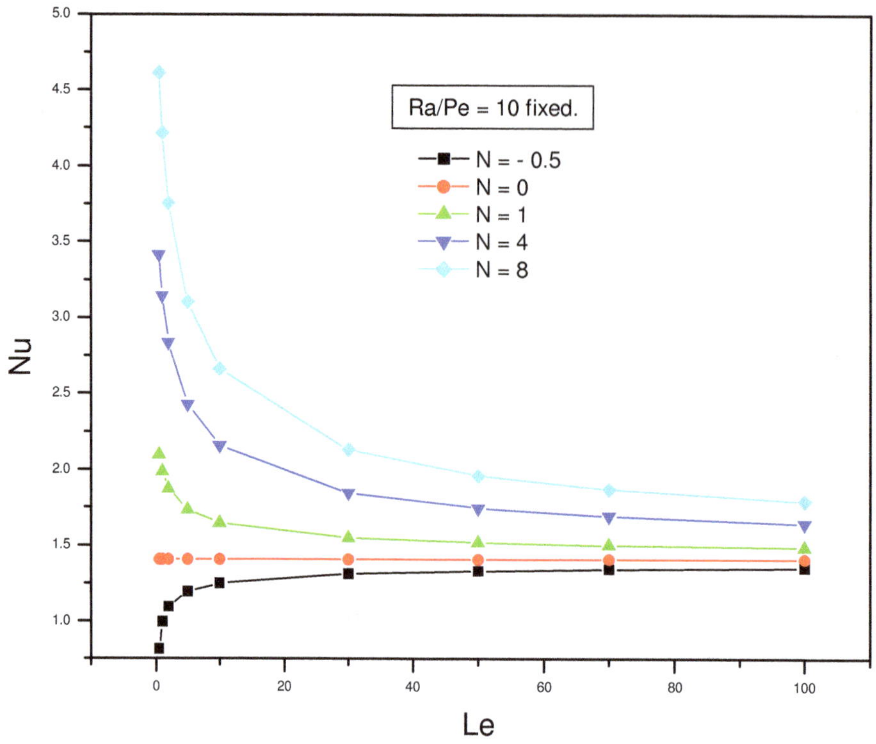

c

Figure 1. (a) Heat transfer results as a function of Lewis number when (Ra / Pe) = 0.1 and N = - 0.5, 0, 1, 4, 8. (b) Heat transfer results as a function of Lewis number when (Ra / Pe) = 1 and N = - 0.5, 0, 1, 4, 8. (c) Heat transfer results as a function of Lewis number when (Ra / Pe) = 10 and N = - 0.5, 0, 1, 4, 8.

dependence of Lewis number to $Le^{1/2}$. As the value of mixed convection parameter increases, the flow approaches the buoyancy-dominated regime. For N > 0, it is observed that the heat transfer is greatly enhanced by the mass buoyancy effect, while it is considerably reduced for N < 0. It is important to note that, the mass transfer coefficient does not have any physical significant meaning for thermally induced flow that is N = 0. The curve N = 0 is included in Figure 1 (a, b and c) only for comparison.

To understand the dependence of the heat and mass transfer coefficients on N, Equations (22) and (23) are plotted in Figures 3(a, b and c) and 4 (a, b and c). For Le < 1, it is observed that the heat transfer coefficient is increased for N > 0 and decreased for N < 0. For Le >1, the situation is reversed and the flow reversal takes place at a larger value of N. In the forced convection dominated regime, both heat and mass transfer coefficients vary linearly with N. To show the mixed convection results, the heat and mass transfer coefficients are plotted in Figures 5 (a, b and c) and 6 (a, b and c), respectively as a function of $(\frac{Ra}{Pe})$. It is clear from these figures that the convection favoring effect of $(\frac{Ra}{Pe})$ is countered by the parameters. For a given Le, it is observed that buoyancy-dominated regime extends toward a smaller value of $(\frac{Ra}{Pe})$ as N increases.

However, for a given N, the buoyancy-dominated regime retreats towards a larger value of $(\frac{Ra}{Pe})$ as Le increases. Therefore, for a given value of $(\frac{Ra}{Pe})$, the heat transfer regime under which the system operation depends on the value of N and Le. Selected values of Nu and Sh are listed in Table (1) for mixed convection. As expected, the heat transfer decreases with Le for opposing buoyancy whereas it increases with Le for aiding buoyancy.

Concluding remarks

Integral solution for mixed convection heat and mass transfer from a vertical surface embedded in a fluid saturated porous medium is reported. The heat and mass transfer in the boundary layer region has been analyzed

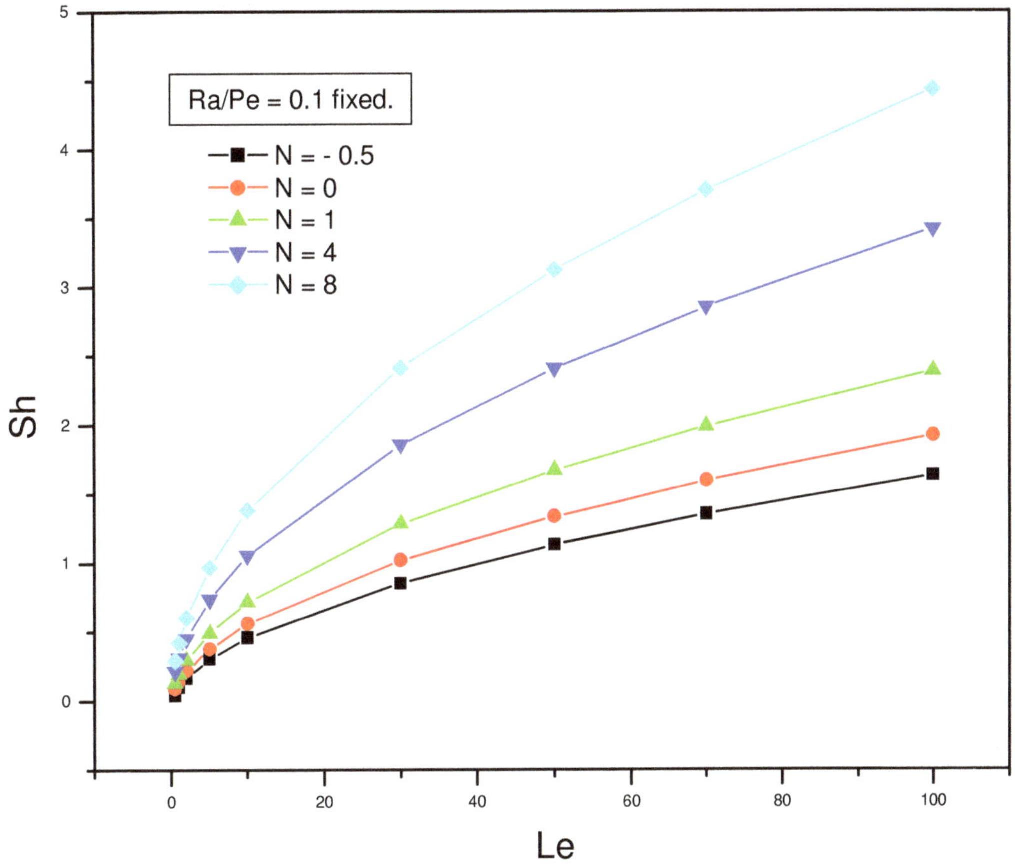

a

b

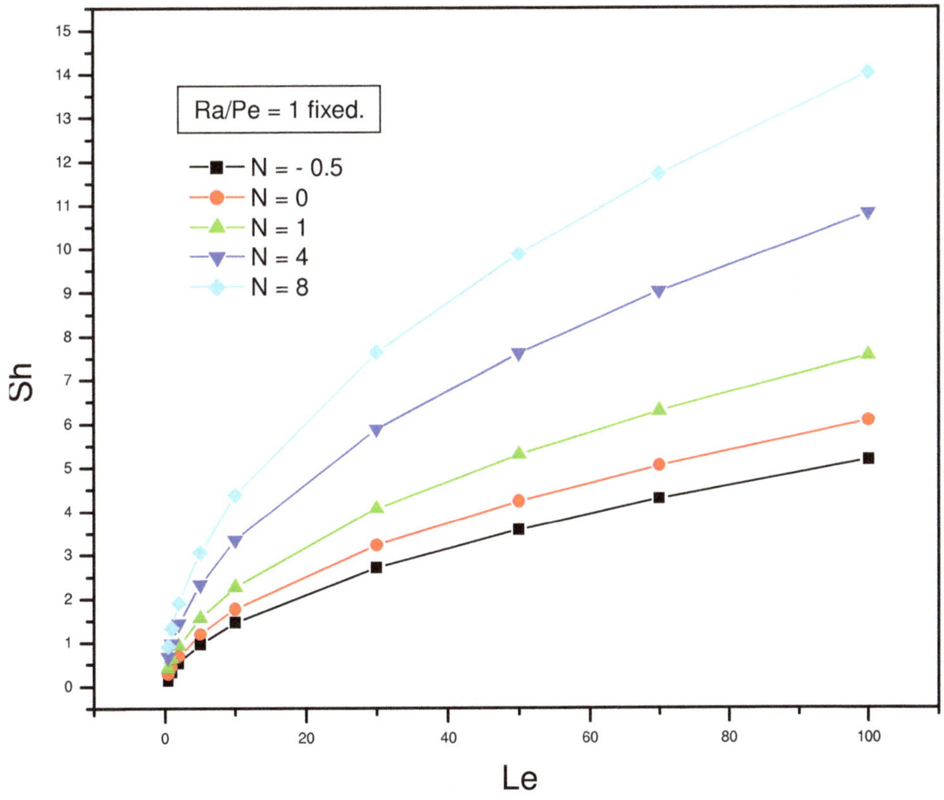

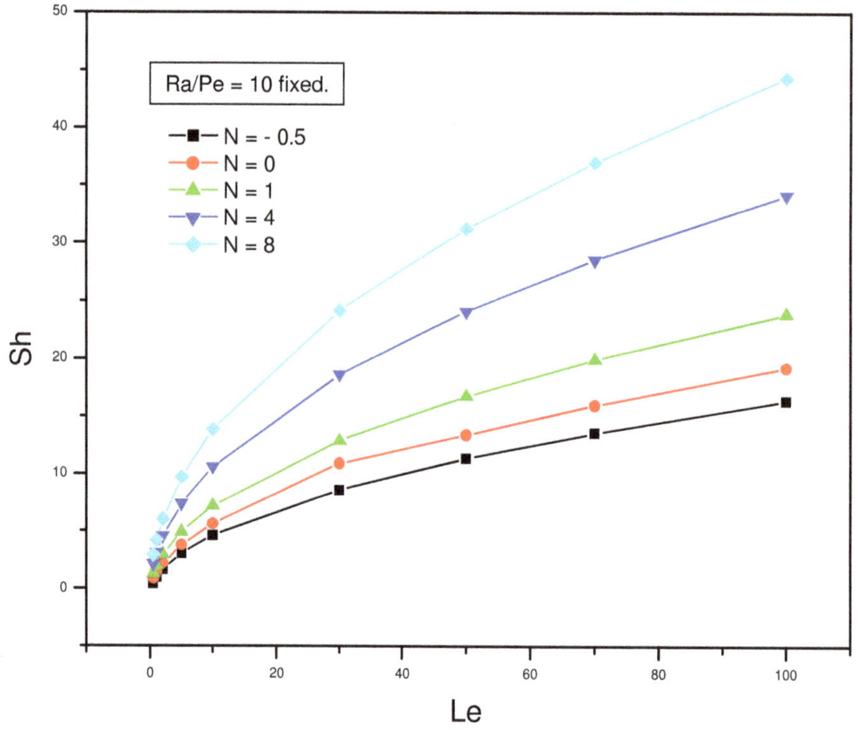

c

Figure 2. (a) Mass transfer results as a function of Lewis number when (Ra / Pe) = 0.1 and N = - 0.5, 0, 1, 4, 8. (b) Mass transfer results as a function of Lewis number when (Ra / Pe) = 1 and N = - 0.5, 0, 1, 4, 8. (c) Mass transfer results as a function of Lewis number when (Ra / Pe) = 10 and N = - 0.5, 0, 1, 4, 8.

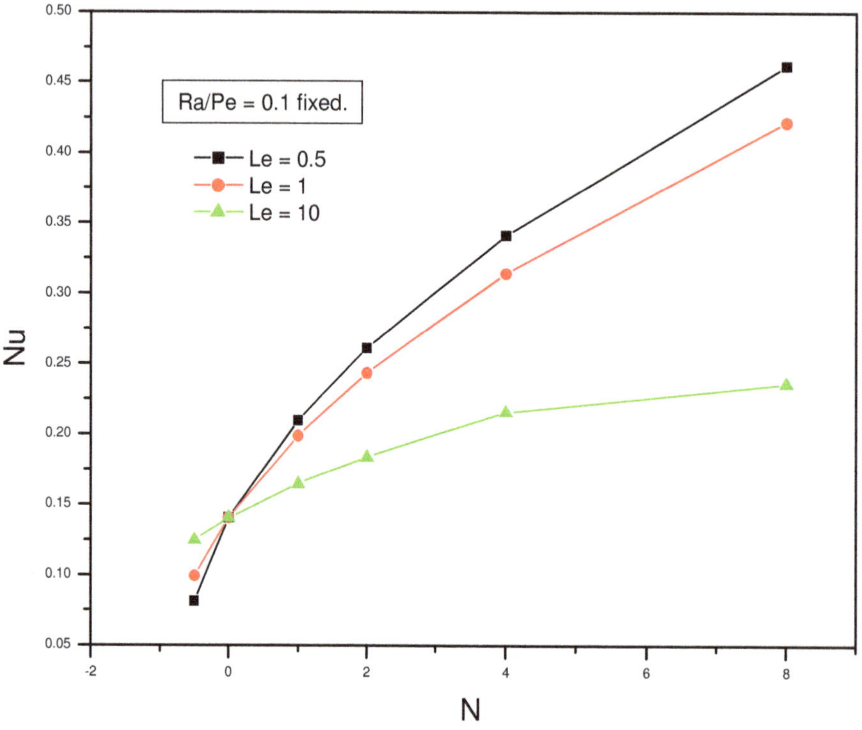

a

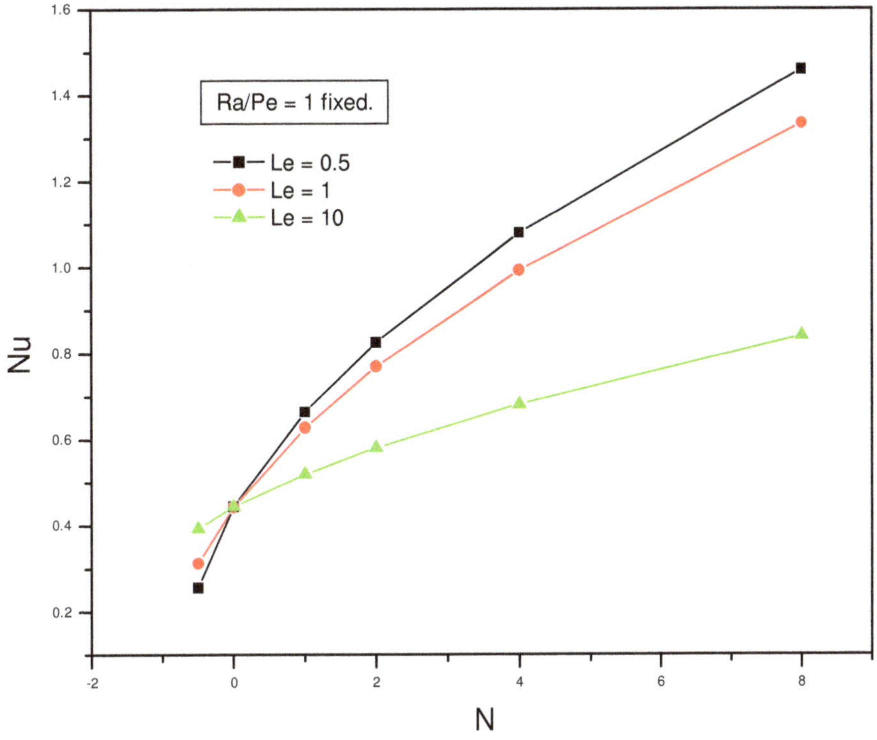

b

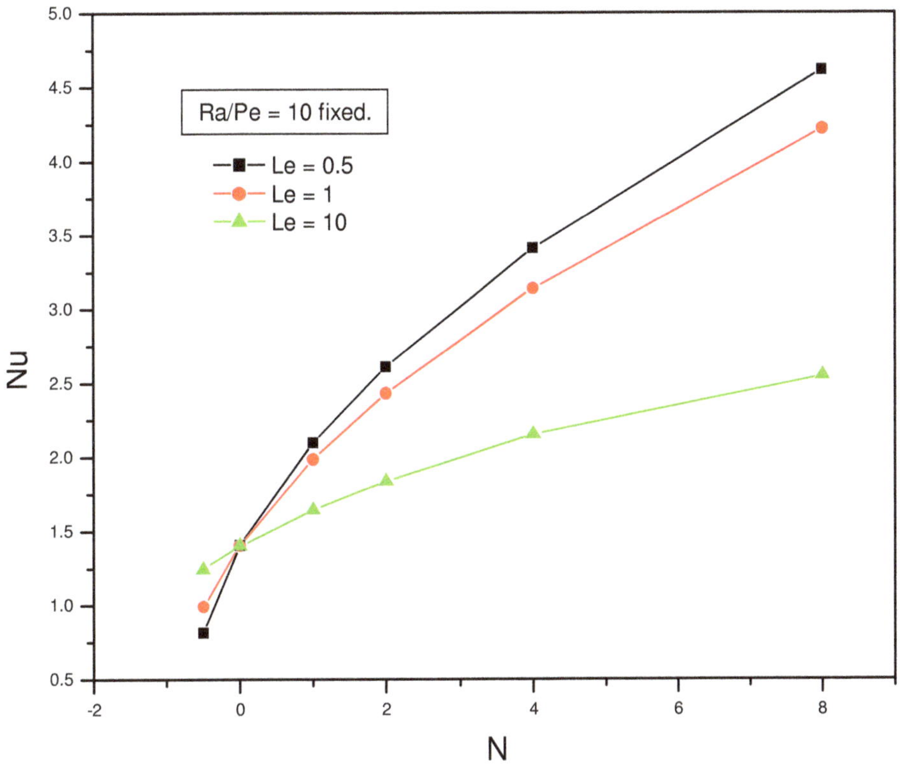

c

Figure 3. (a) Heat transfer results as a function of buoyancy ratio when (Ra / Pe) = 0.1 and N = 0.5, 1, 10. (b) Heat transfer results as a function of buoyancy ratio when (Ra / Pe) = 1 and N = 0.5, 1, 10. (c) Heat transfer results as a function of buoyancy ratio when (Ra / Pe) = 10 and N = 0.5, 1, 10.

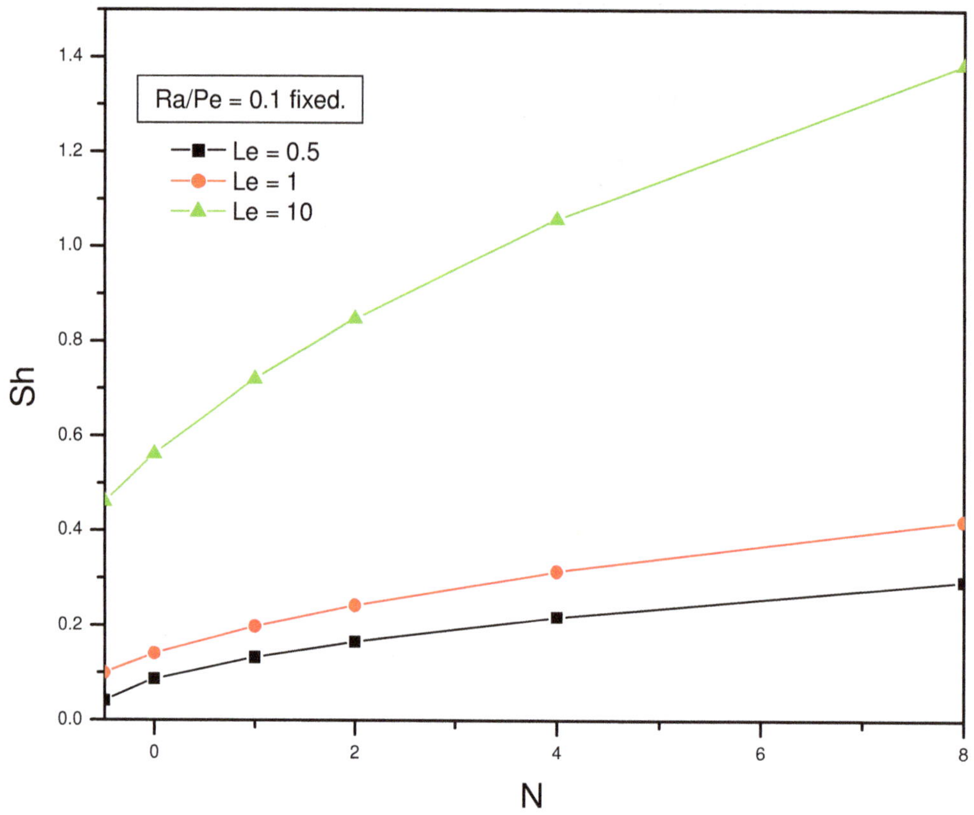

a

b

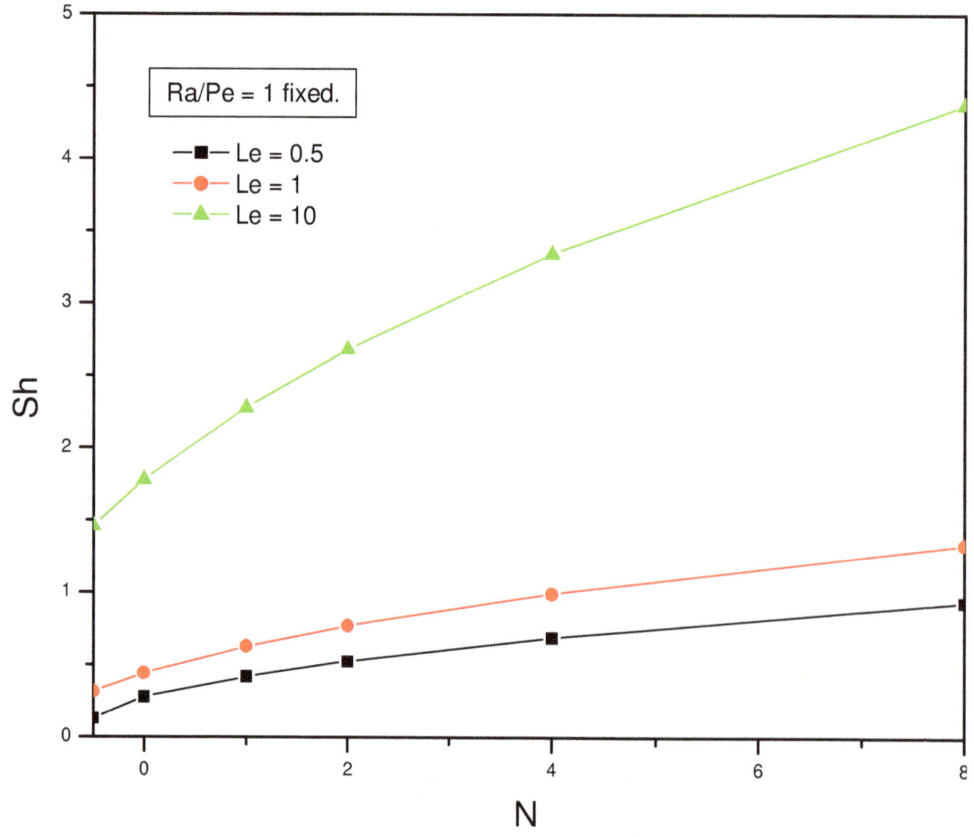

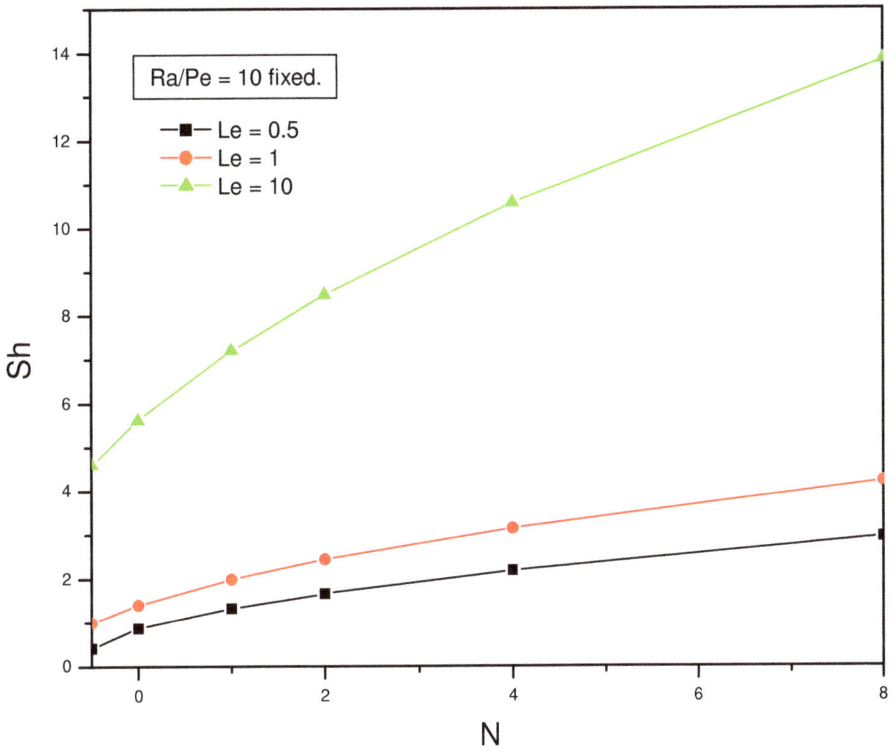

c

Figure 4. (a) Mass transfer results as a function of buoyancy ratio when (Ra / Pe) = 0.1 and N = 0.5, 1, 10. (b) Mass transfer results as a function of buoyancy ratio when (Ra / Pe) = 1 and N = 0.5, 1, 10. (c) Mass transfer results as a function of buoyancy ratio when (Ra / Pe) = 10 and N = 0.5, 1, 10.

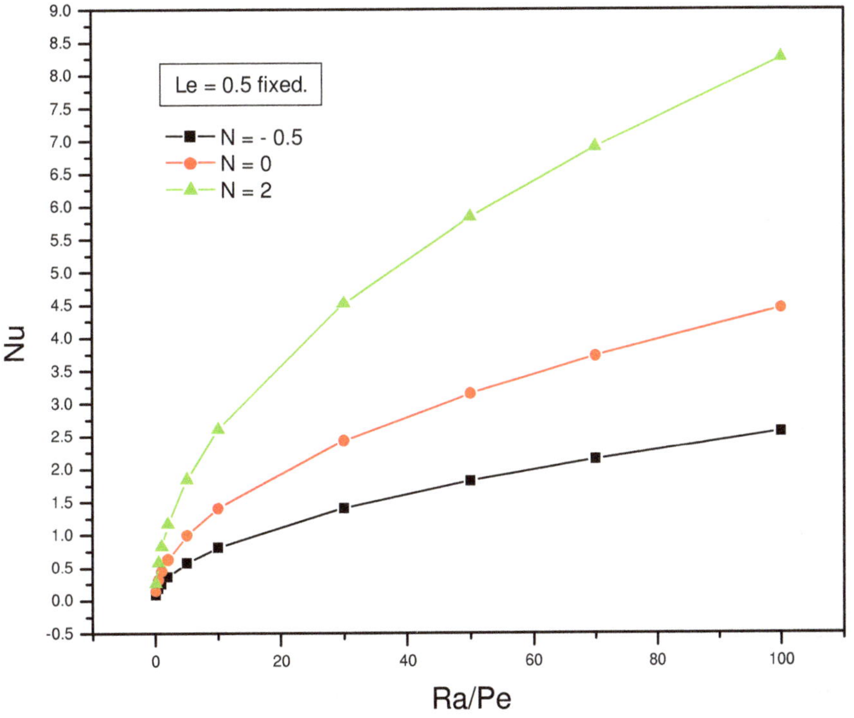

a

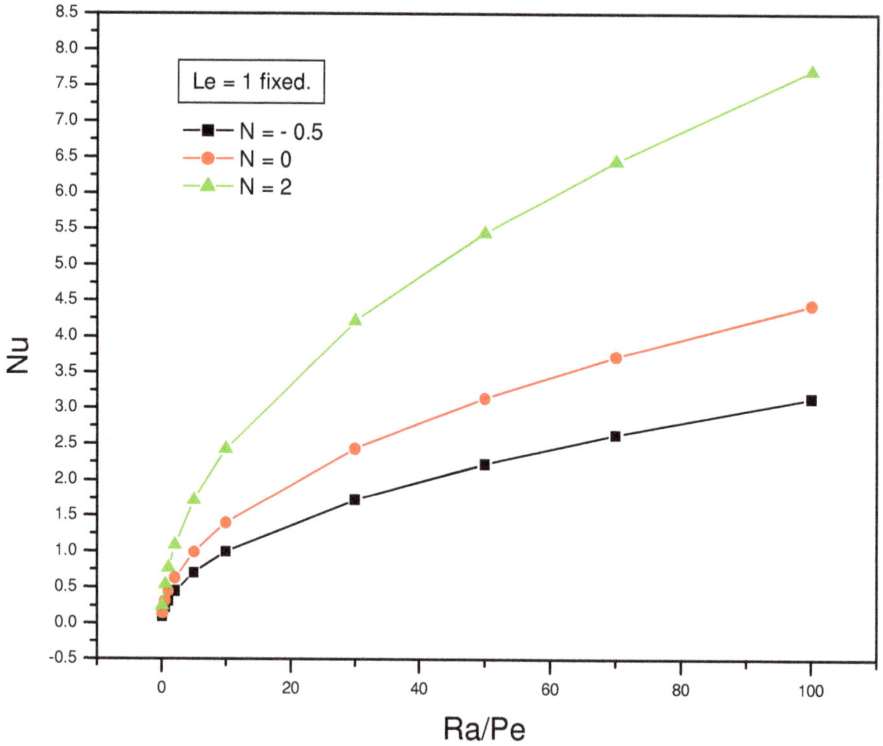

b

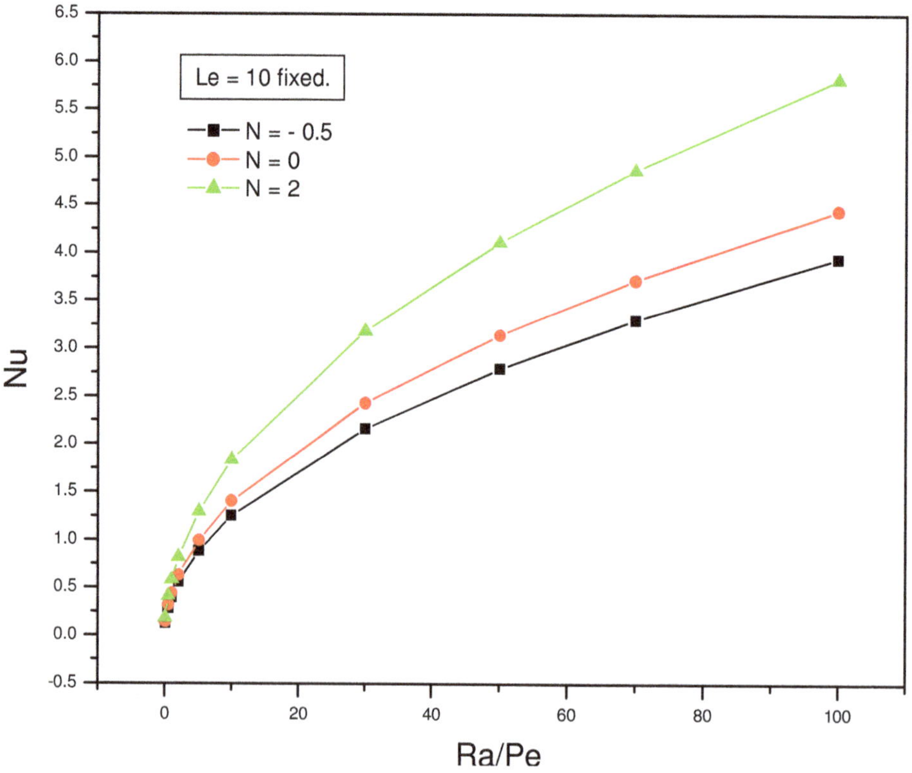

c

Figure 5. (a) Heat transfer results as a function of (Ra / Pe) when Le = 0.5 and N = - 0.5, 0, 2. (b) Heat transfer results as a function of (Ra / Pe) when Le = 1 and N = - 0.5, 0, 2. (c) Heat transfer results as a function of (Ra / Pe) when Le = 10 and N = - 0.5, 0, 2

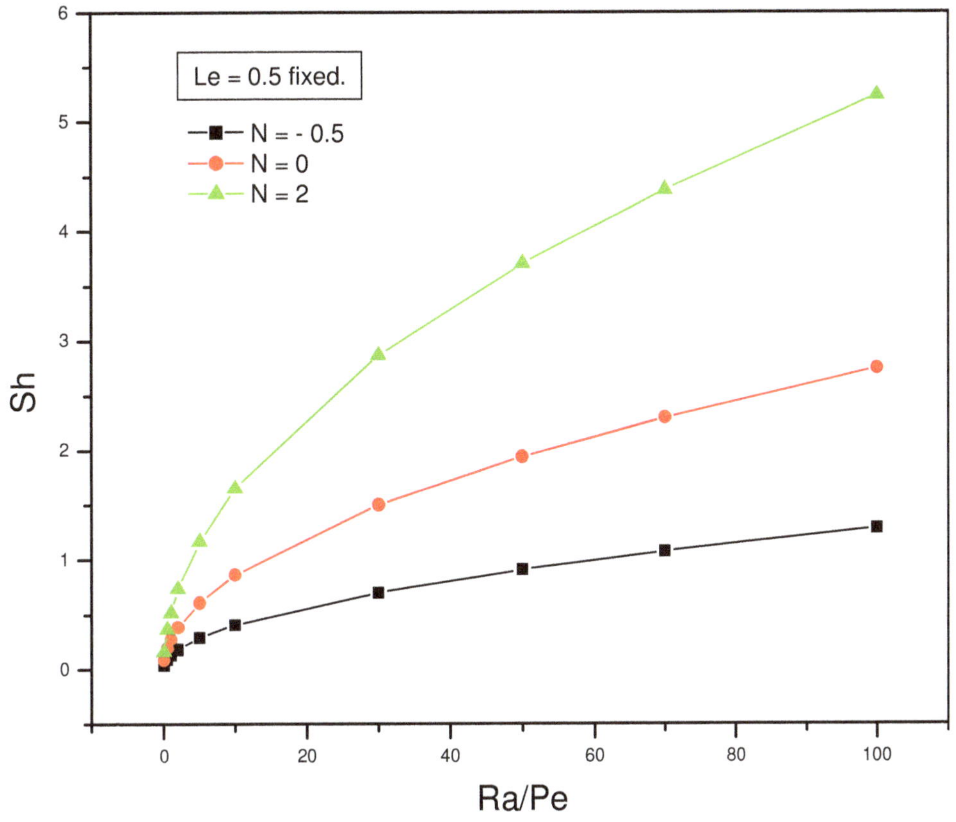
Le = 0.5 fixed.
N = - 0.5
N = 0
N = 2
Sh
Ra/Pe
0
1
2
3
4
5
6
0
20
40
60
80
100

a

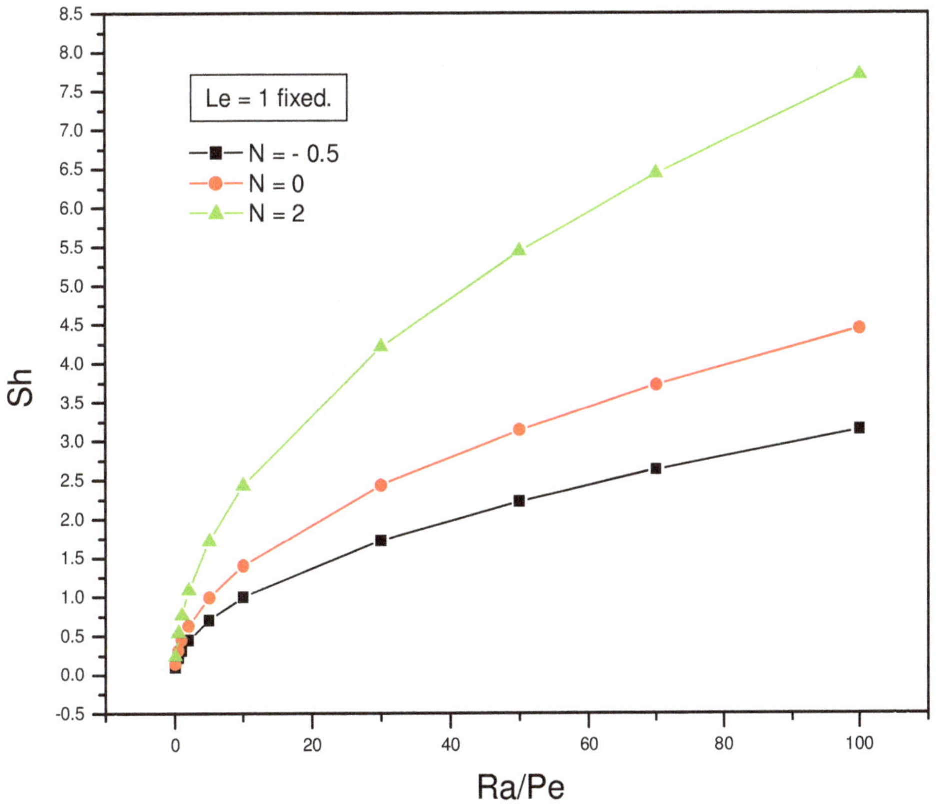
Le = 1 fixed.
N = - 0.5
N = 0
N = 2
Sh
Ra/Pe
-0.5
0.0
0.5
1.0
1.5
2.0
2.5
3.0
3.5
4.0
4.5
5.0
5.5
6.0
6.5
7.0
7.5
8.0
8.5
0
20
40
60
80
100

b

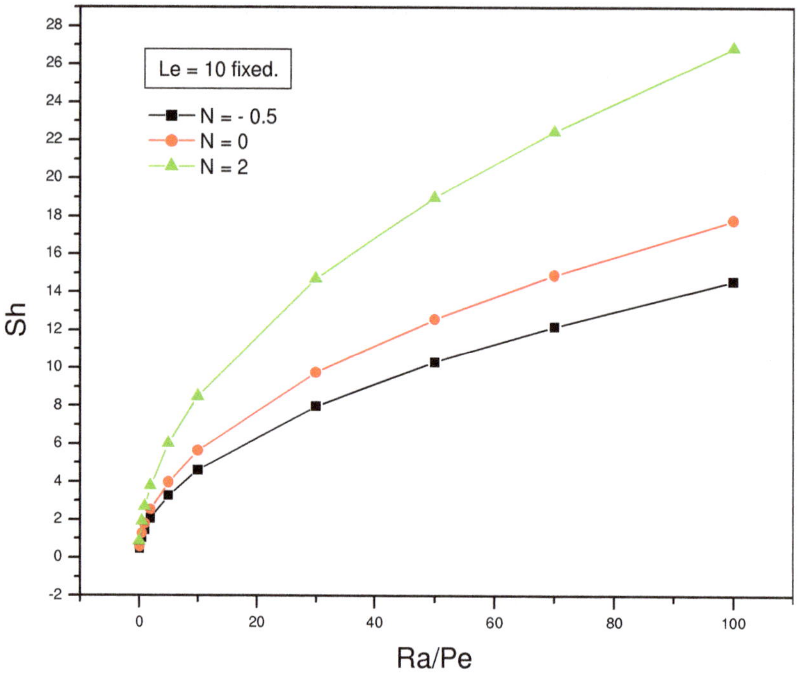

c

Figure 6. (a) Mass transfer results as a function of (Ra / Pe) when Le = 0.5 and N = - 0.5, 0, 2. (b) Mass transfer results as a function of (Ra / Pe) when Le = 1 and N = - 0.5, 0, 2. (c) Mass transfer results as a function of (Ra / Pe) when Le = 10 and N = - 0.5, 0, 2.

Table 1. Integral results for Nusselt and Sherwood number

Le	N	Ra/Pe	Nu	Sh
0.5	-0.5	0.1	0.0812	0.0406
		1	0.2568	0.1284
		10	0.8121	0.4060
	0	0.1	0.1406	0.0869
		1	0.4448	0.2749
		10	1.4067	0.8693
	4	0.1	0.3412	0.2174
		1	1.0793	0.9946
		10	3.4130	2.1742
1	-0.5	0.1	0.0994	0.0994
		1	0.3145	0.3145
		10	0.9946	0.9946
	0	0.1	0.1406	0.1406
		1	0.4448	0.4448
		10	1.4067	1.4067
	4	0.1	0.3145	0.3145
		1	0.9947	0.9947
		10	3.1454	3.1454

Table 1. Contd.

		0.1	0.1247	0.4610
	-0.5	1	0.3946	1.4577
		10	1.2479	4.6098
		0.1	0.1406	0.5626
10	0	1	0.4448	1.7793
		10	1.4067	5.6267
		0.1	0.2156	1.0583
	4	1	0.6827	3.3470
		10	2.1588	10.5839

for aiding and opposing flows. Extensive calculations for a wide range of physical parameters are performed. The heat and mass transfer coefficients increases with the increasing value of flow driving parameter ($\frac{Ra}{Pe}$). The Lewis number has a complex impact on the heat and mass transfer mechanism.

REFERENCES

Bansod VJ (2003). The Darcy model for boundary layer flows in a horizontal porous medium induced by combined buoyancy forces, J. Porous Media, 6: 273-281.

Bansod VJ (2005). The effects of blowing and suction on double diffusion by mixed convection over inclined permeable surfaces, J. Transp. Porous Media, 60: 301-317.

Bansod VJ, Jadhav RK (2009). An integral treatment for combined heat and mass transfer by natural convection along a horizontal surface in a porous medium, Int. J. Heat Mass Trans., 52: 2802-2806.

Bansod VJ, Jadhav RK (2010). Effect of double stratification on mixed convection heat and mass transfer from a vertical surface in a fluid saturated porous medium, J. Heat Transf. – Asian Res., 39(6): 378-395.

Beg O A, Beg TA, Bahier AY, Prasad VR (2009). Chemically-reacting mixing convective heat and mass transfer along inclined and vertical plates with soret and Dufour effects: Numerical solutions, Int. J. Appl. Math. Mech., 5(2): 39-57.

Chamkha AJ, Khaled AR (1999). Non-similar hydro-magnetic simultaneous heat and mass transfer by mixed convection from a vertical plate embedded in a uniform porous medium, Numer. Heat Transfer, 36: 327-344.

Chamkha AJ, Khaled AR (2000). Hydro-magnetic simultaneous heat and mass transfer by mixed convection from a vertical plate embedded in a stratified porous medium with thermal dispersion effect, J. Heat Mass Transfer, 36: 63-70.

Cheng P (1977). Combined free and forced convection flow about inclined surfaces in porous media, Int. J. Heat Mass Transfer, 20: 807-814.

Hassan AM, El-Arabawy (2009). Soret and Dufour effects on natural convection flow past a vertical surface in a porous medium with variable surface temperature, J. Math. Stat., 5: 190-198.

Ingham D, Pop I (2002). Transport Phenomena in Porous Media II, Pergamon Press.

Lai FC (1991). Coupled heat and mass transfer by mixed convection from a vertical plate in a saturated porous medium, Int. Comm. Heat Mass Transfer, 18: 83-106.

Lakshmi NPA, Murthy PVSN (2008). Soret and Dufour effects on free convection heat and mass transfer from a horizontal flat plate in a Darcy porous medium, J. Heat Transfer, 130.

Li CT, Lai FC (1997). Coupled heat and mass transfer from a horizontal surface in saturated porous medium, HTD-Vol. 349, ASME National Heat Transfer Conference, 11: 169-176.

Nield DA, Bejan A (2006). Convection in Porous Media, 3rd Edition, Springer Verlag, New York.

Oladapo OP (2010). Dufour and Soret effects of a transient free convective flow with radiative heat transfer past a flat plate moving through a binary mixture, The Pacific J. Sci. Technol. 11, May (Spring).

Postelnicu A (2007). Influence of chemical reaction on heat and mass transfer by natural convection from vertical surfaces in porous media considering Soret and Dufour effects, Int. J. Heat Mass Transfer, 43: 595-602,

Shateyi S (2008). Thermal radiation and buoyancy effects on heat and mass transfer over a semi-infinite stretching surface with suction and blowing, J. Appl. Math., Article ID 414830, 12 pages,

Singh P, Quenny, Sharma RN (2002). Influence of lateral mass flux on mixed convection heat and mass transfer over inclined surfaces in porous medium, J. Heat Mass Transfer, 38: 233-242.

Tak SS, Mathur R, Gehlot RK, Khan A (2010). MHD free convection-radiation interaction along a vertical surface embedded in a Darcian porous medium in presence of Soret and Dufour effects, J. Therm. Sci., 14: 137-145.

4

Calibrating the multiple orifice mathematical model using physical scale model foam at low Reynolds number

M. Emmanuel Adigio

Department of Mechanical Engineering, Niger Delta University, Wilberforce Island, Amassoma, Bayelsa State, Nigeria.
E-mail: emadigio@yahoo.com.

Recently, gelcast ceramic foams are being considered as potential diesel particulate filter substrates. Consequently, a mathematical model known as the Multiple Orifice Mathematical (MOM) model for the study of fluid flow and the determination of pressure gradients across the foam filters was developed and calibrated by some researchers. However, there was need to establish the model application on a wider range of pore sizes of the foam filters. Hence, this work is to establish the dynamic similarity of the physical scale model used for the calibration and the ceramic foams. Following the conceptual model employed in the development of the MOM model, generic physical scale foam models and a fluid flow rig was fabricated. The pressure drops across the generic physical model foam obtained from experiments over different ranges of low Reynolds number were graph-fitted against the MOM model to determine the kinetic correction factors. The values for the kinetic correction coefficient determined from the generic physical model at low Reynolds number is within the range obtained by other researchers in the calibration of the MOM model, which implies that the MOM model can be applied to a wide range of pore sizes found in gelcast ceramic foam filters.

Key words: Diesel particulate trap, gelcast ceramic foam, kinetic correction coefficient, generic foams, foam filters, pressure gradients.

INTRODUCTION

Engine manufacturers have made progress in the reduction of diesel engine emissions through improved engine design (Swiss Agency for the Environment, Forest and Landscape, 2000a; Mayer, 1998], fuel formulation and improved maintenance practices. Modern diesel engine is reported to have reduced particulate matter (PM) emission by as much as 90% (Mayer, 1998; Nauss, 1997) through the improved engine design and as much as 30% through fuel formulation. Mayer et al. (1998) reported that PM emissions from diesel engines fall in the size range of up to 100 nm and the new engines in particular emit more fine particulates at all operating conditions.

The consequence of improved engine design and fuel formulation is a decrease in PM mass but an increase in fine particulate number which is potentially more hazardous (Mayer, 1998). The key factor for the determination of the effect of diesel particulates on health is their size. Particles that are < 100 nm are invisible to the eyes but can deposit in the bronchial and pulmonary tracts of the respiratory system (Hinds, 1980). It is reported by many researchers (Nikula et al., 1999; Health Effects Institute, 2002; United State Environmental Protection Agency, 2002; Dybdahl et al., 2003; World Health Organization, 2003; Garshick et al., 2004; Warheit et al., 2004; Brown et al., 2004; Arey, 2004) that diesel exhaust emissions affect health and contribute to acid rain and visibility. Siegmann and Siegmann (1997) reported that fine particles from combustion contain thousands of different chemicals that cannot be characterized due to their unstable condition in the atmosphere.

Consequently, there is expected to be a need for after treatment of the exhaust gases to meet future emission limits (Swiss Agency for the Environment, Forest and Landscape, 2000). The diesel particulate filter technology has been proven as a viable option for the effective reduction of PM from diesel engines (Swiss Agency for the Environment, Forest and Landscape, 2000b) and mathematical modelling is increasingly becoming an engineering tool to understand, predict and control the diesel particulate filter (DPF) systems. DPFs consist of a filter designed to collect the PM in the exhaust stream of the diesel engine, while allowing the exhaust gases to pass through the system. The fundamental parameters to assess the quality of the DPF are the filtration efficiency and the pressure drop of the filter. Hence, it is desirable to develop mathematical models to predict these parameters that can be used within given boundary conditions to aid the design of DPFs.

Ceramic foams until recently were mainly used as catalyst supports (Richardson et al., 2000) and molten metal filters (Gabathuler et al., 1991). However, they are now being considered for DPF applications since they exhibit some favorable attributes. Ceramic foams have good filtration in the nano-particle range (Pontikakis et al., 2001). The high porous nature of ceramic foam filters is favorable to the propagation of the combustion zone during regeneration.

The modelling of porous media such as the ceramic foam filters, however, has been of interest to significantly fewer researchers. A work, of interest, reported by Pontikakis et al. (2001) is the development of a mathematical model for the prediction of pressure drop across reticulated foam filters. Pontikakis et al. (2001) assumed that the struts which form the solid frame work of foam filters can be modeled as fiber elements. Other researchers (Adigio et al., 2008) reported the development of a mathematical model referred to as "Multiple Orifice Mathematical (MOM) model" for understanding fluid flow through the filters and an aid for filter design. This MOM model was developed by applying the fluid flow theory on a simplified conceptual model, where the ceramic foam was represented with rows of cells across the filter, connected by openings called the windows. The resultant mathematical model was calibrated by fixing the viscous correction coefficient to determine the kinetic correction coefficient, β by "graph fitting" the mathematical model on a graph developed from experimental data of fluid flow on a generic physical scale model foam filter.

This report presents the calibration of the MOM model using experimental data from generic physical scale model foam of external diameter 60 mm and lengths of 100 and 125 mm at low flow rate, thus, Reynolds number ranging from 35 to 890. This was necessary because it was observed that the flow rates through the ceramic foam filters samples are low and the corresponding Reynolds number is within the above range. The aim of this study is to establish the dynamic similarity of the physical scale model foam and the ceramic foam filter samples used for the model validation, thus, confirming the general application of the MOM model on gelcast ceramic foam filters.

MATERIALS AND METHODS

The generic physical scale model foam samples used for the model calibration are based on a combination of five pieces of foams of length 25 mm and diameter 60 mm, which are reproduction of the conceptual model used for the development of the MOM model, where the cells are arrayed across the length of the foam with connecting windows. The method used in the manufacture of the model foam is a rapid manufacturing process known as stereolithography (SL). The rapid manufacturing refers to a class of technologies that can automatically produce physical models from computer assisted design (CAD) data. The structure is illustrated by a Solid Edge (Version 15) as shown in Figure 1.

The advantage of using the SL method to produce the generic physical scale model foam lies in the accuracy of the process and the ability to produce complex geometries without the need to resort to mould tooling, therefore, the relatively complex structure of the filter could be manufactured comparatively easily. This would have been difficult to achieve with other manufacturing approaches, or indeed on a real ceramic foam sample.

An experimental rig was designed using the Solid Edge CAD Package and constructed to measure the pressure drop across the model foam filter samples, the flow rates through them, and the temperature of the fluid is shown in Figure 2. A flow conditioner was mounted on the rig to straighten the swirling air flow and reduce the pulsating effect from the centrifugal blower that generates the air flow. The distance from the conditioner to the orifice plate was more than ten times the pipe diameter to allow the full development of the fluid flow before the orifice plate. Flow rates were measured using a calibrated orifice flow meter designed and assembled in accordance to the ISO 5167 standard (BS EN ISO 5167, 2003). Using the Reader-Harris/Gallagher Equation (Reader-Harris and McNaught, 2005) the values of the orifice discharge coefficients (C_D) corresponding to the 15 mm orifice diameter plate was calculated to be 0.6285. The absolute pressure and temperature were measured before the filter holder to determine the density of the air flowing through the rig. The experiments on the model foam samples were repeated three times to assure repeatability of the results obtained.

The data was collected from filters of lengths 100 and 125 mm which are made up of 25 mm × 4 pieces and 25 mm × 5 pieces of generic physical model foam samples respectively. The samples have cell size of 7.0 mm and porosity of 80%. The data was collected at three ranges of Reynolds number by adjusting the flow rates across the filters accordingly; 148 to 890, 86 to 269 and 35 to 99 respectively, and graphs corresponding to each range were produced.

RESULTS

Figures 3, 4 and 5 show the MOM model calibration graphs of pressure gradients vs. fluid mass flow rate developed from the experimental data presented in Tables 1, 2 and 3 respectively. The graphs demonstrate the graph-fitting of the model where the value of the kinetic correction coefficient β in the MOM model (Equation 3) was adjusted until the model fits the graphs

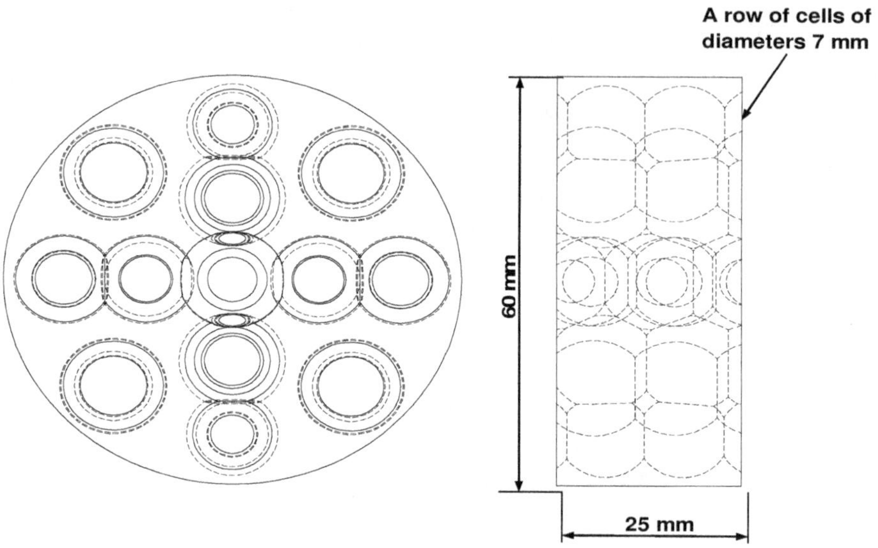

Figure 1. A generic multiple orifice physical scale model designed using a CAD package.

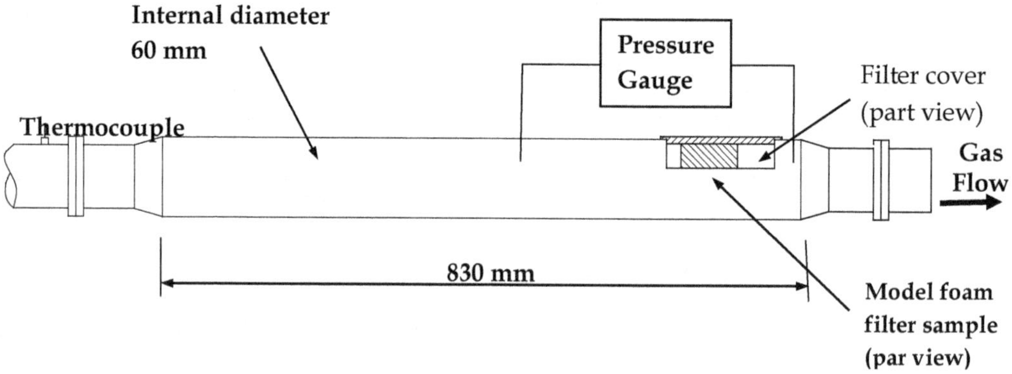

Figure 2. Schematic diagram of a flow rig model foam sample holder.

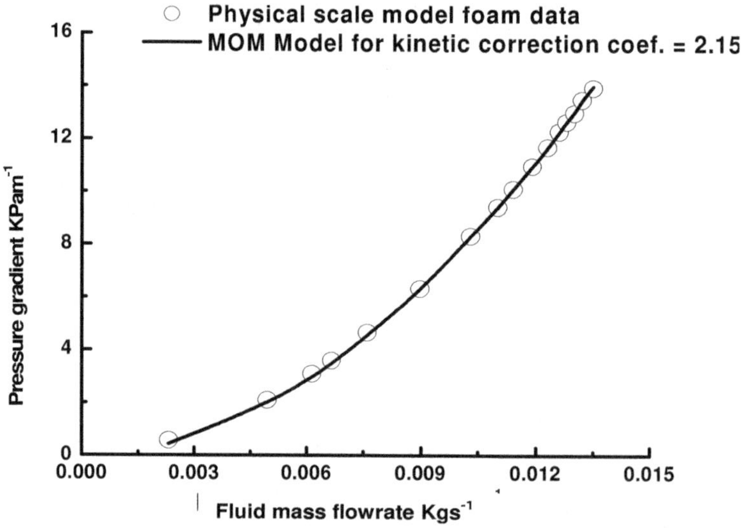

Figure 3. Graph of pressure gradient vs. fluid flow rate in a physical scale model foam sample, for the calibration of the MOM model. Reynolds number is from 148 to 890.

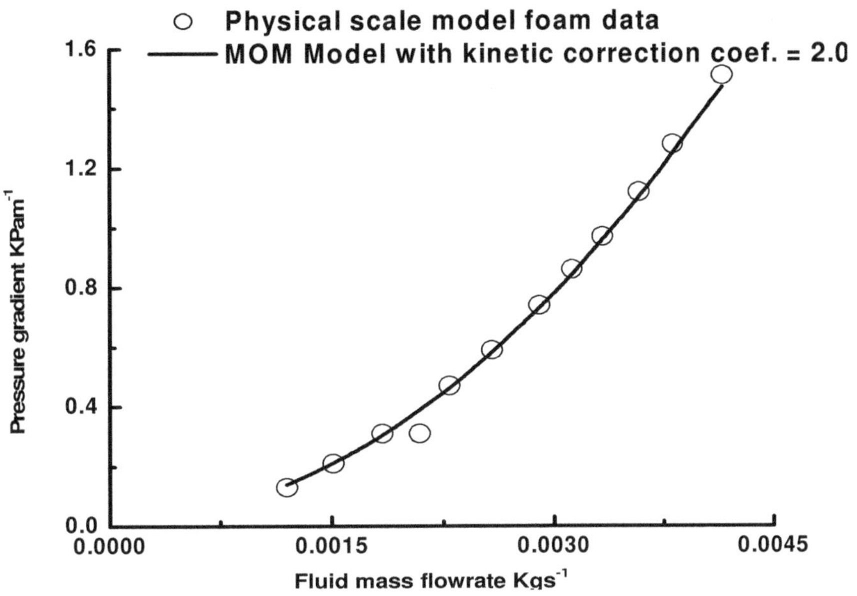

Figure 4. Graph of pressure gradient vs. fluid flow rate in a physical scale model foam sample, for the calibration of the MOM model. Reynolds number is from 86 to 269.

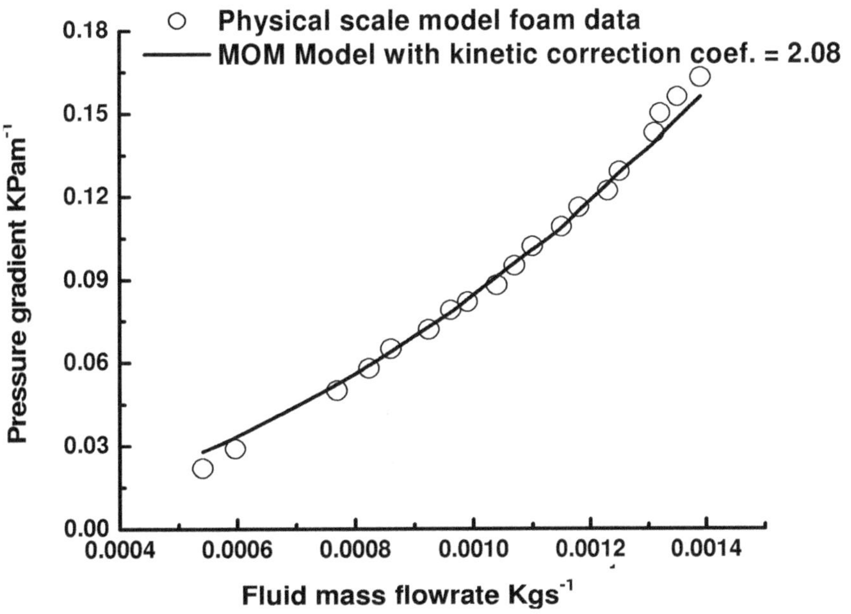

Figure 5. Graph of pressure gradient vs. fluid flow rate in a physical scale model foam sample, for the calibration of the MOM model. Reynolds number is from 35 to 99.

developed from the experimental data. The corresponding kinetic correction coefficients are indicated on the graphs.

DISCUSSIONS

In computing the experimental data to calculate the mass flow rate of the fluid, the densities were determined from the ideal gas law, $\rho = p/RT$, where R is the universal gas constant equal to 287 $Jkg^{-1}K^{-1}$, T is the absolute temperature in Kelvin, K, and p is the gas pressure in Pascal, Pa. The mass flow rate is the product of the volumetric flow rate, Q (m^3s^{-1}) and the fluid density, that is,

$$\text{Mass flow rate} = \rho Q \quad (1)$$

Consequently, the fluid mass flow rate is determined by

Table 1. Physical scale model foam of length 100 mm, cell diameter = 7.0 mm and porosity = 80%, Reynolds number is from 148 to 890.

P_{orif} (Pa)	P_{filt} (Pa)	T (°C)	P_{abs} (Pa)	Air mass flow rate (kgs^{-1})	Pgrad. ($kPam^{-1}$)	Pgcal. ($kPam^{-1}$)
21	59	26.9	56	2.32E-03	0.59	0.44
96	211	27.3	200	4.95E-03	2.11	1.94
147	310	27.6	290	6.13E-03	3.10	2.95
173	360	28	338	6.65E-03	3.60	3.46
226	466	28.2	436	7.60E-03	4.66	4.51
315	634	28.8	594	8.97E-03	6.34	6.26
417	832	29	779	1.03E-02	8.32	8.27
473	941	29.5	884	1.10E-02	9.41	9.37
509	1010	29.6	948	1.14E-02	10.10	10.07
552	1097	29.8	1029	1.19E-02	10.97	10.92
589	1169	29.9	1097	1.23E-02	11.69	11.64
619	1228	29.9	1152	1.26E-02	12.28	12.23
640	1265	30	1190	1.28E-02	12.65	12.64
657	1298	30	1220	1.30E-02	12.98	12.98
681	1348	30	1266	1.32E-02	13.48	13.45
708	1393	29.9	1310	1.35E-02	13.93	13.98

Table 2. Physical scale model foam of length 100 mm, cell diameter = 7.0 mm and porosity = 80%, Reynolds number is from 86 to 269.

P_{orif} (Pa)	P_{filt} (Pa)	T (°C)	P_{abs} (Pa)	Air mass flow rate kg s^{-1}	Pgrad. kPa m^{-1}	Pgcal. kPa m^{-1}
50	13	20.7	15	1.20E-03	0.13	0.14
79	21	20.7	23	1.51E-03	0.21	0.21
118	31	20.7	33	1.84E-03	0.31	0.30
151	31	20.7	43	2.09E-03	0.31	0.39
181	47	20.7	51	2.29E-03	0.47	0.46
231	59	20.8	65	2.58E-03	0.59	0.58
292	74	20.9	81	2.90E-03	0.74	0.73
337	86	21	94	3.12E-03	0.86	0.84
384	97	21.1	107	3.33E-03	0.97	0.96
444	112	21.2	123	3.58E-03	1.12	1.10
504	128	21.3	139	3.81E-03	1.28	1.25
597	151	21.4	165	4.15E-03	1.51	1.47

developing a relationship between flow rate and pressure difference across the orifice plate, thus, applying the Bernoulli equation across the orifice plate then simplifying gives the following relationship,

$$\rho Q = C_D \frac{\pi}{4} D_O^2 \sqrt{\frac{2\rho\Delta p}{(1-(\frac{D_O}{D_I})^4)}} \qquad 2$$

where C_D is the orifice plate discharge coefficient, D_o is the orifice diameter, ρ is the fluid density, D_I is the pipe diameter and Δp is the pressure difference measured across the orifice plate.

The first four columns of Tables 1, 2 and 3 (pressure difference across orifice Porif, pressure difference across filter P_{filt} temperature T and absolute pressure P_{abs}) were measurements from the experimental rig while the remaining columns were calculated from appropriate relationships. The atmospheric pressure was also measured during the experiments to enable the calculation of the air mass flow rate. The air mass flow rate was calculated from Equation 2. The pressure gradients across the filter from the experimental data were the ratio of the pressure difference across the filter and the filter length.

The last column is the calculated pressure gradients. Across the filters using the MOM model (Equation 3) (Adigio et al., 2008), by primarily adjusting the Kinetic correction coefficient β for a given foam window and cell size and filter length, until the model graph fits to the graphs developed from the experimental data.

Table 3. Physical scale model foam of length 125 mm, cell diameter = 7.0 mm and porosity = 80%, Reynolds number is from 35 to 99.

P_{orif} (Pa)	P_{filt} (Pa)	T (°C)	P_{abs} (Pa)	Air mass flow rate (kgs^{-1})	Pgrad. (kPm^{-1})	Pgcal. ($kPam^{-1}$)
10	3	22.1	4	5.41E-04	0.022	0.028
12	4	22.1	6	5.97E-04	0.029	0.033
21	6	22.4	10	7.69E-04	0.050	0.052
24	7	22.4	12	8.23E-04	0.058	0.059
26	8	22.4	13	8.60E-04	0.065	0.064
30	9	22.5	14	9.24E-04	0.072	0.073
32	10	22.5	16	9.61E-04	0.079	0.078
34	10	22.6	17	9.90E-04	0.082	0.083
38	11	22.6	18	1.04E-03	0.088	0.091
40	12	22.7	19	1.07E-03	0.095	0.096
43	13	22.7	21	1.10E-03	0.102	0.101
46	14	22.8	22	1.15E-03	0.109	0.108
49	14	22.9	24	1.18E-03	0.116	0.115
53	15	23	25	1.23E-03	0.122	0.124
55	16	23.1	27	1.25E-03	0.129	0.129
60	18	23.1	29	1.31E-03	0.143	0.139
61	19	23.3	30	1.32E-03	0.150	0.142
64	20	23.5	31	1.35E-03	0.156	0.148
68	20	23.5	33	1.39E-03	0.163	0.156

$$\left.\begin{aligned} &\frac{\Delta p}{L} = \frac{\alpha\mu A(1-\varepsilon)^2}{D^2\pi\varepsilon^3} S_v^2 Q + \frac{(1-w^4/d_o^4)}{\sqrt{d^2-w^2}}\left[\frac{d_o^2}{\beta\pi\varepsilon D^2 w^2}\right]^2 8\rho Q^2 \\ &\text{where} \quad d_o = \left[\frac{d^3(2-3B(3k^2+B^2))}{3\sqrt{d^2-w^2}}\right]^{0.5} \\ &B = 1-\sqrt{1-k^2} \quad \text{and } k = w/d \end{aligned}\right\} \qquad 3$$

α is the viscous pressure loss correction coefficient which was chosen and fixed at 5 as suggested by MacDonald et al. (1979), w and d are the window and cell size respectively. The specific surface, S_V was calculated from the expression below developed by Adigio et al. (2008);

$$S_v = \frac{12[1-6B+5k^2/2]\varepsilon}{d[2-3B(3k^2+B^2)][1-\varepsilon]} \qquad 4$$

The graphs (Figures 3, 4 and 5) shows that the kinetic correction coefficients obtained from the three ranges of Reynolds numbers varied from 2.0 to 2.15. These values of kinetic correction coefficients determined from the use of the multiple orifice physical scale model corroborates the results offered by Adigio et al. (2008).

Conclusion

A cost effective, accurate and rapid methodology was developed to evaluate the general use of the MOM model in the study of fluid flow through gelcast ceramic foam filters. The kinetic correction coefficient β of the MOM model has been determined by graph fitting the mathematical model to experimental data obtained from generic physical scale model foams. The corresponding values of kinetic correction coefficient are 2.15, 2.0 and 2.08 respectively, which lie within the range established by Adigio et al. (2008). This implies that the kinetic correction coefficient for the MOM model is independent of the Reynolds number. Hence, this research work has established the use of the MOM model as a potential tool in the study of fluid flow and diesel particulate filter design in a wide range of filter cell sizes found in gelcast ceramic foam filters.

The application of the mathematical model can be extended to other types of foams, including reticulated foams. The model can also be extended to the development of filtration efficiency modelling in foam filters which are also a tool in filter design and the understanding of fluid flow in porous materials.

ACKNOWLEDGEMENTS

The author acknowledges the Niger Delta University Bayelsa State, Nigeria for financial support. The author also thanks Loughborough University, UK for technical support. Particular thanks must go to the author's research supervisor Colin Garner for his untiring encouragement and Head of Department, P. P. Jombo for his understanding.

REFERENCES

Adigio EM, Binner JGP, Garner CP, Hague RJ, Williams AM (2008). Modeling gas flow pressure gradients in gelcast ceramic foam diesel particulate filters. Proc. IMechE Part D: J. Automobile Eng., DOI: 10.1243/09544070JAUTO508 17. 222: 1471-1487.

Arey J (2004). A tale of two diesels. Environ. Health Perspect., DOI:10.1289/ehp.7031,112: 812-813.

Brown DM, Donaldson K, Borm PJ, Schins RP, Dehnhardt M (2003) Calcium and ROS-mediated activation of transcription factors and TNF-α cytokine gene expression in macrophages exposed to ultra-fine particles. Am. J. Phys. Lung Cell Mol. Phys., http://cat.inist.fr/?aModele=afficheN&cpsidt=15421220,

BS EN ISO 5167: Part 2: Measurement of fluid flow by means of pressure differential devices inserted in circular cross-section conduits running full. 286: 344-353.

Dybdahl M, Risom L, Mùller P, Autrup H, Wallin H (2003). DNA adduct formation and oxidative stress in colon and liver of big blue rats after dietary exposure to diesel particles. Carcinogenesis, DOI: 10.1093/carcin/bgg147. 24: 1759-1766.

Gabathuler JP, Mizrah T, Eckert L, Fischer A, Kaser P, Maurer A (1991). "New Developments of Ceramic Foam as a Diesel Particulate Filter", SAE Paper no. 910325.

Garshick E, Laden F, Hart JE, Rosner B, Smith TJ, Dockery DW, Speizer FE (2004). Lung cancer in railroad workers exposed to diesel exhaust. Environ. Health Perspect., DOI: 10.1289/ehp.7195. 112: 1539-1543.

Health Effects Institute, HEI, Perspectives, (2002) Understanding the health effects of components of the particulate matter mix: Progress and next steps. http://rds.yahoo.com/_ylt=A0geu4_NJOhJn9cAiadXNyoA;_ylu=X3oDMTE1dDhwNGY0BHNlYwNzcgRwb3M heffects.org/Pubs/annualreport01-02.pdfDNwRjb2xvA2FjMgR2dGlkA01BUDAwN 18xMDY-/SIG=12bijfbsq/EXP=1240036941/**http%3a//www.healt.

Hinds WC (1980). "The drug and the Environment", Sem. Medic. Respir., 1: 197-210.

Macdonald IF, El-Sayed MS, Mow K, Dullien FAL (1979). "Flow through Porous Media – the Ergun Equation Revisited", Industrial Engineering Chemistry Fundamentals, 18: 198-208.

Mayer A (1998). Selection Criteria for Diesel Particulate Trap Systems: VERT Experience. http://www.dieselnet.com/papers/9812mayer.html.

Mayer A, Czerwinski J, Matter U, Wyser M, Scheidegger, Kieser D, Weidhofer (1998). "VERT: Diesel Nano-Particulate Emissions: Properties and Reduction Strategies", SAE No. 980539.

Nauss K (1997). Diesel exhaust: A critical analysis of emissions, exposure and health effects. DieselNet technical report. http://www.dieselnet.com/papers/9710nauss.html.

Nikula KJ, Finch GL, Westhouse RA, Seagrave J, Mauderly JL, Lawson DR, Gurevich M (1999). Progress in understanding the toxicity of gasoline and diesel engine exhaust emissions. Society of Automotive Engineers. Technical paper series. http://www.osti.gov/bridge/servlets/purl/771098g01olN/native/771098.pdf.

Pontikakis GN, Koltsakis GC, Stamatelos AM (2001). Dynamic Filtration Modelling in Foam Filters for Diesel Exhaust, Chemical Engineering Comm., 00: 1-26.

Reader-Harris MJ, McNaught JM (2005). Best Practice Guide Impulse Lines for Differential-Pressure Flowmeters, NEL No.: 2005/224. http://rds.yahoo.com/_ylt=A0geu8g6H.hJYw4APNFXNyoA;_ylu=X3oDMTE2bWZ2MzcwBHNlYwNzcgRwb3MDMTEEY29sbwNhYzlEdnRpZANNQVAwMDdfMTA2/SIG=1442s42ps/EXP=1240035514/**http%3a//www.idconline.com/technical_references/pdfs/instrumentation/Differential%2520pressure%2520meters_nel.pdf.

Richardson JT, Peng Y, Remue D (2000). "Properties of Ceramic Foam Catalyst Support: Pressure Drop", Applied catalysis A: General 204: 19-32.

Siegmann K, Siegmann HC (1997). "The Formation of Carbon in Combustion and How to Quantify the Impact on Human Health, Euro physics news 28.

Swiss Agency for the Environment, Forest and Landscape, SAEFL, 3oo3 Berne, (2000) Particulate Trap for Heavy Duty Vehicles, Environmental Documentation. No. 130. http://rds.yahoo.com/_ylt=A0geu7hnDehJywMANThXNyoA;_ylu=X3oDMTE1ZGw0b24zBHNlYwNzcgRwb3 MDMQRjb2xvA2FjMgR2dGlkA01BUDAwN18xMDY-/SIG=16oq3agei/EXP=12400 30951/**http%3a//www.ba fu.admin.ch/luft/00596/00597/00609/index.html%3flang=en%26 download=NHzLpZig7t,lnp6I0NTU042l2Z6ln1ad1IZn4Z2qZpnO2Yuq2Z6gpJCDeIB7fmym162dpYbUzd,Gpd6emK2Oz9aGodetmqaN19XI2ldvoaCVZ,s-.pdf.

Swiss Agency for the Environment, Forest and Landscape, SAEFL, 3oo3 Berne (2000) "Particulate Trap for Heavy Duty Vehicles", Environmental Documentation No. 130 http://www.umweltschweiz.ch/imperia/md/content/luft/fachgebiet/e/industrie/Partikelfilter_UM130_e.pdf for further details (last accessed 17th November 2004).

United State Environmental Protection Agency, (2002) Health Assessment Document for Diesel Engine Exhaust. US Environmental Protection Agency, Office of Research and Development, National Center for Environmental Assessment, Washington Office, Washington, DC, EPA/600/8-90/057F. http://oaspub.epa.gov/eims/eimscomm.getfile?p_download_id=36319

Warheit DB, Laurence BR, Reed KL, Roach DH, Reynolds GAM, Webb TR (2004). Comparative pulmonary toxicity assessment of single-wall carbon nanotubes in rats. Toxicol. Sci., DOI: 10.1093/toxsci/kfg228. 77: 117-125.

World Health Organization (2003). Health Aspects of Air Pollution with Particulate Matter, Ozone and Nitrogen Dioxide. Report on WHO Working Group, Bonn, Germany. EUR/03/5042688 http://rds.yahoo.com/_ylt=A0geu9SpCehJvQoAbRZXNyoA;_ylu=X3oDMTE1ZGw0b24zBHNlYwNzcgRwb3MDMQRjb2xvA2FjMgR2dGlkA01BUDAwN18xMDY-/SIG=11v59svcc/EXP=1240029993/**http%3a//www.euro.who.int/document/e79097.pdf.

Effect of heat treatment on dry sliding wear of titanium-aluminum-vanadium (Ti-6Al-4V) implant alloy

B. K. C. Ganesh[1]*, N. Ramaniah[2] and P. V. Chandrasekhar Rao[3]

[1]Department of Mechanical Engineering, Narsaraopeta Engineering College, Narsaraopeta, A. P. India.
[2]Department of Mechanical Engineering, College of Engineering (Autonomous), Andhra University, Visakhapatnam, India.
[3]Department of Mechanical Engineering, L. B. R. College of Engineering, Mylavaram. A. P. India.

Titanium and its alloys have very attractive properties that enable them to be used in the fields of aero space, biomedical, marine and also in many corrosive environments. The application of these alloys is more attractive today in the field of biomedical implant materials due to their superior biocompatibility and strength. These alloys have high coefficient of friction and poor abrasive wear resistance which results in the wear of the implant during its fixation in the body. Implant wear is a common phenomenon which results due to high friction between artificial implant materials when in contact with natural bone - this is much higher than healthy and natural joints that can withstand cyclic loads acting on them. The corresponding wear of the implant results in the accumulation of wear debris in the body tissues which results in inflammation, pain and loosening of implant resulting in shorter life period of the implant. Heat treatment of the alloy is one of the important techniques to improve the sliding wear properties of the alloy. The property of poor abrasive resistance can be altered by changing the microstructure of the alloy where the formation martensitic structure (acicular α or retained β) is resulted. The formation of martensitic structure in titanium alloy results in improved hardness value with a subsequent improvement in its sliding wear behavior. In this work, the implant material is subjected to heat treatment above its transformation temperature followed by slow cooling in furnace, air and water. These specimens were further aged and tested for dry sliding wear properties against hardened steel disc using a pin-on-disc apparatus using weight loss method with an optimal load of 50 N and a sliding distance of 500 m. An improvement of wear rate is reported under different heat treatment condition and an analysis of wear track of the specimens is done using scanning electron micrographs (SEM).

Key words: Biomaterial, heat treatment, martensitic structure, dry sliding wear.

INTRODUCTION

Titanium and its alloys have been extensively used today to replace the natural bone joint with artificial joints. The important correlation of bone substitute materials include biocompatibility with hard and soft tissue, an elastic modulus near that of bone and tensile ,compressive strength and fracture toughness equal to or greater than that of toughness equal to or greater than that of bone. In addition to fatigue strength, the wear resistance of the material should guarantee a safe operation of the implant during the expected period of use (Boehlart et al., 2008).

Degradation of artificial implant materials due to high wear rates can lead to unfavorable biological effects on bone density and implant fixation, resulting in shorter lifetime. Therefore, wear resistance is an important property while considering a suitable implant material. The wear resistance is directly proportional to the contact stress and friction at the part of contact and the type of heat treatment the material is subjected to. The type of heat treatment in a particular solution treatment of the implant results in the formation of acicular martensitic structure which greatly improves the hardness and wear resistance of implant.

*Corresponding author. E-mail: bkcganesh@yahoo.com.

Wear resistance plays an important role whenever a material (a bone plate for fracture fixation) is attached to fractured bones of a different stiffness values and modules of elasticity; relative movements between the two different materials (bone and implant) and between the parts of multi-component systems must occur when a cyclic load is applied to the system. These relative movements causes a wear stress on the attachment devices (bone screws as well as on the eyelets). Therefore, high wear resistance is required for orthopedic implants to obtain biocompatibility and acceptability (Thomann, 2000).

Titanium exists in various allotropic forms. At low temperatures, it has a closed packed hexagonal crystal structure known as α, whereas above 883°C, it has a body centered cubic structure known as β. The α to β transformation temperature of pure titanium either increases or decreases based on the nature of the alloying elements. The alloying element such as aluminum, oxygen, nitrogen, etc., that tend to stabilize are called α stabilizers and the addition of these elements increase beta transus temperature, while elements that stabilize β phase are known as β stabilizers such as vanadium, molybdenum, niobium, iron, etc., and addition of these elements depress the β transus temperature. Some of the elements such as zirconium and stantium which do not have marked effect on the stability of either of the phase but form solid solutions with titanium are termed as neutral elements (Geetha et al., 2009).

Mitsuo (2008) reported that the application of stress by rapid quenching results in the formation of martensitic structure in steels which contain residual austenite in their microstructures. This phenomenon is called stress or strain induced martensitic transformation which enhances ductility or fracture toughness of steel. Deformation induced martensitic transformation also occurs in titanium where unstable β phase is retained at room temperature by rapid cooling such as water quenching from a high temperature near the β transformation temperature.

The draw back of extensive use of titanium alloys in hip replacement and other artificial joints is poor tribological properties such as poor abrasive wear resistance, poor fretting behavior and high coefficient of friction. The improvement of the above properties can be done with the help of four main mechanisms as suggested by Zhecheva et al. (2005). They are as follows:

i. To induce a compressive residual stress.
ii. To decrease the coefficient of friction.
iii. To increase the hardness.
iv. To increase the surface roughness.

The surface modification and change in microstructure can be obtained by heat treating the various samples at beta transus temperature (transformation temperature) where primary α changes from hexagonally closely packed crystallographic structure (α) to body centered cubic crystallographic structure (β). These specimens were aged to complete the transformation of retained beta in order to achieve strengthening of the alloy.

Wear rate calculation by pin and rotating disc machine using weight loss method is one of the common techniques to evaluate dry sliding wear behavior of Titanium implant materials. Molinari et al. (1997) investigated on dry sliding wear mechanism of titanium-aluminum-vanadium (Ti-6Al-4V) alloy. In their experimental work, it has been found that the wear volume of the rotating specimens is reported as a function as sliding speed and the load applied on the pin. An increase in wear volume results with an increase in applied load. Alam et al. (2002), in their experimental work of dry sliding wear, reported that under a constant load of 45 N applied on pin, an increase of wear rate was identified up to a sliding distance of 500 m. There after, a steady state is attained; In this condition no appreciable change in the wear rate behavior of the alloy was observed.

Venkatesh et al. (2009) reported an improvement in various mechanical properties when α + β alloys were heat treated above the beta transus temperature and cooled by water quenching, air cooling and furnace cooling. Ajel et al. (2007) have studied the influences of heat treatment on Ti-6Al-7Nb implant alloy, where strength of the alloy is reported due to change in the alloy microstructure.

Loads acting on human joint vary considerably from joint to joint. For a particular joint, it varies with time during the loading cycle (Majumdar et al., 2008). It has been reported that stresses in the living area are of the order of 1 MPa. Gispert et al. (2007) used a normal pressure of 0.88MPa.In this work a load of 50 N with a contact pin diameter of 8mm is used to obtain a pressure of 1 Mpa which is considered to be a safe stress acting on the joint during the loading conditions.

It is evident that from this literature that the effect of various types of heat treatment influences the various changes in the respective microstructures which further results in obtaining tailor made mechanical properties and tribological properties of the implant alloy. In this work, dry sliding wear tests are conducted on Ti-6Al-4V implant alloy subjected to various heat treating conditions. These tests were conducted by considering an optimal load of 50 N with sliding distance of 500 m under a sliding velocity 1 m/s Microstructure evaluation along with Scanning electron microscope (SEM) analysis is done in support of the experimental work.

MATERIALS AND METHODS

The chemical composition (% by weight) of the alloy is as follows: 89.6% titanium, 6.29% aluminum, 3.95% vanadium, 0.09% iron, and 0.029% carbon.

The pins were cut according to the standard dimensions as shown in Figure 1 by wire EDM process employing a brass wire of nominal diameter of 0.3 mm. For Microstructure analysis all the cut specimens were mechanically polished via a standard

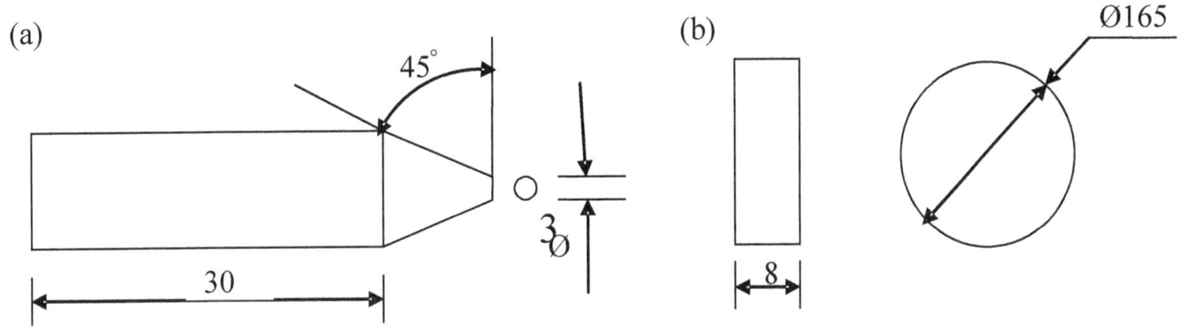

Figure 1. Dimensions of the (a) friction pin; (b) rotating disc in millimeters.

metallographic procedure to a final level of 0.3 μm alumina powder and etched with a solution of water, nitric acid and hydrofluoric acid (80:15:5 in volume). Scanning electron micrograph (SEM) analysis was also conducted to study the wear track behavior of various heat treated specimens. Hardness values were measured by Vickers micro hardness testing machine with a constant load of 0.5 kg. Average surface roughness (Ra) value of the wear track was measured by surface roughness testing instrument.

In this work, three samples of each condition were heat treated at 950°C for one hour in an argon controlled atmosphere for a period one hour. It is then aged at 550°C for a period of three hours in an air controlled furnace and subsequently cooled under various cooling environments such as cooling in a furnace, air and quenching in water. These specimens were then tested by DUCOM wear testing machine and the results are reported.

RESULTS AND DISCUSSION

Microstructure

Figure 2a represents the microstructure of the 'as received' alloy at room temperature showing equiaxed α+β matrix. Figure 2b reveals the formation of lamellar microstructure of α plates surrounded by prior β grain boundary which is due to the influence of slow cooling in the furnace. Figure 2c represents more number of primary α grains surrounded by needle like transformed or retained beta which is formed as a result of moderate cooling of the alloy in the air. Figure 2d represents acicular or needle like α in the matrix of primary α and β. Further, the scanning electron micrograph of quenched specimen shown in Figure 3 represents the presence of acicular (needle like) martensitic structure (ά) in white globular primary α in α + β matrix. Equiaxed microstructures shown in Figure 2a often have high ductility as well as fatigue strength and are preferred for super plastic deformation, while lamellar microstructure shown in Figure 2b have high fracture toughness and show superior resistance to creep and fatigue crack growth .The formation of bimodal microstructure as shown in Figures 2d and 3 combine the advantage of both equiaxed and lamellar microstructures. The presence of acicular martensitic structure greatly improves the hardness values and also its ultimate tensile strength (Leyens and Peter, 2003). The microstructures have been consistent with respect to the work done on this alloy by Jha et al. (2010) and Molinari et al. (2010).

Wear rate analysis

Figure 4 shows the wear rate behavior of the alloy under different heat treating conditions. Table 1 shows various properties associated with sliding wear of the implant alloy. Presence of high amount of wear is reported from the wear testing of 'as received' material. This corresponds to low hardness value which is Vickers hardness number (VHN) 311 as compared to quenched specimen which is having a hardness value VHN value of 380. High amount of hardness values in quenched specimen is due to the presence of acicular ά (martensitic structure) which was formed due to heat treatment above beta transus temperature followed by water quenching and aging. The wear rate as shown in Table 1 is also high as in the case of 'as received' material when compared to air cooling or water quenched specimens. The wear rate of furnace cooling specimen is greater than the water quenched and air cooled specimens due to the formation of lamellar α plate like structure where there is no presence acicular ά or retained beta as shown in Figure 2b. This is due to slow cooling of the specimen where complete transformation of body centered cubic crystallographic (β) structure to hexagonally closely packed (α) structure has taken place. This transformation to hexagonally packed structure limits plastic deformation of the furnace cooled specimen due to which higher wear rate is obtained

Weight loss and wear track analysis

Maximum amount of weight loss has been reported from 'as received' material as shown in Figure 5, where as less amount of wear is reported from both water quenched and air cooled specimens. The results commensurate with the hardness values and microstructure behavior of all the heat treated specimens.

Surface roughness values measured perpendicular to

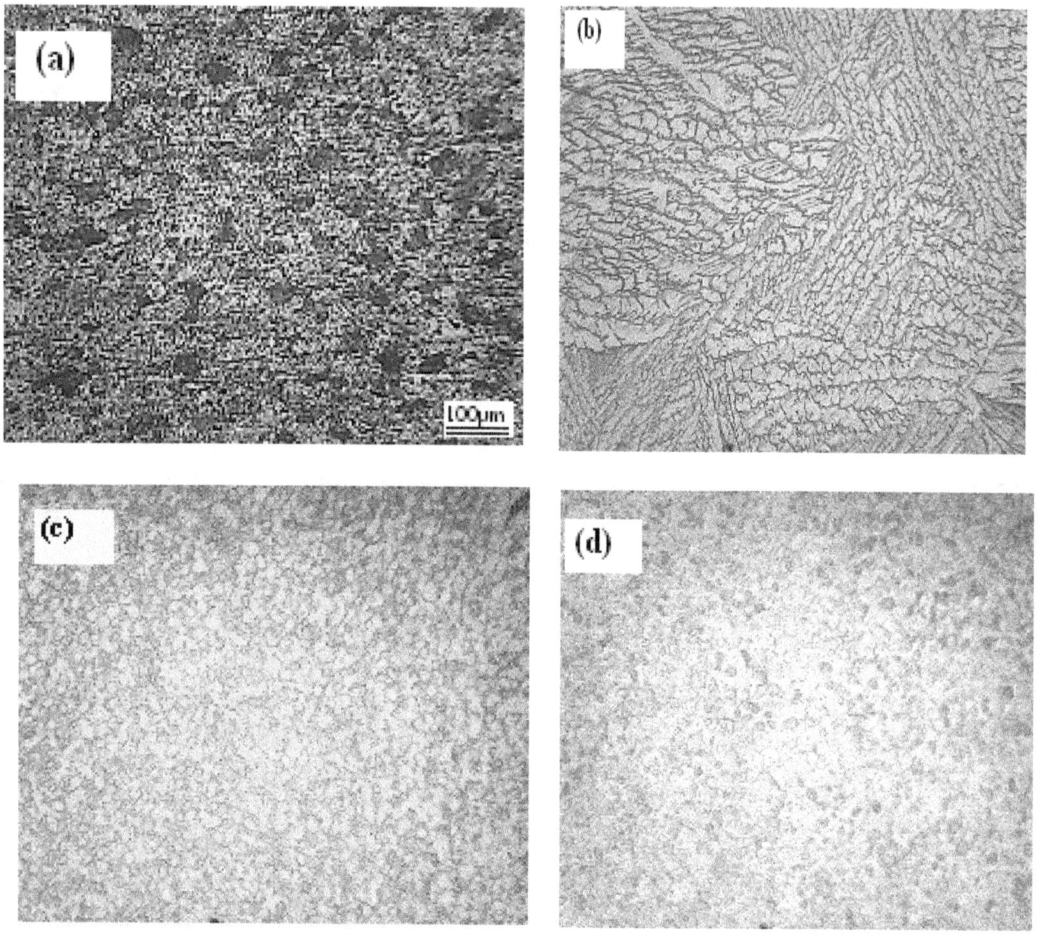

Figure 2. (a) Microstructure of 'as received' material; (b) furnace cooled and aged; (c) air cooled and aged; (d) water quenched and aged specimen at a Magnification of 100X.

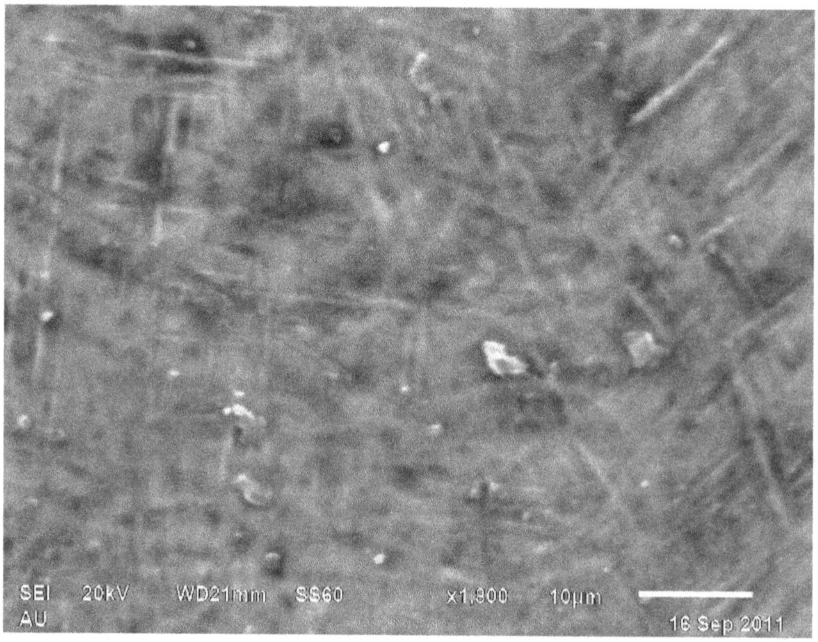

Figure 3. Scanning electron micrograph of quenched specimen.

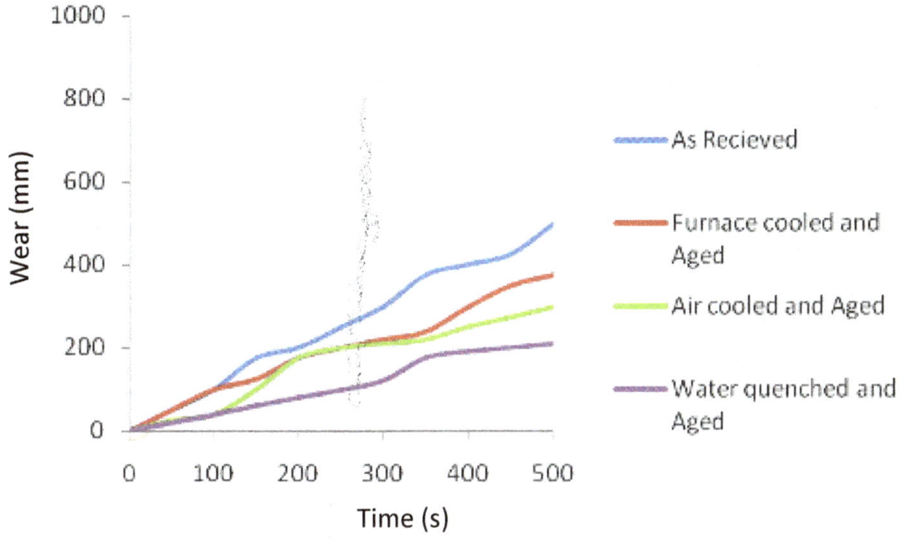

Figure 4. Wear behavior of the alloy under various heat treatment conditions.

Table 1. Wear properties under various heat treating conditions.

Heat treated condition	Wear track surface roughness (Ra)	Micro hardness ($HV_{0.5}$)	Wear rate x $10^{-11} m^3/m$
As received	2.11	311	1.954
Furnace cooled and aged	1.29	351	0.93
Air cooled and aged	0.858	340	0.186
Water quenched and aged	1.411	380	0.139

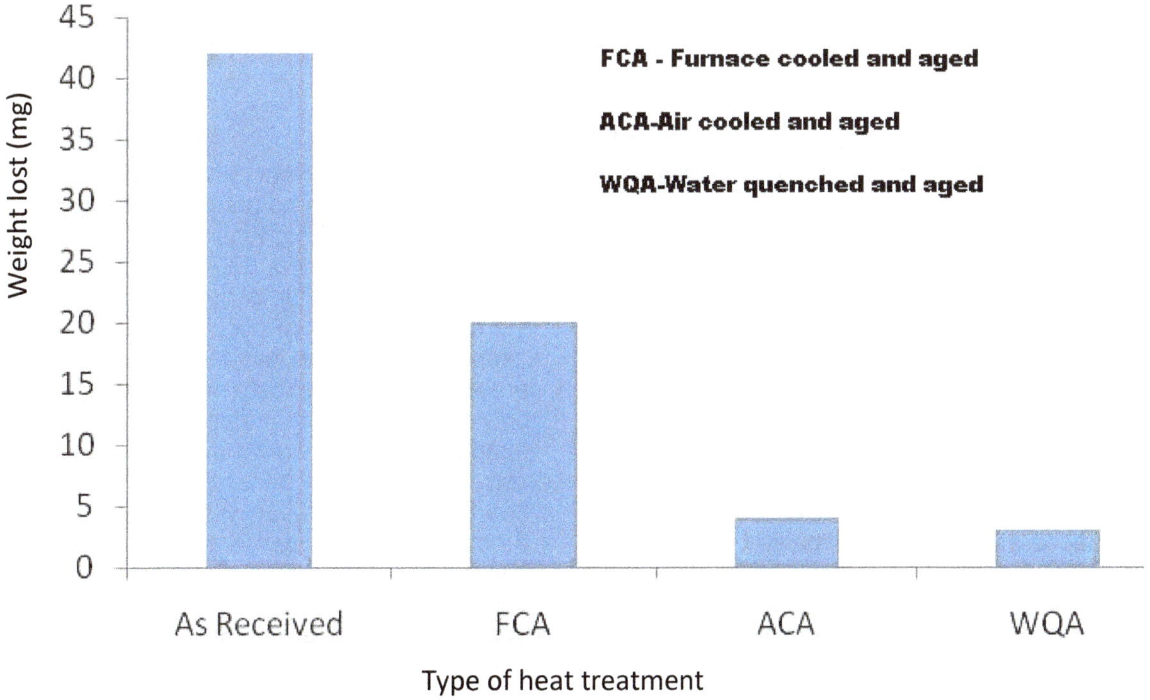

Figure 5. Weight loss of the alloy after heat treatment.

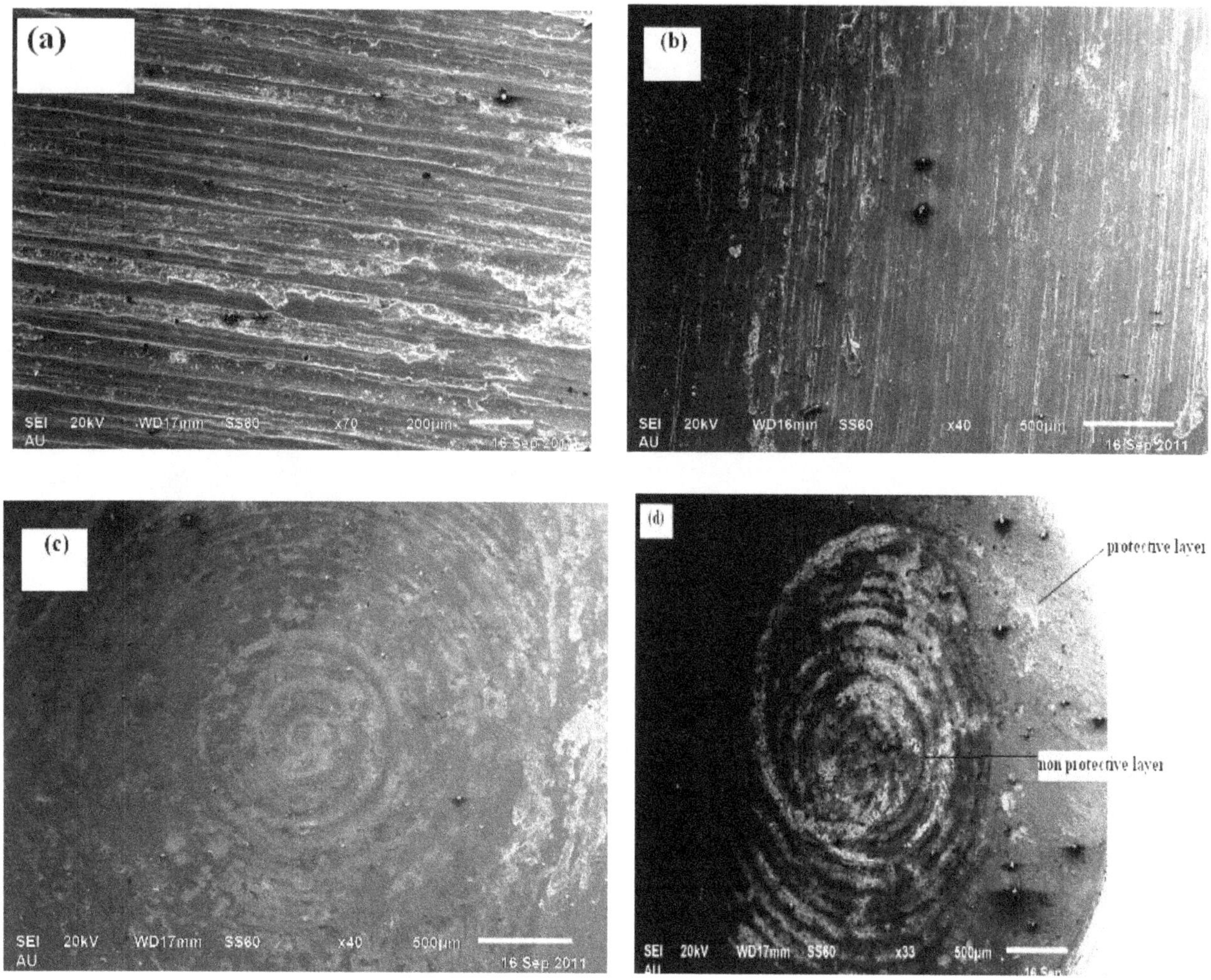

Figure 6. (a) Wear track of 'as received' material; (b) furnace cooled and aged; (c) air cooled and aged; (d) water quenched and aged specimen.

the wear track are given in Table 1. The wear rate of as received material is high due to low hardness of the specimen. This is due to coarse wear track developed during the testing procedure. The surface roughness value of water quenched specimen is moderately higher than that of the other heat treated specimens due to formation martensitic structure by heat treatment which resulted in an increase in the hardness value.

The wear track of 'as received' specimen as observed in scanning electron microscope (SEM) is coarse and thick with the presence of white titanium oxide globules as shown in Figure 6a. The wear track of furnace cooled specimen is moderately fine as compared to base metal specimen. Both the air cooled and water quenched specimens shown in Figure 6c and d accounts for the presence of protective layer coating. The coating is only formed due to faster rate of cooling, where the specimen is subjected to water quench or cooling in air. This coating, along with the martensitic (ά) in the microstructure of the quenched specimen, play an important role in limiting the wear rate when subjected to the above conditions. There is no presence of protective layer which can be observed from Figure 7a and b, when the implant alloy is heat treated at 600°C and simultaneously cooled by air and water. This is due to the fact that oxidation of the specimen has not occurred below the transformation temperature of the implant alloy. It is also clear that presence of retained beta or acicular alpha will take place only during heat treatment above transformation temperature with cooling in water and air medium. The corresponding wear track of these specimens shows very fine track layer of wear with no protective layer present. From the energy dispersive spectrometry analysis (EDS) shown in Figure 8a, a clear picture of the constituent elements is available. The protective layer mainly consists of carbon along with significant amount of oxygen which indicates that the specimen has undergone oxidation. It is evident from the literature that when titanium and its alloys are exposed to oxygen containing atmosphere, it results in the formation of an oxide

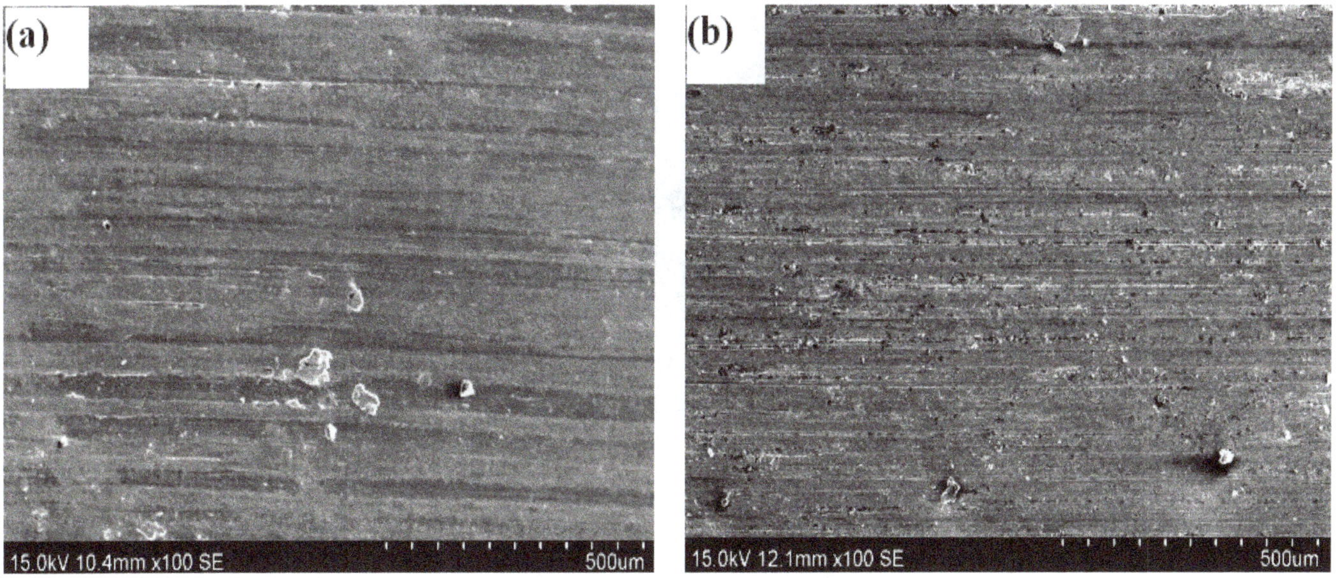

Figure 7. (a) Wear track of water quenched specimen heat treated at 600°C; (b) furnace cooled specimen.

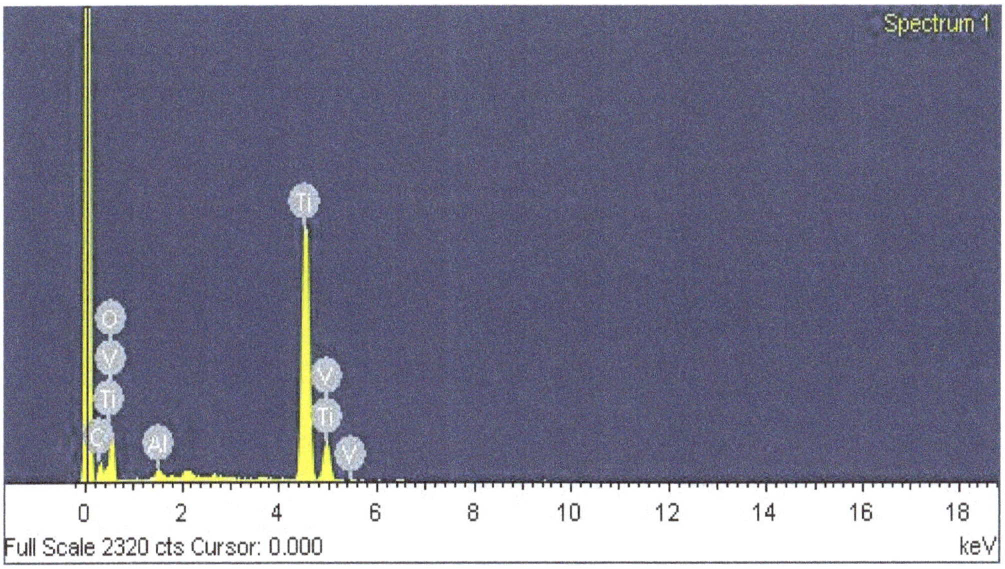

Element	Weight (%)	Atomic (%)
C K	1.71	5.33
O K	11.29	26.45
Al K	0.46	0.63
Ti K	83.26	65.17
V K	3.29	2.42
Total	00	100

Figure 8a. EDS analysis of protective layer.

layer on its surface with an oxygen diffusion zone beneath it (Guleryuz and Cimenglu, 2008). This formation is more pronounced when the alloy is heat treated above transformation temperature and cooled rapidly in air or quenching by water. This formation further plays an important role in developing a protective layer which promotes remarkable advantage of the alloy while working in a friction and wear environment. The presence of protective layer is crucial in improving corrosive and wear resistance. The presence of carbides which would have formed during heat treatment also plays an important role in minimizing the weight loss as well as the wear rate of the quenched specimen when in contact with the rotating disc. On the other hand, more amount of iron is detected from the non protective layer shown in Figure 8b, indicating adhesive wear of the alloy specimen while in contact with the sliding steel disc.

Conclusions

i. The wear rate of quenched specimen is very low due to the presence of protective oxide coating layer formed during heat treatment and also due to the presence of

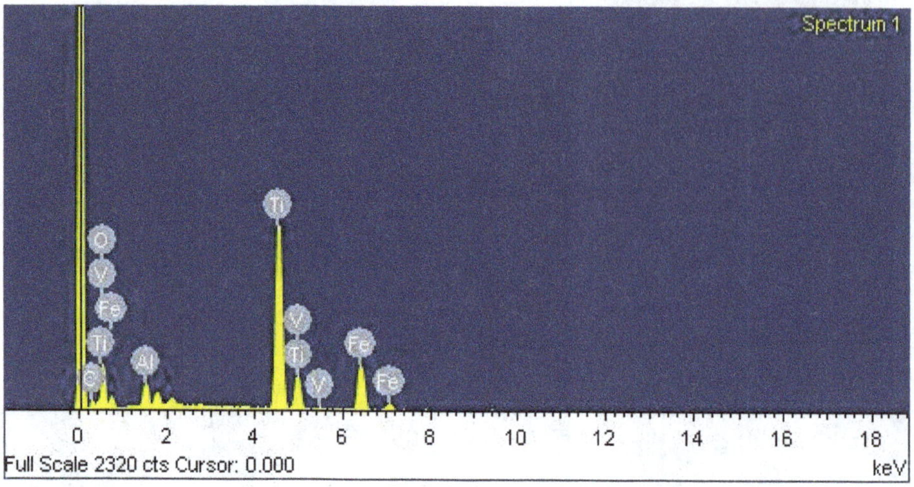

Element	Weight (%)	Atomic (%)
C K	1.11	3.49
O K	12.14	28.57
Al K	1.60	2.24
Ti K	71.47	56.19
V K	4.30	3.18
Fe L	9.38	6.32
Total	100	100

Figure 8b. EDS analysis of non protective layer.

acicular martensitic structure (retained β) in its microstructure.

ii. Finer wear tracks were observed when the specimen hardness is increased during rapid quenching and air cooling of the specimen below the transformation temperature

iii. Finer α grains in α+β matrix were observed in all the microstructures which are formed as a result of aging where complete transformation of retained β has taken place to α.

iv. Formation of protective oxide has taken place only in quenched and air cooled specimens, which indicates the presence of oxide layer at times of faster rate of cooling the alloy below the transformation temperature

v. Formation of a protective oxide layer plays an important role in improving corrosive resistance and wear resistance of the implant alloy.

vi. Ti-6Al-4V implant alloy due to its α+β binary alloy composition results in obtaining various dry sliding wear properties when it is heat treated above its transformation temperature and cooled through various medium such as furnace, air and water..

ACKNOWLEDGEMENTS

The authors would like to thank Prof. B. V. RamaMohan Rao, Prof. M. Koteswara Rao and Prof. Ramesh and Prof. K. Varaprasada Rao for their kind support in extending their support in utilizing the institute laboratory facilities. Technical guidance by Prof. T. R. Parthasarathy of METMECH Laboratories, Chennai, and Dr. S. V. N. Pammi Scientist, DST-PURSE, Advanced Analytical Laboratory, Andhra University, Dr. K. SivaPrasad, Metallurgical and Materials Engineering Department, NIT Trichy, Dr. A. Krishanaih, Department of Mechanical Engineering, Osmania University and Mr. Ganesh Puthucode, Ultra Surface Finishers, Chennai is greatly acknowledged.

REFERENCES

Ajeel SA, Thair AL, ASK (2007). 'Influence of heat treatment conditions on microstructure of Ti-6al-7Nb alloy as used surgical implant materials'. J. Eng. Technol., 25: 431 442.

Alam O, Haseeb ASMA Include names of other authors (2002).'Response of Ti-6Al-4V and Ti-24Al-11Nb alloys to dry sliding wear against hardened steel'. J. Tribol. Int., 35: 357-362.

Boehlart CJ, Cowen CJ, Quast JP, Akahori T, Niinomi M (2008). ' Fatigue and wear evaluation of Ti-Al-Nb alloys for biomedical applications'. J. Mater. Sci. Eng., 28: 323-330.

Geetha M, Singh AK, Asokamani R, Gogia AK (2009). 'Ti based biomaterials, the ultimate choice for orthopedic implants' - A Review . J. Prog. Mater. Sci., 54: 397-425.

Gispert MP, Serro AP, Colaco R, Botelho Do Rego, Tribological AM (2007). Behavior of Cl-Implanted TiN Coatings for Biomedical Applications. J. Wear., 262: 1337-1345.

Guleryuz H, Cimenglu H (2008). Article in Press, DOI: 10.1016/j.jallcom.2008.04.024.

Jha AK, Singh SK, Kiranmayee MS (2010). 'Failure analysis of titanium alloy (Ti6Al4V) fastener in aerospace application'. J. Eng. Fail. Anal., 17: 1457-1465.

Leyens C, Peters M (2003). "Titanium and Titanium Alloys. Fundamentals and Applications", Wienham Germany: Wiley-Vch Publishers, Chapter 1.

Majumdar P, Singh SB, Chakraborty M (2008). Wear response of Heat Treated Ti-13 Zr-13Nb Alloy in Dry Condition and Simulated body Fluid. J. Wear., 264: 1015-1025.

Mitsuo Ni (2008). 'Mechanical biocompatibilities of titanium alloys for biomedical applications'. J. Mech. Behav. Biomed. Mater., 1: 30-42.

Molinari A, Straffelini G, Tesi B, Bacci T (1997). 'Dry sliding wear mechanisms of the Ti6Al4V alloy, J. Wear., 208: 105-112.

Thomann Ul, Petar UJ (2000). 'Wear Corrosion Behavior of biocompatible Austenitic stainless steel', J. Wear., 239: 48-58.

Venkatesh BD, Chen DL, Bhole SD (2009). 'Effect of heat treatment on mechanical properties of Ti-6Al-4Valloy', J. Mater. Sci. Eng., 506: 117-124.

Zhecheva A, Sha W, Malinov S, Long A (2005). 'Enhancing the microstructure and properties of titanium alloys through nitriding and other surface engineering methods'. J. Surf. Coat. Technol., 200: 2192-2207.

Experimental determination of entropy generation in flat heat pipes

Maheshkumar P.* and Muraleedharan C.

Department of Mechanical Engineering, National Institute of Technology, Calicut NIT Campus (P.O), Kerala, India 673601.

Heat pipe is a thermodynamic device which transfers heat over considerable distances with a very small temperature drop. Entropy generation can be considered as a significant parameter on heat pipe performance. Major reasons for entropy generation in a heat pipe system are temperature difference between cold and hot reservoirs, frictional losses in the working fluid flows and vapor temperature/pressure drop along heat pipe. The present objective is to estimate the entropy generation in a two dimensional flat heat pipe. An experimental set up is fabricated for the analysis of the transient operation of a flat heat pipe. The entropy generation depends on both temperature and velocity distributions. The analysis involves the measurement of vapor temperature and vapor velocity and using the obtained temperature and velocity distributions, entropy generation rate is calculated.

Key words: Heat pipe, entropy generation, transient analysis, second law of thermodynamics.

INTRODUCTION

A heat pipe is a simple device of very high thermal conductance. It can transmit heat at high rate over considerable distance with extremely small temperature drop (Bejan, 1982). The design of heat pipes is simple and it is easy to manufacture. Its maintenance is also easy. Heat pipes have found various applications including energy conversion systems, cooling of nuclear reactors and electronic components, etc.

A conventional heat pipe has three sections; (1) the evaporator where heat is added to the system, (2) the condenser where heat is rejected from the system and (3) the adiabatic section which connects the evaporator and condenser, serving as a flow channel. The working fluid inside the heat pipe undergoes a thermodynamic cycle which generates entropy. The entropy generation in a heat pipe is due to frictional losses in the flow of working fluid and heat transfer across a finite temperature difference. The entropy generation rate can be used to quantify the irreversibility of the system which is directly related to the lost work during any process.

A large number of theoretical and experimental studies on heat pipes have been reported over past few decades. Cotter (1965) analyzed the laminar, steady, incompressible one-dimensional vapor flow in a cylindrical heat pipe. Bankston and Smith (1973) presented the solutions for the axisymmetric Navier-Stokes equations for steady laminar vapor flow in circular heat pipes with various evaporator and condenser lengths. Numerical calculations of the vapor flow in a flat heat pipe were presented by Van Ooijen and Hoogendoon (1981). The pressure profiles along the vapor channel of a flat heat pipe were experimentally found by Van Ooijen and Hoogendoon (1981). Numerical analysis of the vapor flow in a double walled concentric heat pipe was presented by Faghri (1986). Chen and Faghri (1990) studied the overall performance of the heat pipe with single or multiple sources of heat. A transient two dimensional analysis of the vapor core and wick regions of a flat heat pipe was performed by Unnikrishnan and Sobhan (1997).

Vasilev and Konev (1990) presented a thermodynamic analysis based on the assumption of constant vapor pressure along heat pipe. Rajesh and Raveendran (1994) developed an optimum design of heat pipe using nonlinear programming technique. Khalkali et al. (1999) presented the entropy generation in a heat pipe system. They developed a thermodynamic model of conventional

*Corresponding author. E-mail: nandumahesh03@yahoo.co.in.

heat pipe based on the second law of thermodynamics. A detailed parametric analysis was presented in which the effects of various heat pipe parameters on entropy generation were examined.

This paper aims at finding the entropy generation developed by thermodynamic irreversibility for a two dimensional flat heat pipe. It is seen that the entropy generation can be quantified in terms of velocity and temperature distributions of both liquid and vapor flow.

The three major factors causing entropy generation in a heat pipe are (1) temperature difference between hot and cold reservoirs (attached to the evaporator and condenser outer surfaces), (2) temperature drop in the vapor flow and (3) frictional losses associated with the vapor and liquid flows of working fluid in the heat pipe.

ANALYSIS OF HEAT PIPE

The heat pipe consists of an evacuated chamber, the interior of which is lined by a capillary structure saturated with a working fluid. The heat is essentially transferred as latent heat by evaporating the liquid working substance in a heating zone called evaporator and condensing the vapor in a cooling region called condenser. The circulation is completed by the return flow of the condensate to the evaporator through the wick under the driving action of capillary forces. This process will continue as long as the flow passage for the working fluid is not blocked and a sufficient capillary pressure is maintained. Due to this heat transfer and fluid flow between the reservoirs, entropy is generated and is formulated thus.

Entropy generation

The volumetric rate of entropy generation in a convective heat transfer problem is given as:

$$S^{III}_{gen} = \left(\frac{K}{T^2}\right)(\nabla T)^2 + \left(\frac{\mu}{T}\right)\phi \tag{1}$$

where the first term represents the entropy generation due to heat transfer and second term, entropy generation due to fluid flow friction.

For a two dimensional flow, the entropy generation equation becomes:

$$S^{III}_{gen} = \left(\frac{K}{T^2}\right)\left(\left(\frac{\partial T}{\partial x}\right)^2 + \left(\frac{\partial T}{\partial y}\right)^2\right) + \left(\frac{\mu}{T}\right)\left\{2\left(\left(\frac{\partial u}{\partial x}\right)^2 + \left(\frac{\partial u}{\partial y}\right)^2\right) + \left(\left(\frac{\partial u}{\partial y} + \frac{\partial v}{\partial x}\right)^2\right)\right\} \tag{2}$$

So, for obtaining the entropy generation rate, the velocity and temperature distributions of both vapor and liquid flow in a heat pipe are required.

THE PHYSICAL MODEL

An experimental model for analyzing the transient and steady state performance of a flat heat pipe system is presented. An experimental set up of heat pipe was fabricated. The experimental set up consists of two stainless steel flat plates as container material and three layers of stainless steel wire mesh as wick structure. The aim is to measure the temperature and pressure of the working fluid at different locations along the axial direction of heat pipe for different heat inputs. The temperature was measured at 10 min time interval until steady state is achieved. Experiments were carried out with water and acetone as working fluids. Other details of the tested heat pipe are: length of evaporator = 85 cm, length of adiabatic section = nil, length of condenser section = 15 cm (overall length = 100 cm) width = 5 cm and thickness = 1 cm. Calibrated thermocouples (8 numbers) are used to measure the temperature of the vapor along the axial direction of heat pipe. Heat pipe is provided with solar heating with the help of a parabolic collector. A pyranometer is employed to measure the solar radiation.

A wick porosity of 0.65 is used in the present analysis with the voids assumed to be saturated with water. The temperature and pressure characteristics of the heat pipe were obtained by using both the working fluids in the heat pipe system. Using the obtained velocity and temperature distributions, entropy generation rate due to vapor and liquid flows were estimated.

RESULTS AND DISCUSSIONS

Transient and steady state results from the experimental analysis are discussed here. The important results of analysis are the distributions of the velocity components, pressure, temperature, and entropy generation rate in the heat pipe. Figures 1, 2, 3 and 4 represent the axial distribution of vapor temperature along the centerline of vapor core at different time instants for heat inputs of 40, 50, 60 and 70 W, respectively when water is the working fluid. As expected, the temperature increases at the beginning portion of the evaporator and then decreases. This is because the solar radiation is more concentrated on the beginning of the evaporator section. The variation of vapor temperature along the centre line of the vapor core with acetone as the working fluid for heat inputs of 50, 60 and 70 W are shown in Figures 5, 6 and 7. The nature of variation is similar to that of water but magnitude of temperature is less. Figures 8, 9, 10, 11 and 12 show the steady state vapor temperature variation for different heat inputs with water and acetone were the working fluids. Figures 13, 14, 15 and 16 represent the variation of vapor velocity along vapor core for heat inputs of 40, 50, 60 and 70 W with water as working fluid. The steep increase in velocity along the evaporator section is due to the mass addition into the vapor core. At

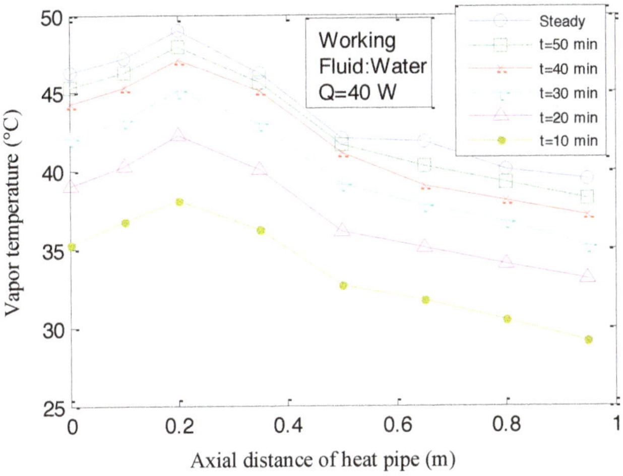

Figure 1. Variation of vapor temperature with axial length of heat pipe.

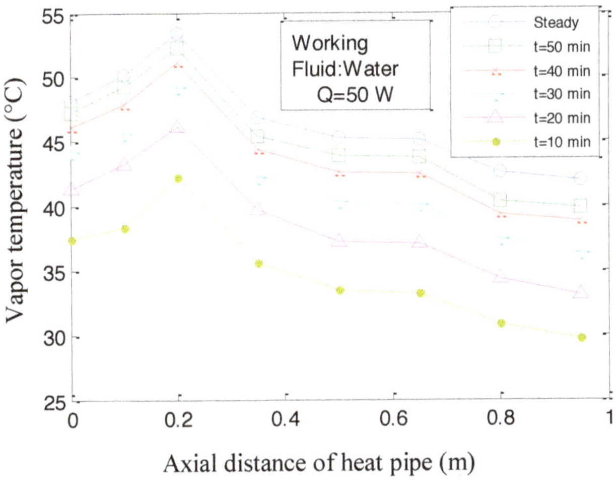

Figure 2. Variation of Vapor temperature with axial length of heat pipe.

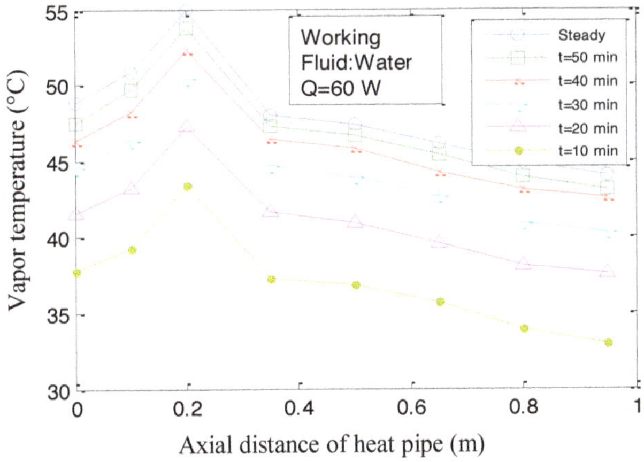

Figure 3. Variation of Vapor temperature with axial length of heat pipe.

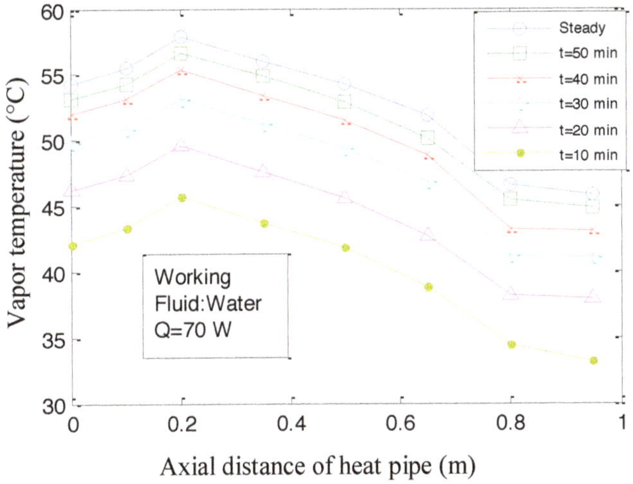

Figure 4. Variation of vapor temperature with axial length of heat pipe.

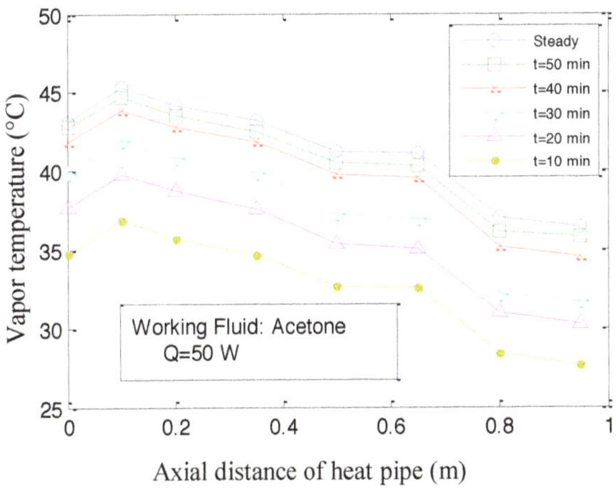

Figure 5. Variation of Vapor temperature with axial length of heat pipe.

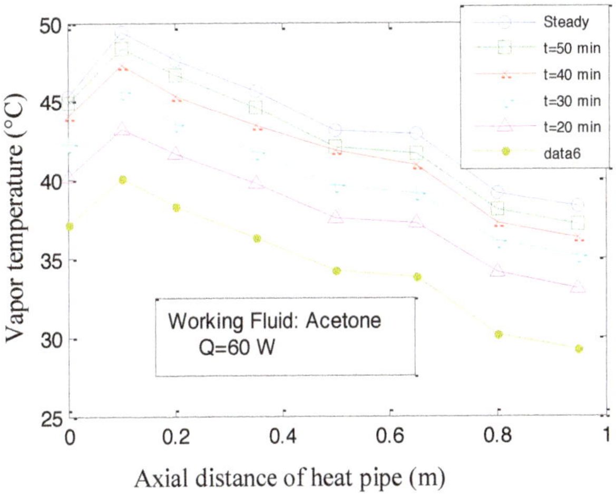

Figure 6. Variation of Vapor temperature with axial length of heat pipe.

Figure 7. Variation of Vapor temperature with axial length of heat pipe.

Figure 10. Variation of Steady State Vapor temperature with axial length of heat pipe.

Figure 8. Variation of Steady State Vapor temperature with axial length of heat pipe.

Figure 11. Variation of Steady State Vapor temperature with axial length of heat pipe.

Figure 9. Variation of Steady State Vapor temperature with axial length of heat pipe

Figure 12. Variation of Steady State Vapor temperature with axial length of heat pipe.

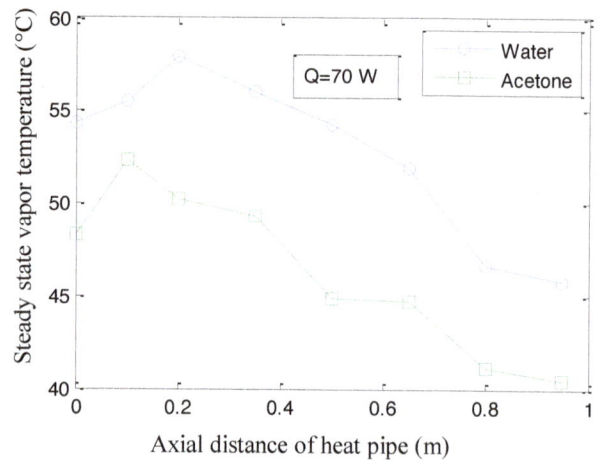

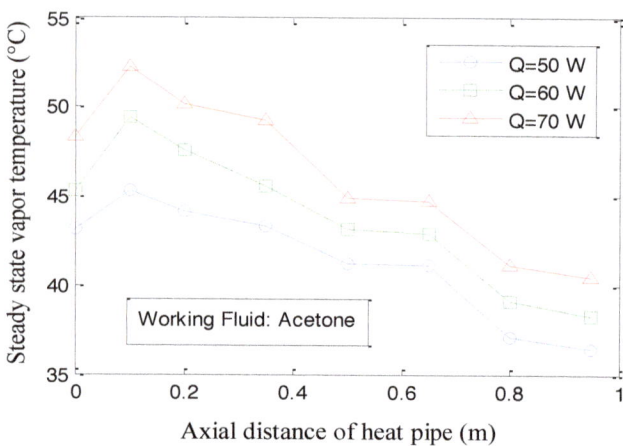

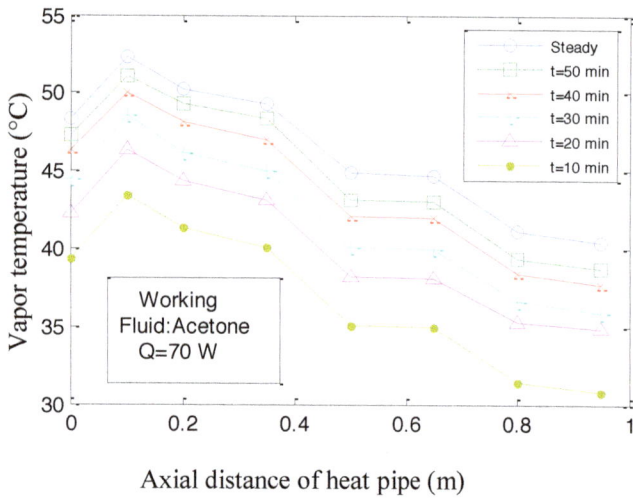

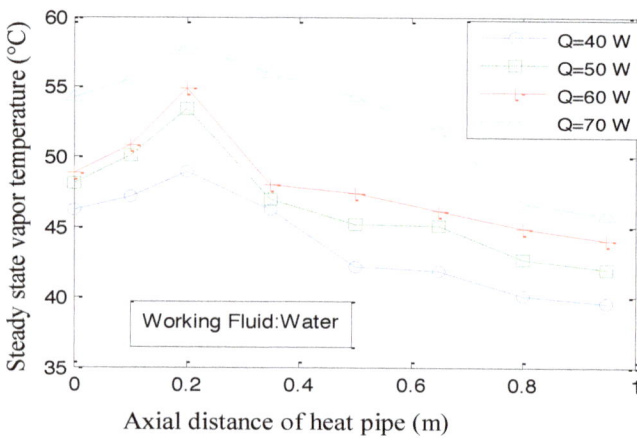

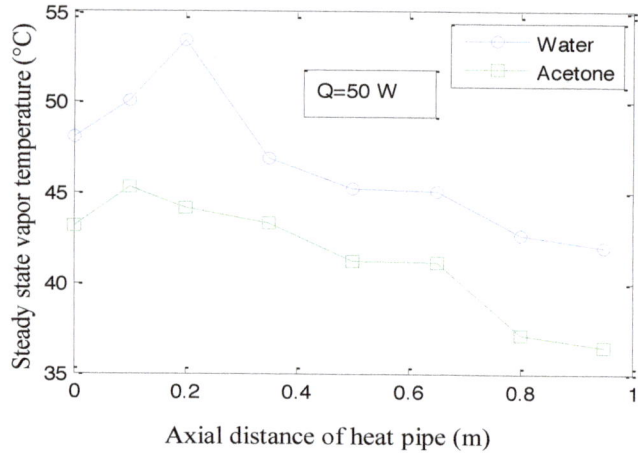

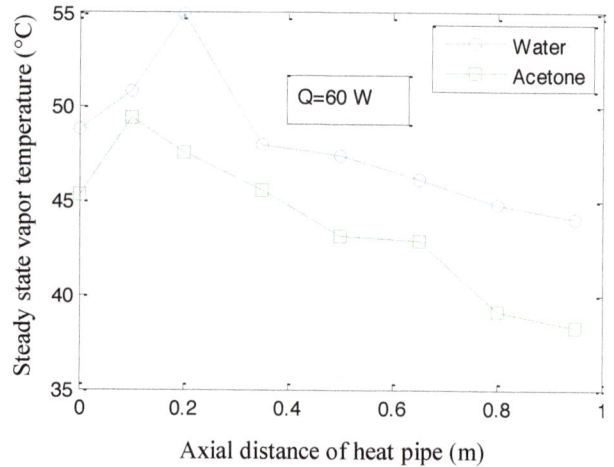

Figure 13. Variation of vapor velocity with axial length of heat pipe.

Figure 16. Variation of vapor velocity with axial length of heat pipe.

Figure 14. Variation of vapor velocity with axial length of heat pipe.

Figure 17. Variation of entropy generation rate with axial length of heat pipe.

Figure 15. Variation of vapor velocity with axial length of heat pipe.

the adiabatic and condenser sections the velocity is found to be decreasing. The rate of decrease is less at the adiabatic section. The steep decrease in velocity at the condenser section is due to high rate of mass transfer from vapor core into the wick as a result of condensation. Figures 17 and 18 show the variation of entropy generation due to vapor flow when water and acetone are the working fluids; it shows an irregular variation of entropy generation rate, since the temperature gradient along the length of the heat pipe is not varying uniformly. Figures 19 and 20 show the variation of thermal conductance with heat input for different working fluids, that is, water and acetone. It is found that the thermal conductance increases with heat load, as a result of more rate of increase in heat input compared to rate of

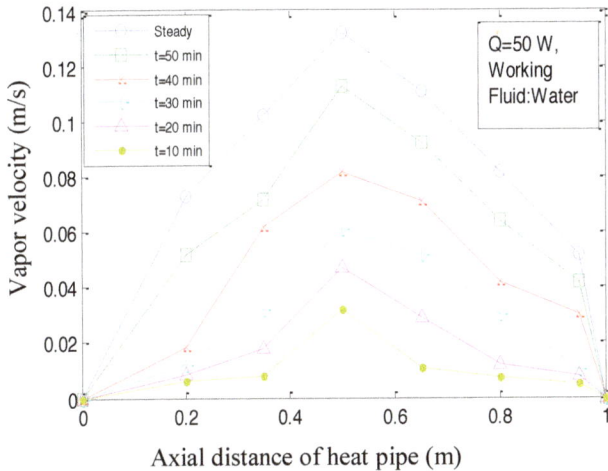

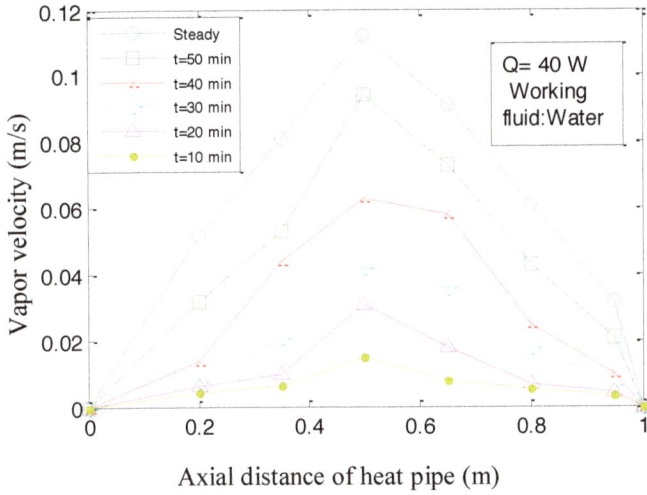

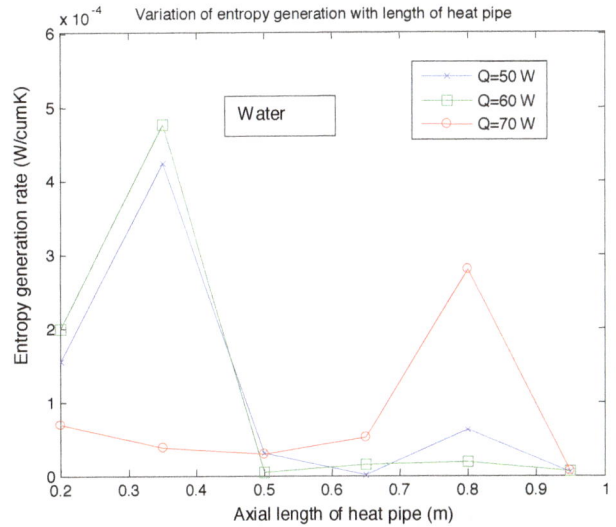

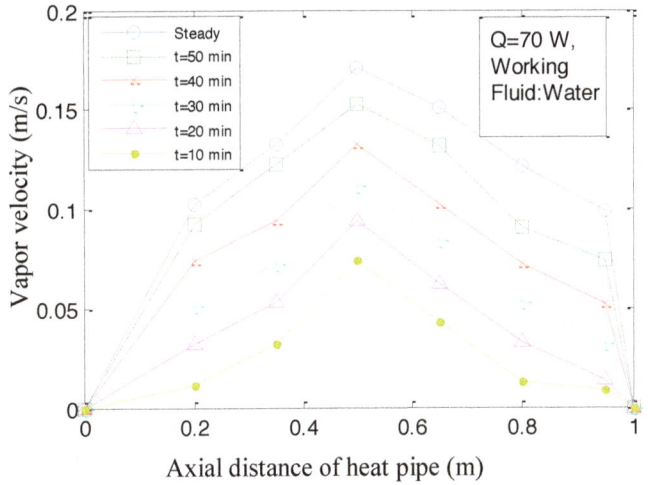

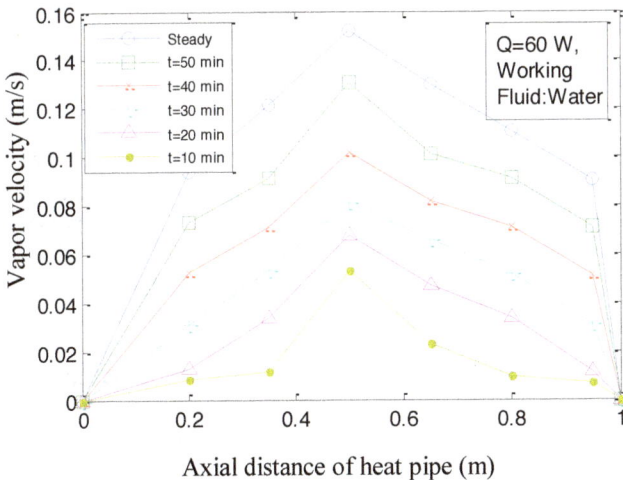

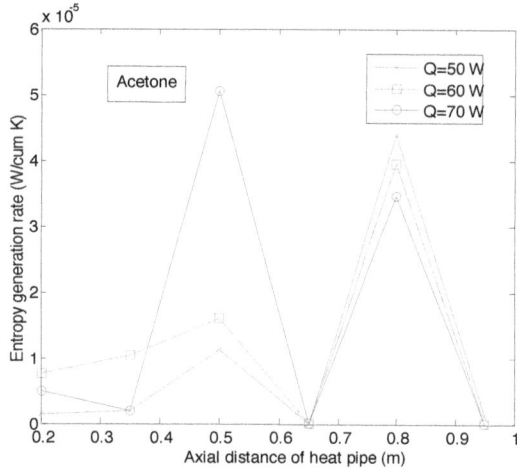

Figure 18. Variation of entropy generation rate with axial length of heat pipe.

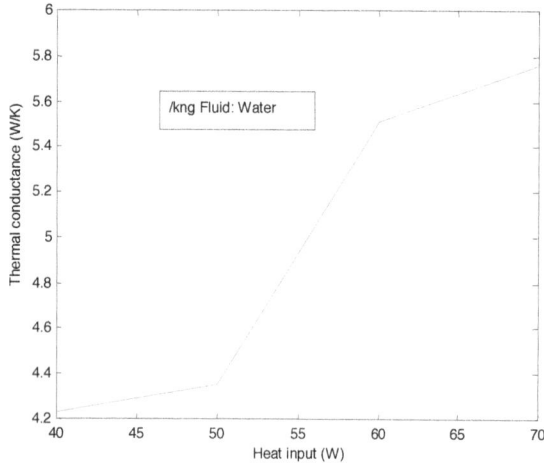

Figure 19. Variation of thermal conductance with axial length of heat pipe.

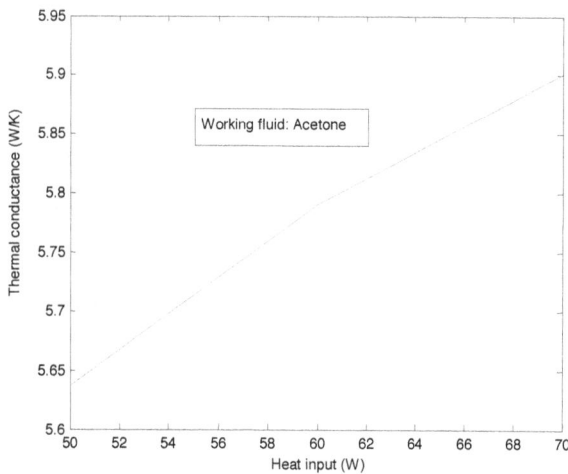

Figure 20. Variation of thermal conductance with axial length of heat pipe.

Figure 21. Comparison of experimental and numerical methods (Vapor temperature).

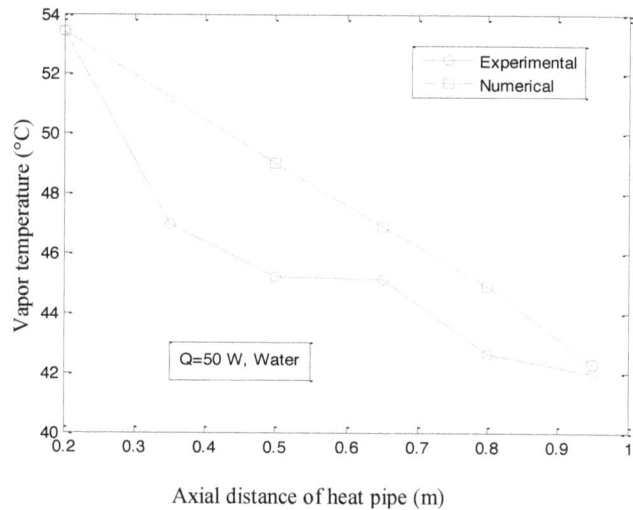

Figure 22. Comparison of experimental and numerical methods (Vapor temperature).

increase in temperature gradient. Figures 21 and 22 show the comparison of vapor temperature obtained by experimental and numerical methods for water for different heat inputs. It is found that both the results are matching. Figures 23 and 24 show the comparison of vapor temperature for acetone for different heat inputs; Figure 25 shows comparison of vapor velocity obtained by experimental and numerical methods for water. Figure 26 shows comparison of entropy generation rate obtained by experimental and numerical methods for water as working fluid. It is found that the results obtained by the numerical method are not much varying with those results obtained by the experimental method.

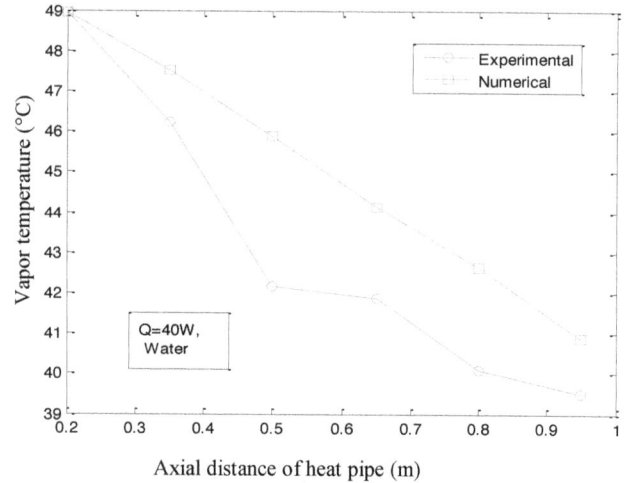

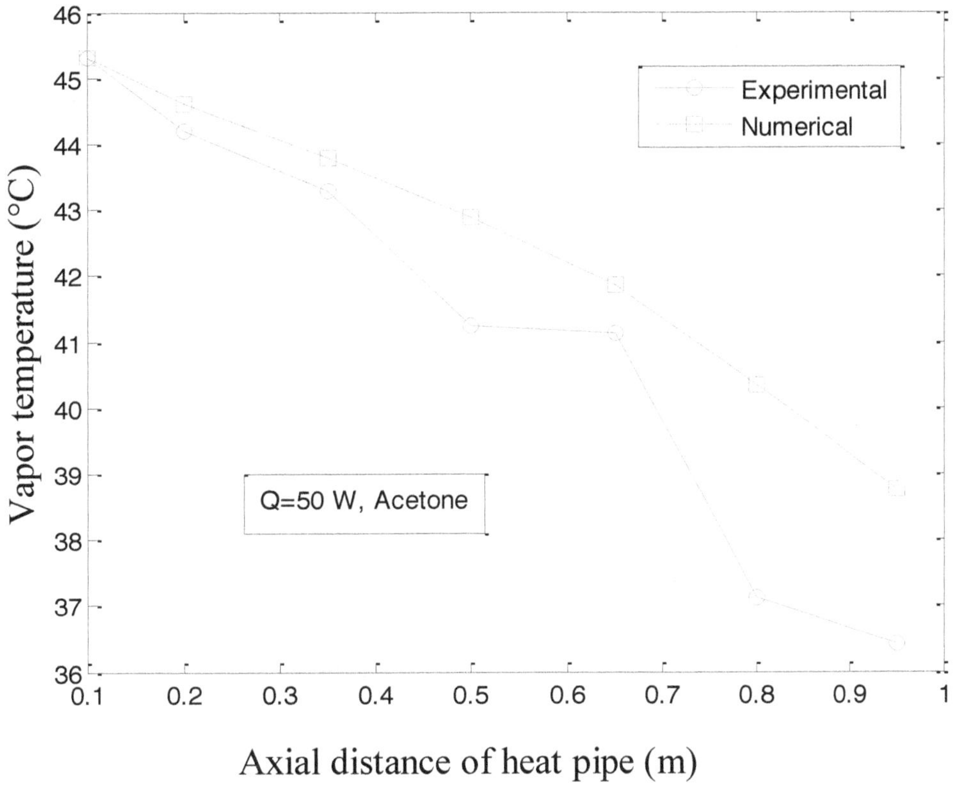

Figure 23. Comparison of experimental and numerical methods (Vapor temperature).

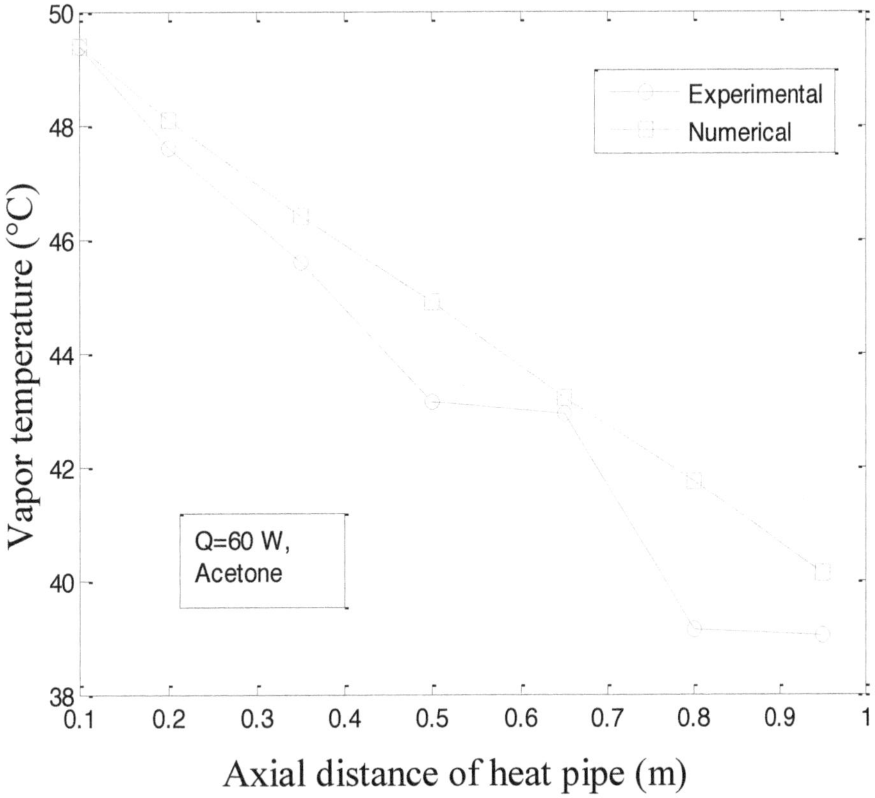

Figure 24. Comparison of experimental and numerical methods (Vapor temperature).

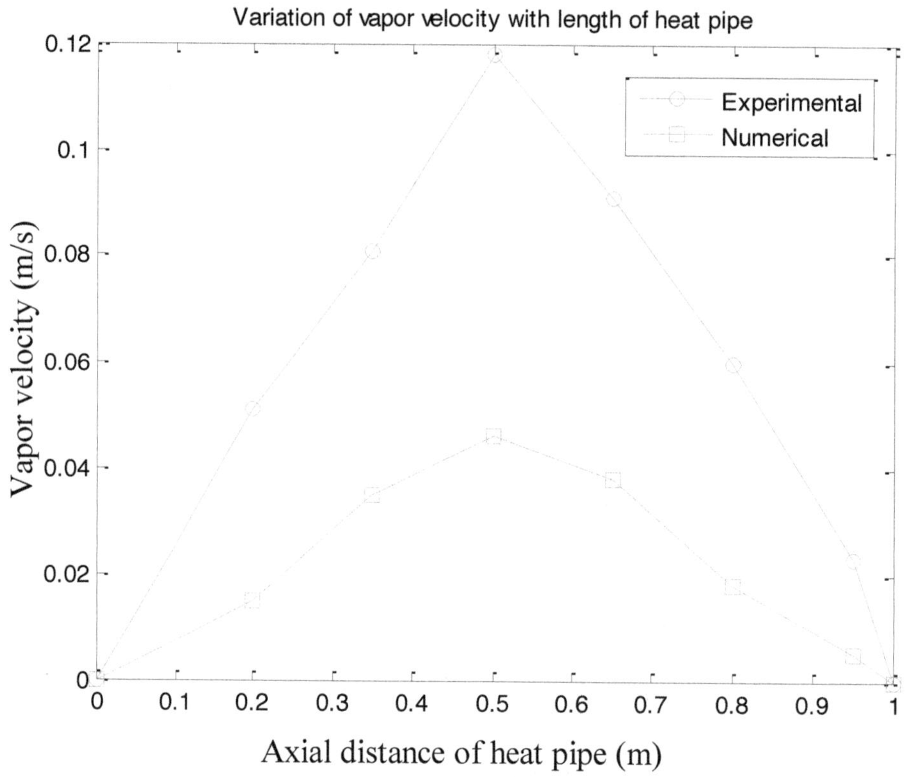

Figure 25. Comparison of experimental and numerical methods (Vapor temperature).

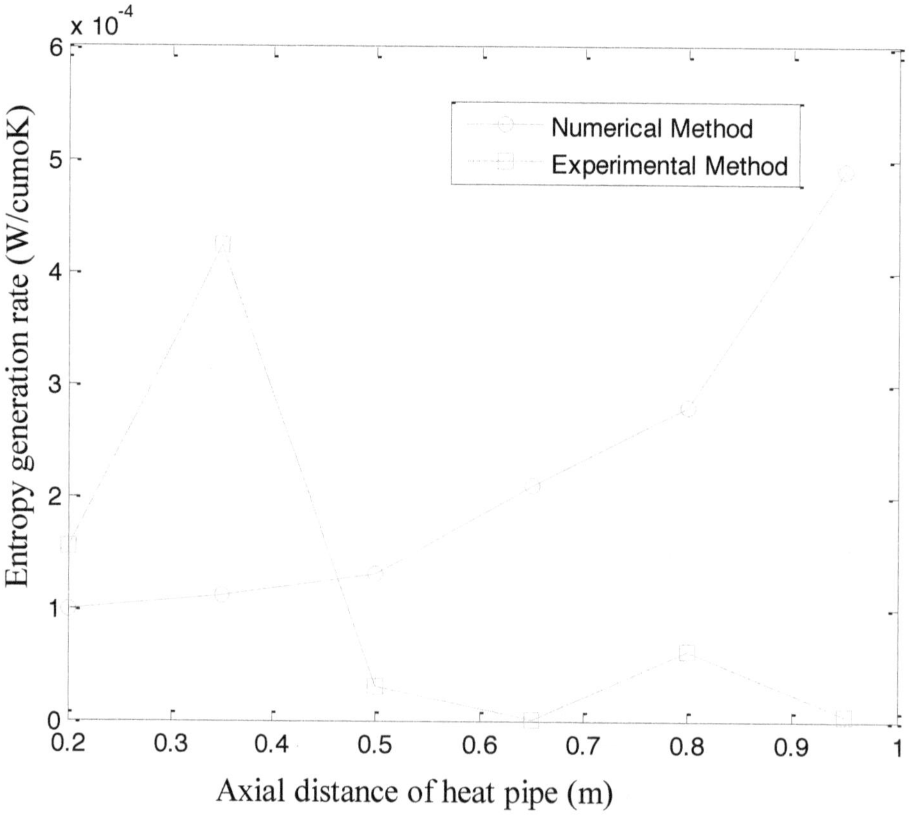

Figure 26. Comparison of experimental and numerical methods (Vapor temperature).

Nomenclature: **K,** Thermal conductivity, W/mK; **S,** entropy, J/kgK; $\mathbf{S}^{111}{}_{\mathbf{gen}}$**,** volumetric entropy generation, W/m^3K; **T,** temperature, K; **u,** longitudinal velocity, m/s; **v,** transverse velocity, m/s; **x,** axial distance, m; **y,** transverse distance, m.

Greek symbols: **μ,** Dynamic viscosity, Ns/m^2; ϕ, viscous dissipation.

REFERENCES

Bankston CA, Smith HJ (1973). Vapor Flow in Cylindrical Heat Pipes, J. Heat Transfer, 95: 371-376.

Bejan A (1982). "Entropy Generation Through Heat and Fluid Flow", Wiley, New York.

Chen MM, Faghri A (1990). "An Analysis of Vapor Flow and Heat Conduction Through the Liquid Wick in a Heat Pipe with Single or Multiple Heat Sources". Int. J. Heat Mass. Transfer, 33: 1945-1955.

Cotter TP (1965). Theory of Heat Pipes, Los Alamos Scientific Laboratory Report No. LA-3246-MS,

Faghri A (1986). Vapor Flow Analysis in a Double walled Concentric Heat Pipe, Numerical. Heat Transfer, 10: 583-593.

Khalkali, Faghri A, Zuo ZJ (1999). "Entropy Generation in a Heat Pipe System". Appl. Thermal. Eng., 19: 1027-1043.

Rajesh VG, Raveendran KP (1994). "Optimum Heat Pipe design: a non linear programming", M. Tech thesis, NIT Calicut.

Unnikrishnan VV, Sobhan CB (1997). "Finite Difference Analysis of the Transient Performance of a Flat Heat Pipe" Int. Conference on Numerical Methods in Thermal Problems", Swansea, UK, pp. 391-400.

Van Ooijen H, Hoogendoon CJ (1981). Experimental Pressure Profiles Along the Vapor Channel of a Flat Heat Pipe, Procs. International Heat Pipe Conference, London, pp. 415-426.

Van Ooijen H, Hoogendoon CJ (1981). Vapor Flow Calculation in a Flat Heat Pipe, AIAA J, 17: 1251-1259.

Vasilev LL, Konev SV (1990). "Thermodynamic Analysis of Heat Pipe", Proceedings of seventh International Heat Pipe Conference, Minsk,

7

Experimental approach to study friction factor and temperature profiles of fluid flow in circular microchannels

Mohd Nadeem Khan[1]*, Mohd Islam[2] and M. M. Hasan[2]

[1]Department of Mechanical Engineering, Krishna Institute of Engineering and Technology, Ghaziabad, India.
[2]Department of Mechanical Engineering, Jamia Millia Islamia, New Delhi, India.

Heat dissipation in electronic components becomes an important issue in efficiency promotion and stable operation. Microchannel heat exchanger plays the major role for heat dissipation from such high heat generating electronic components. In this connection an experimental investigation was conducted to explore the validity of classical correlations of friction factor based on conventional sized channels for predicting the fluid behavior in single-phase water flow through circular microchannels. The microchannels under investigation have the hydraulic diameter of 279 µm and 45 mm long. Test piece was made of stainless steel and the test section contained a total of seventy nine microchannels arranged in circumferential manner. The experiments were conducted with deionized water of Reynolds number ranging from approximately 300 to 3000. Pressure drop and flow rates were measured to analyze the flow characteristics. The results show good agreement between the classical correlations of friction factor and the experimentally measured data. The temperature profiles along as well as across the channels shows that channel length and channel diameter play the major role on its behavior.

Key words: Microchannels, friction factor, temperature profiles.

INTRODUCTION

The development of micro heat exchanger becomes an important area of interest in many fields of developing technology that requires compact high heat energy removal such as micro miniature refrigerators, micro heat pipe spreader, microelectronics, biomedical, fuel processing and aerospace etc. It is necessary to study not only the theoretical aspects but also experimental investigation of fluid flow and heat transfer in micro channel heat sink. In the beginning of the 1980s, Tuckerman and Pease (1981) conducted initial experiments on water flow and heat transfer characteristics in microchannel heat sinks that demonstrate the cooling of electronic components by the use of forced convective flow of fluid through microchannels. This opened a wide area in the field of electronics cooling and heat transfer in microscale geometries. The first study concerning circular microtube configurations was reported by Choi et al. (1991). They conducted experiments for evaluating the heat transfer and pressure drop characteristic of circular silica microtubes with inside diameters 3, 7, 10, 53 and 81.2 µm, and length 24 to 52 mm using nitrogen gas as a working fluid. They found that the Nusselt number in both laminar and turbulent flow depends on Reynolds Number in a different manner compared to macroscale theory. Based on their measurements the critical Reynolds number equals to 2000, which is concurrent with microtubes.

Yu et al. (1995) conducted experiments with liquid water flowing in circular tubes having inner diameters 19.2, 52.1 and 102 µm, measuring the heat transfer coefficient. They found that the heat transfer data at low Reynolds number agreed with those for microtubes, and diverged as Reynolds number increased. Adams et al. (1998,1999) investigated the heat transfer coefficient of turbulent water flowing in circular channels with inner diameters 0.76 and 1.09 mm, and found that the Nusselt number for these channels were higher than those for

*Corresponding author. E-mail: khanrkgit@rediffmail.com

microchannels. Mala and Li (1999) experimentally tested water flow through microtubes with inner diameter from 50 to 254 μm. Their results show that the deviation from the macro scale theory increases as the Reynolds number increases, and diameter decreases. They also concluded that there is an earlier transition from laminar to turbulent flow in microchannels. Celata et al. (2000) experimentally investigated liquid R114 flowing through six parallel tubes with inner diameter 30 μm and the length 90 mm. The transition from laminar to turbulent flow was reported in the range 1881<Re<2479, and they found a large discrepancy between the experimental heat transfer coefficient in microchannels and those predicted by classical correlations, particularly at high Reynolds numbers. Experimental analysis of the flow and heat transfer characteristics of water flowing through the microchannel made of stainless steel was done by Peng et al. (1994) and their fluid flow results were found to deviate from the values of classical correlation and the transition was observed to occur at the Reynolds number in the range of 200 to 700. These results were contradicted by Xu et al. (2000). Chein and Chuang (2007) experimentally studied the performance of microchannel heat sink using nanofluids. They found that nanofluid cooled microchannel heat sink could absorb more energy than water-cooled microchannel heat sink when the flow rate is low. For high flow rates, the heat transfer was dominated by the volume flow rate and nanoparticles did not contribute to the extra heat absorption. Their measured microchannel heat sink wall temperature variations agreed with the theoretical prediction for low flow rate.

For high flow rate, the measured microchannel heat sink wall temperatures did not completely agree with the theoretical prediction due to the particle agglomeration and deposition. It was concluded that the literature is inconclusive concerning prediction about heat transfer and fluid flow in microchannels. Therefore carefully designed experiments are necessary before final conclusions can be drawn. The objective of this work is to investigate experimentally the characteristics of water flow and heat transfer in microchannels and attempt to explain the obtained results.

EXPERIMENTAL SETUP AND PROCEDURES

A schematic diagram of the experimental setup used in this investigation for measurements of pressure and temperature difference at inlet and outlet of the test section is shown in Figure 1. Deionized water from a holding tank is driven through the flow pump, which provided smooth and steady flow over a wide range of flow rates that corresponds to a Reynolds number ranging from 300 to 3000. The fluid then passes through a 0.1 μm filter before entering the microchannel test section. The test section was enveloped by the heated oil in the oil bath. Details of the microchannel test section are shown in Figure 2. The test section consists of total seventy nine stainless steel tubes having the inner diameter 279 and 45 mm long arrange in the circumferential manner. Three copper–constantan (Type-T) thermocouples were used to measure the temperatures at the inlet and outlet of the test section as well as of the oil of the oil bath and hand operated digital RS232 manometer (has a pressure range of 0 to 100 Psi with an accuracy of ±0.3% of its full scale at 25℃) was used to measure the differential pressure between the inlet and outlet of the test section as shown in Figure 2. HJ-123 heater of 500 W was placed in the oil bath to heat the oil and the connection of this heater through the blind temperature controller (BTC) which cut the electric supply of the heater when the temperature of the oil reaches the desired temperature. The first step in conducting the experiment was to fill the water tank with Deionized water and note down the initial as well as final reading of measuring scale of water tank before and after filling. This gives the volume of water contained in the water tank.

Once the water was filled, heater is turned on the heater and wait until the temperature of the oil reaches the desired temperature. Once the oil in the oil bath beaker attends this temperature, open the valve of water tank and motor is switch on. This allowed water to flow through the Microchannel Heat Exchanger. A set flow rate was established with the help of controlled valves by monitoring the digital manometer and setting the valve at a position where a predetermined pressure was measured on the digital manometer. After a steady state was reached, temperatures of water at inlet and outlet of the test section were recorded from the monitor of temperature indicator keeping in mind that the temperature of the oil in the oil bath remained constant with time during measurement. Once the temperature measurements were completed, water is collected in the volumetric beaker from the exit section for predetermined period of time and measured the flow rate. This procedure was repeated for several times for various differential pressure readings. The experiments cover the Reynolds numbers from 304 to 2997.

RESULTS AND DISCUSSION

Friction factor

Yang et al. (2003) measured the friction factors of water flow in tubes with diameter ranging from 0.5 to 4.0 mm. They found that there is no significant discrepancy for water flow in these tubes in comparison with large tubes. Phillips (2008) studied the fluid flow and heat transfer characteristic in rectangular microchannels having the hydraulic diameter 234 μm using water as the working fluid. Their results agree well with the conventional correlations in the laminar region. The present study measured experimentally the friction factors of Deionized water flows in microchannels having the hydraulic diameter 279 μm and 45 mm long shown in Figure 3. The results of experimental data are plotted and shown in Figure 3. It was found that the experimental data agree very well with the developing laminar flow equation and Churchill Equation (13) in laminar and turbulent flow regime, respectively.

Developing laminar flow equation:

$$f_{lam,dev} = \frac{16}{\mathrm{Re}} + \frac{1.28D^2\,\mathrm{Re}}{4L} x10^{-9}$$

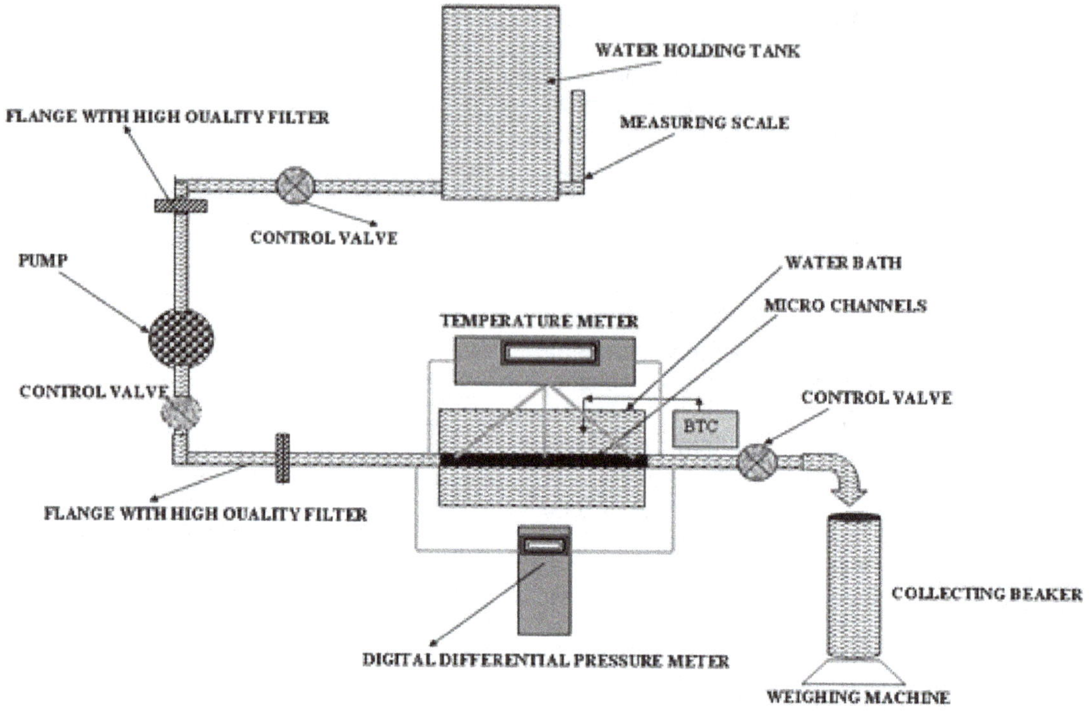

Figure 1. Schematic of the experimental system.

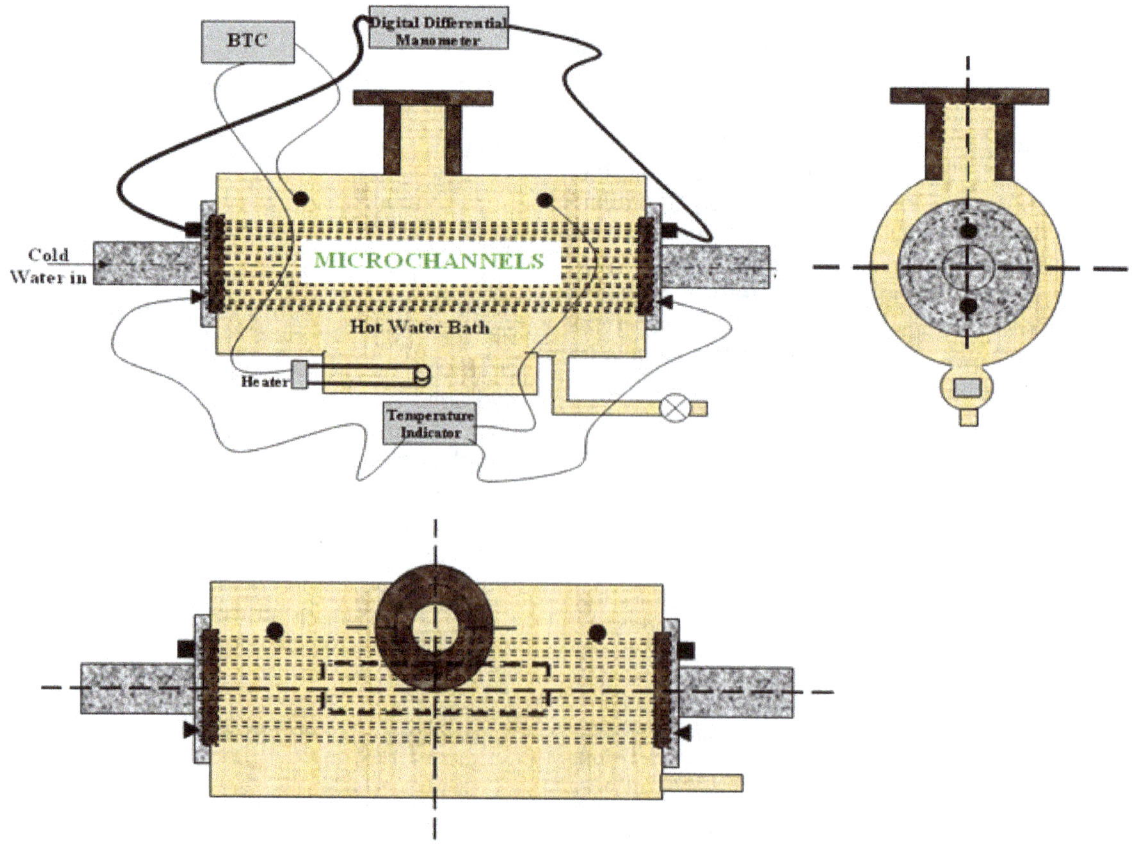

Figure 2. Detail of heat exchanger system.

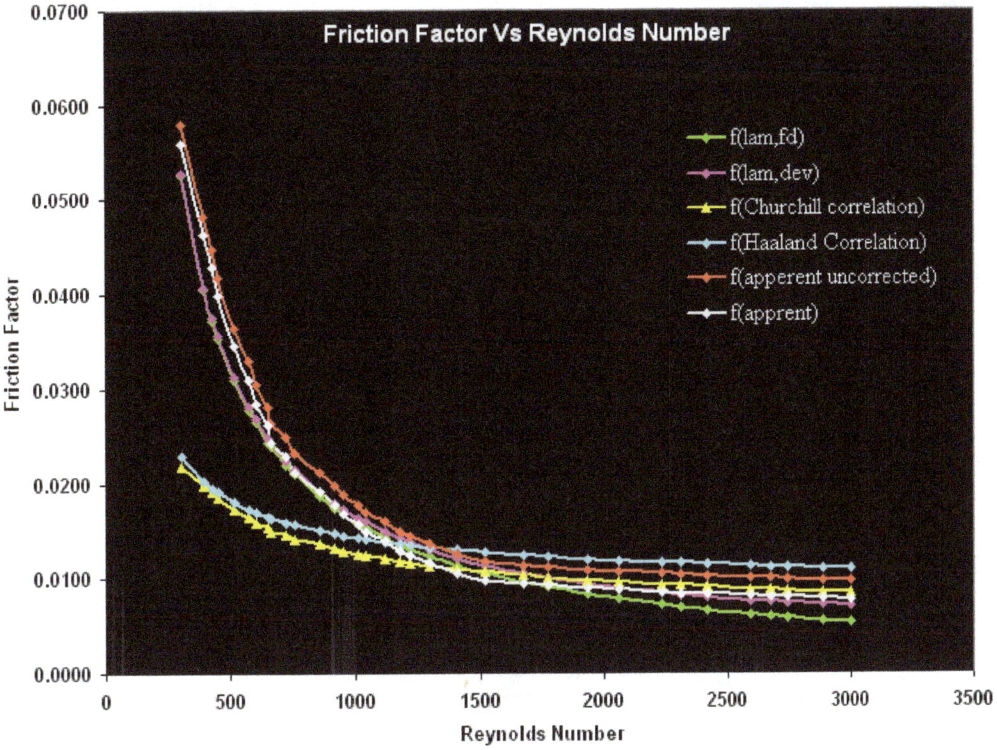

Figure 3. Comparative plots of friction factor with the experimental results.

Churchill equation: $f_{ch} = 8\left\{ \frac{1}{\left[\left(\frac{8}{Re}\right)^{10} + \left(\frac{Re}{36500}\right)^{20}\right]^{1/2}} + \left[2.21 ln\left(\frac{Re}{7}\right)\right]^{10} \right\}^{-1/5}$

Figure 3 shows that for Reynolds number greater than 2000, the experimental data follow the Churchill equation more exactly and in the range of 500<Re<1000, the experimental friction factor is close to developing laminar flow equation. This indicates that the in range 1000<Re<2000 flow is in the transition region. This early transition from laminar to turbulent seen in the microchannels tested here is due to the reason of inlet condition at the entrance and hydrodynamic entrance length. Figure 3 also shows the comparison of theoretical friction factor with the experimental friction factor. The plots show that there is no single equation that agrees well with the experimental as well as theoretical data over the entire range of Reynolds number under study. Using curve fitting method the best equation that describes the whole range of experimental data is f = 2.336 $Re^{-0.6886}$ and Figure 4 shows the comparison of proposed equation for friction factor with the experimental results.

The results show that the proposed equation of friction factor gives very close results with the experimental data. Therefore this equation help to predict the friction factor with the variation of Reynolds number ranging of this study without performing the experiment.

Temperature profiles

Experiments were conducted for single-phase water flow on the same experimental setup in order to find out the characteristics of temperature profiles in microchannels. The thermocouples inserted at the inlet and exit section of the test piece gives the inlet and exit temperature of water at different flow rates. The total heat applied to the working fluid to all the channels is given by $q_{tot} = \dot{m} c_p (T_{out} - T_{in})$ where the mass flow rate, $\dot{m}$, was measured from the volume of water collected from the outlet of the channels over a specified period of time. The mass flow rate in term of measured quantities of experiment is reduce as $\dot{m} = \rho A_c V_{avg}$. It is noted that the boundary condition of the experiment was that of a constant surface temperature for the walls of the microchannels. Under this assumption, the temperature profiles are plotted w.r.t. Reynolds number as shown in Figure 5. Figure 5 shows that in the low Reynolds number range exit temperature as well as difference of exit to inlet temperature decreases more rapidly than in the high Reynolds temperature range and this might be due to that in low Reynolds number range fluid in the microchannels get more time for gaining more heat as compared to in high Reynolds temperature range. Figure 5 also indicates that all the temperature profiles are more

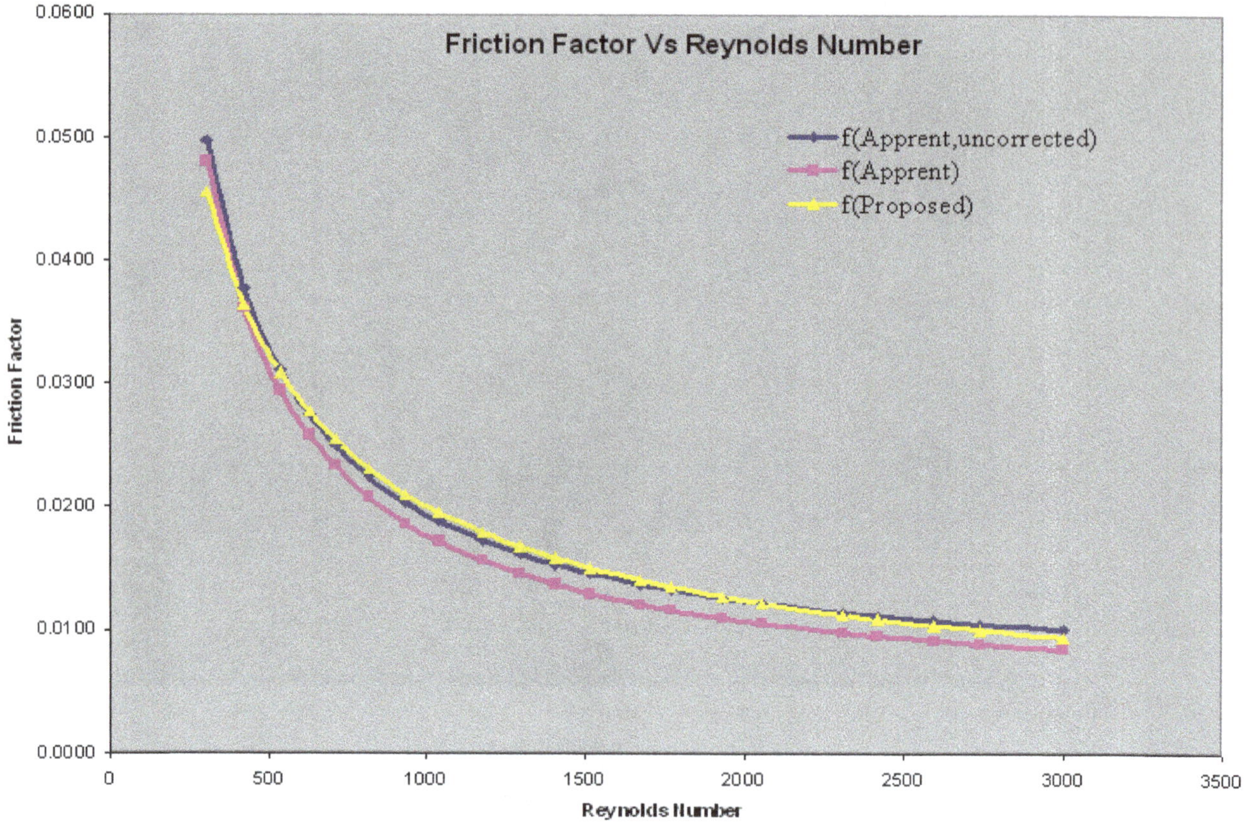

Figure 4. Comparative plots of experimental friction factor with proposed equation.

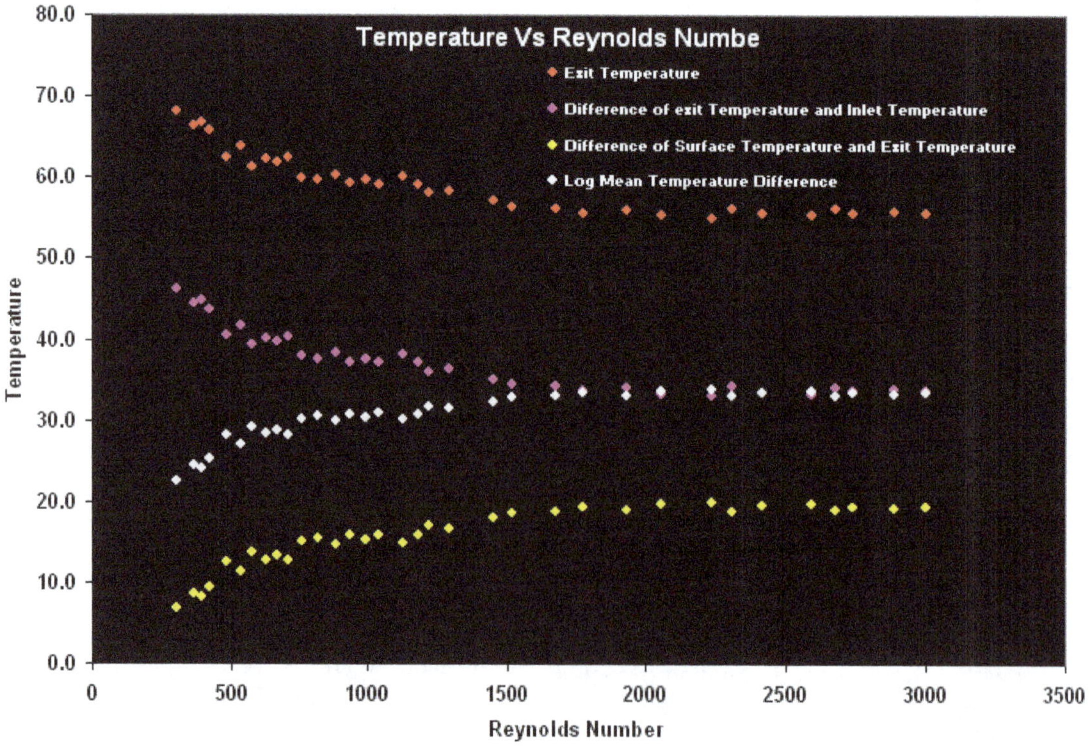

Figure 5. Temperature profiles of w.r.t. Reynolds number.

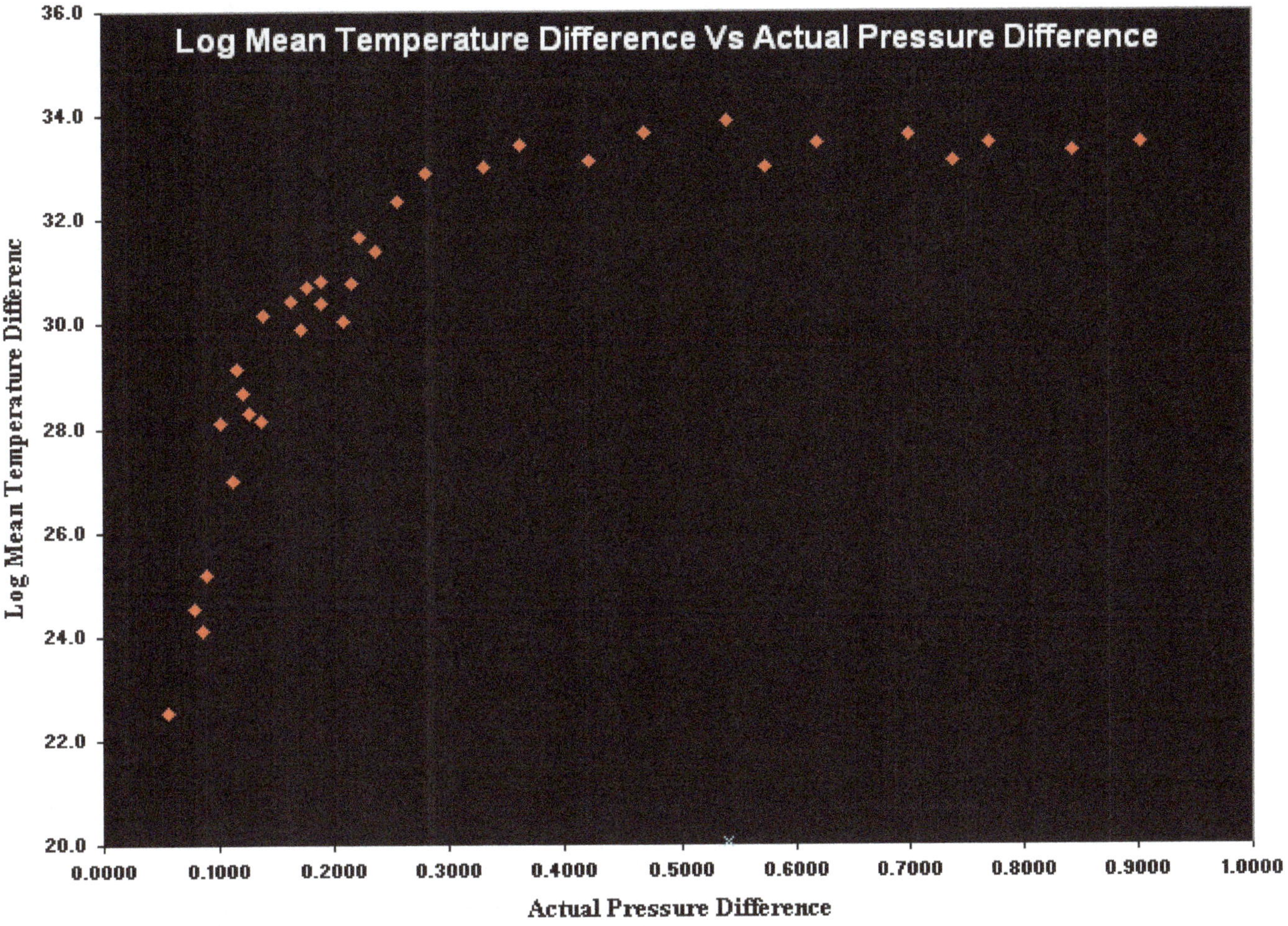

Figure 6. Temperature profile of "log mean temperature difference" w.r.t. Reynolds number.

or less constant with increase of Reynolds number for Reynolds number greater that 2000, this also indicates that for high Reynolds number range the heat gain by the working fluid is constant.

Figure 6 is the plot between log mean temperature difference (LMTD) Vs pressure difference at the inlet and exit section of the test piece. This figure indicates that as the pressure difference across the channels increases "log mean temperature difference" also increases but it achieves it constant value after the particular value of pressure difference. Again the vertical rise of "log mean temperature difference" at low values of pressure difference is due to more heat gain by the working fluid. Figures 5 and 6 represents the temperature profiles with respect to the Reynolds number and pressure difference respectively here it is not clear what sort of profile of working fluid along as well as across the channel for a particular Reynolds number. So in order to find out the temperature profiles along as well as across the channels curves are plotted as shown in Figures 7 and 8. Figure 7 is the plot of temperature profile of the working fluid along the channel length for various Reynolds number without varying the channel diameter. Figure 7 indicates that for the particular channel length the temperature of the working fluid increases with decrease of Reynolds number and this results match with Figure 5. Also for the particular Reynolds number the temperature of the working fluid increases with increase of channel length but the rate of this increase of temperature decreases with the channel length that is in the last portion of the channel the temperature of working fluid is more or less constant. Figure 8 is the plot between the temperature profiles of the working fluid across the channel that is these plots shows the temperature variation of the working fluid in the radial direction. Figure 8 also indicates the same results as in case of Figures 7 and 5 that is the temperature of the working fluid increases with decrease of Reynolds number.

It is also clear from this figure that for a particular value of Reynolds number, the major portion of the temperature of the working fluid is constant in radial direction except that part, which is very close to the channel walls.

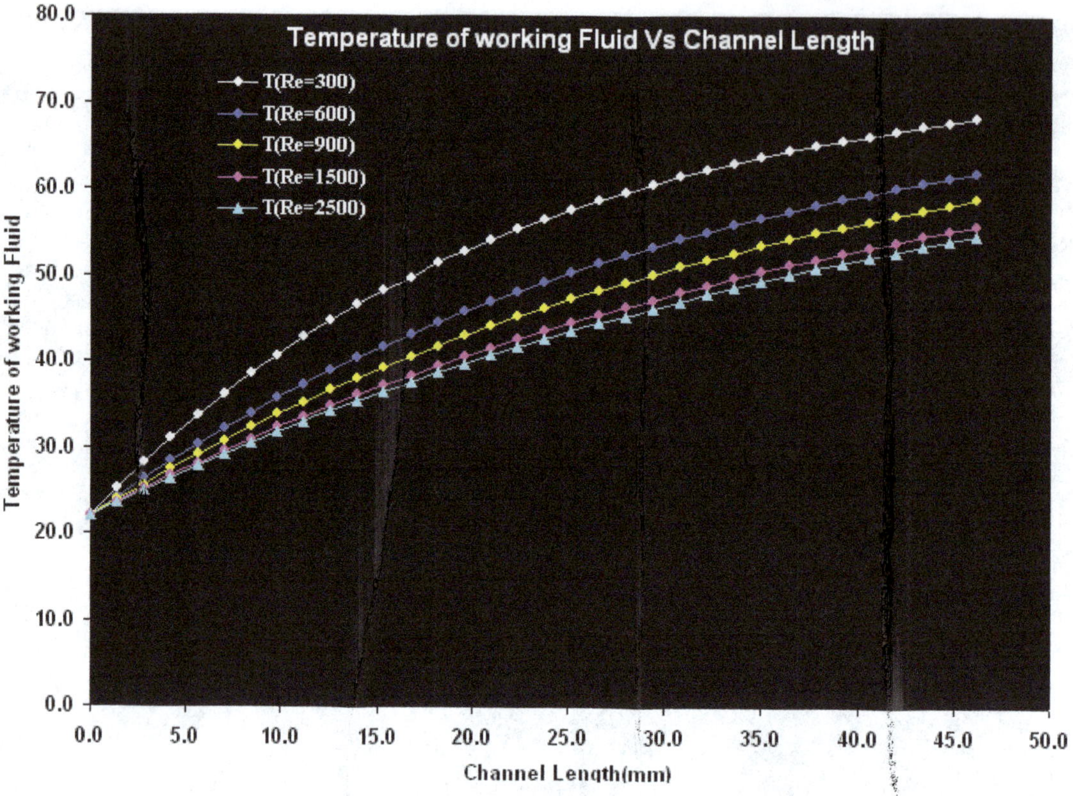

Figure 7. Temperature profile of working fluid along the channel length for different Reynolds number.

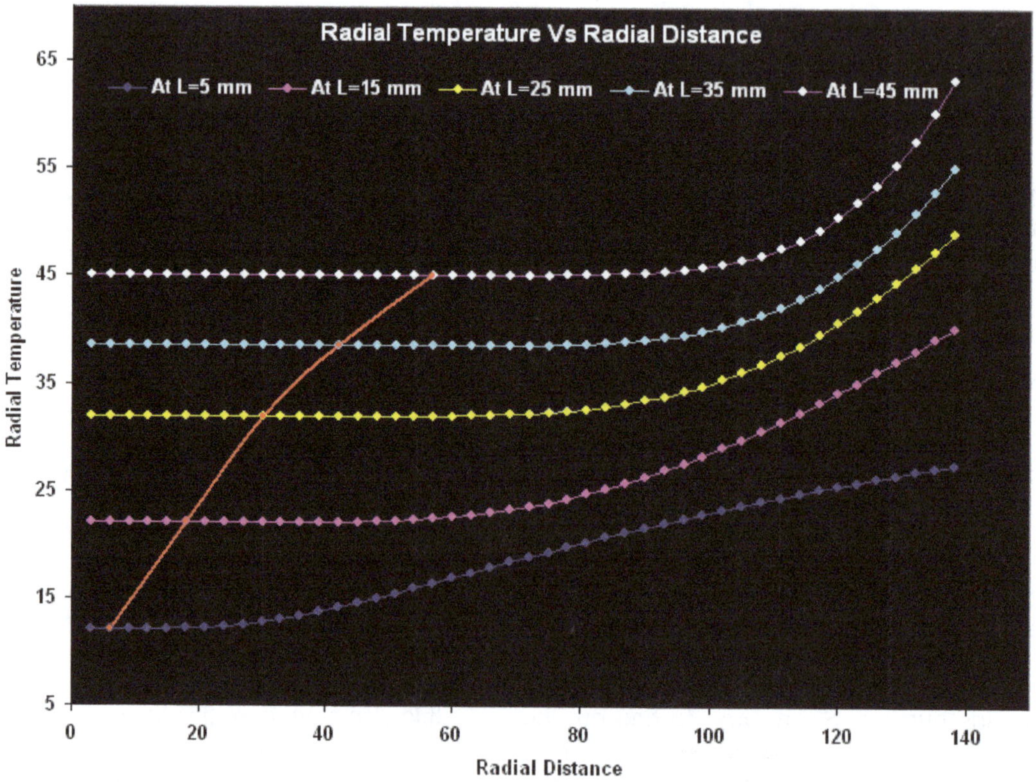

Figure 8. Temperature profile of working fluid along the channel diameter for different Reynolds number.

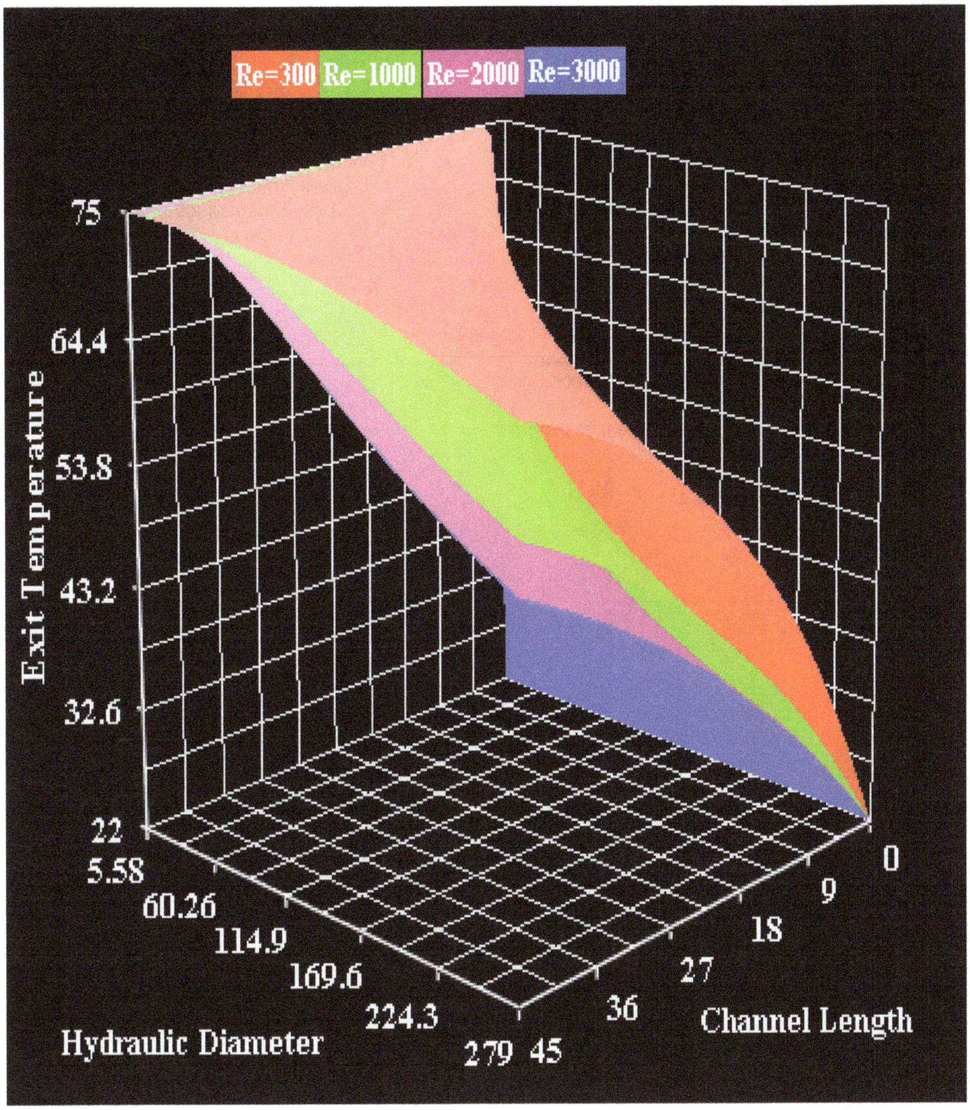

Figure 9. Surface plots of temperature profile of working fluid along as well as across the channel for different Reynolds number.

Figure 9 shows the combine effect of channel length as well as channel diameter on the temperature profiles of the working fluid.

Conclusion

Fluid flow and temperature variation in microchannels was investigated. Experiments were conducted on Microchannel Heat Exchanger. On the basis of certain measurements friction factor were calculated and temperature profiles are plotted for a Reynolds number range of 304 to 2997. Experimental friction factor matched reasonably well with the theoretical equations. On the basis of experimental data the correlation for friction factor was proposed in term of Reynolds number. The proposed correlation for friction factor agrees very well with the experimental data as well as existing correlations. As summary the proposed correlations for friction factor can be used for predicting the behavior of fluid flow in microchannels and temperature profile help to understand the variation of temperature along as well as across the channel.

REFERENCES

Adams TM, Abdel-Khalik SI, Jeter SM, Qureshi ZH (1998), An experimental investigation of single-phase forced convection in microchannels. Int. J. Heat. Mass. Transfer ., 41: 851-857.

Adams TM, Dowling MF, Abdel-Khalik SI, Jeter SM (1999), Applicability of traditional turbulent single-phase forced convection correlations to non-circular microchannels. Int. J. Heat. Mass. Transfer., 42: 4411-4415.

Celata GP, Cumo M, Gugleilmi M, Zummo G (2000), Experimental Investigation of Hydraulic and Single Phase Heat Transfer in 0.130

mm Capillary Tube", International Conference on Heat transfer and Transport Phenomena in Microscale, Banff, Canada, pp.108-113.
Chein R, Chuang J (2007). Experimental Microchannel Heat Sink Performance Studies using Nanofluids. Int. J. Thermal. Sci., 46(1): 57-66.
Choi SB, Barron RF, Warrington RO (1991). Fluid flow and heat transfer in microtubes, in: Micromechanical Sensors, Actuators and Systems, in: D. Choi, et al. (Eds.), ASME DSC, 32: 123-128.
Mala Gh, Mohiuddin Li D (1999). Flow characteristics in microtube. Int. J. Heat Fluid Flow, 20: 142-148.
Peng XF, Wang BX, Peterson GP, Ma HB (1994). Experimental investigation of heat transfer in flat plates with rectangular microchannels. Int. J. Heat .Mass. Transfer ., 38(1): 127-137.
Phillips WA, Jr (2008). Experimental and Numerical Investigation of Fluid Flow and Heat Transfer in Microchannels, Thesis of Master of Science in Mechanical Engineering, Department of Mechanical Engineering, B.S., Louisiana State University.
Tuckerman DB, Pease RFW (1981). High-performance heat sinking for VLSI", IEEE Electron Dev. Lett. EDL- 2: 126-129.
Xu B, Ooi KT, Wong NT, Choi WK (2000). Experimental Investigation of Flow Friction for Liquid Flow in Microchannels. Int. Comm. Heat Transfer, 27: 1165-1176.
Yang CY, Wu JC, Chien HT, Lu SR (2003). Friction characteristics of water, R-134a, and air in small tubes', Microscale. Thermophy. Eng., 7: 335-348.
Yu D, Warrington R, Barron R, Ameen T (1995). An Experimental and Theoretical Investigation of Fluid Flow and Heat Transfer in Microtubes, Proceedings of the ASME/JSME Thermal. Eng. Confer., 1: 523-530.

8

Integration of thermo mechanical strains into optimal tolerance design of mechanical assembly using NSGA II and FE simulation

Jayaprakash G.[1]*, Sivakumar K.[2] and Thilak M.[3]

[1]Department of Mechanical Engineering, Shivani Engineering College, Tiruchirapalli, India.
[2]Department of Mechanical Engineering, Bannari Amman Institute of Technology, Sathiyamangalam, India.
[3]Department of Mechanical Engineering, P.A.B.I.I.T, Tiruchirapalli, India.

In the design of mechanical assembly, the dimension-chain tools take into account the manufacturing dispersion of the parts and assembly defects. This ensures the interchangeability of the different components and guarantees that an assembly can carry out different service functions, as it is modeled in infinitely rigid solids. However, this approach does not take thermo-mechanical effects and deformation due to inertia effects like gravity, angular velocity etc., into account. Most materials change length as they change temperature. As a result of this change, the dimensions and tolerances of a product become at variance with the design values. Hence, thermal effects must be taken into account when designing a product that will undergo temperature cycling and yet, the different operating regimes of an assembly make it indispensable that the effects caused by the thermodynamic cycle should be integrated. In this regard, a finite element model of a machine assembly is created in order to determine the deformation due to change in temperature and inertia effects. The aim of this article is to include the deformation determined by Finite element analysis in the dimension chain thereby controlling clearances in the mechanical assembly. The approach first generates a Cost-tolerance model using neural network where the inputs are parameters and tolerance levels. Then, Finite element analysis of the machine assembly is carried out. The deformation obtained by FEA is then included in the dimension chain. Finally, optimization is done using Non-dominated sorting genetic algorithm II (NSGA II). The results provide designers with optimal component parameters and tolerance values, and the critical components and the manufacturing cost. The approach can also guarantee that the parameter and tolerance values found remain within tolerance for the temperature variation. Then, the product can function as intended under a wide range of temperature conditions for the duration of its life.

Key words: Dimension chain, thermo mechanical tool, finite element analysis, neural network and NSGA II.

INTRODUCTION

Both the performance and reliability of products are strongly influenced by temperature (Bejan et al., 1996; Sergent and Krum, 1998). Hence, exposure to a temperature that is higher or lower than the product is designed to withstand, may result in the failure of the product to perform to specification, or in total failure. As demands for product quality continually increase, the problem of temperature impact becomes an important and challenging issue. A survey shows that the impact of temperature on a product contributes to a substantial portion of product failures. Unfortunately, the effects of temperature impact are often ignored during the design process or are considered too late; consequently, design changes are limited and become very costly. Hence, integration of thermal impact into the early design cycle will ultimately lead to more robust and reliable products which undergo temperature changes during their application. The detrimental effects of excessive

*Corresponding author. E-mail: jpjaya_74@yahoo.co.in.

temperatures may be divided into two categories. The first category is soft failures which are caused by the tendency of the parameters and tolerances of components to exhibit a degree of sensitivity to temperature variation.

As the temperature increases or decreases, the cumulative effects of component parameter drift and tolerance variation may eventually cause the output variables of interest to deviate from the design specifications. The second category is hard failures, which occur as a result of component breakdown resulting from temperature variation. This deviation of temperature from the acceptable operating temperature established by the design specification creates mechanical stresses in components, which may cause fatigue, cracking, fracture, or displacement. One possible way to reduce temperature impact is through thermal control by reducing or adding heat to the product, so that the temperature remains within an acceptable range.

However, thermal control results in a dramatic increase in cost owing to the required cooling or heating equipment and the increase in size, weight, and cost. Robust design should be employed to find the appropriate parameter and tolerance values for each component of a product, so that the output quality of interest is resistant to extreme temperature variation. As is known, quality engineering uses robust design to improve product or process quality by reducing the effects of variation. The variation of output can be reduced by two methods (Kacker, 1985; Nair, 1992; Padke, 1989). One is parameter design which adjusts the parameter value so that the output is less sensitive to causes of variation. The other is tolerance design which reduces the tolerance value to control the variation (Evans, 1974). There is usually no cost associated with changing parameter values. However, reducing tolerance values always leads to additional costs. Hence, parameter design is normally carried out prior to tolerance design for economic reasons.

Generally, there are two types of input variables in product or process design: 1) those with a tolerance requirement and 2) those without a tolerance requirement. As a result of the fact that the values of these input variables do not influence product application and manufacturing operation, only the nominal values need be determined. Hence, for the first type of input variable in product or process design, if the quality determined by measuring the output variables which result from parameter and tolerance design have the same unit, optimization of the parameter and tolerance design may be completed in one step (Jeang, 2001; Jeang and Chang, 2001). For product or process design, the quality function describing the relationship between the output quality of interest and the input variable of the design may or may not be available. For the former situation, well-known methods such as root-sum- square (RSS), worse-case (WC), and numerical simulation can be applied directly for design analysis; however, the latter situation requires a physical experiment (Nigam and Turner, 1995). This paper considers a situation where the quality function is known. A distinct advantage over the situation where the quality function is not available is that costly physical experimentation can be replaced by numerical simulation (Welch et al., 1990).

Furthermore, the computation of design analysis will be more accurate because the quality function is not estimated in this case. A measurement score which is converted from the values found through numerical simulation will be used. The measurement score (or total cost) includes quality loss, manufacturing cost, and failure cost. This score is also called the response value in the statistical analysis presented. Normally, it is efficient to proceed with the design activities if a functional relationship (or response function) exists between the measured score (or response value) and the set of design input variables. In addition to finding the best component design values during the design process, it is necessary to locate the critical components of product or process design, particularly by a repeated application under uncertain design conditions.

Traditional tolerance analysis methods assume that all objects have rigid geometry. The variance is increasingly stacked up as components are assembled. The geometric variation of assembly is always assumed to be larger than those of its subassemblies and components. This rigid body analysis overlooks the role of deformation of flexible parts of the assembly due to inertia effects like gravity, angular velocity, etc. The conventional addition theorem of tolerances has to be suitably modified to accommodate deformation due to the inertia effects. Several studies have been carried out to manage compliant structure (Jack and Camelio, 2006; Stewart and Chase, 2005; Soderberg et al., 2006; Xie et al., 2007). The finite element (FE) simulation is used to predict the influence of geometric tolerances on the part distortions for complex part-forms and assembly designs (Manarvi and Juster, 2004). Tolerance analysis of hull is done considering thermo mechanical effect (Pierre, 2009), where the effect of thermal flux in modifying the contacts and distortion the geometry of parts are studied. In this paper, a finite element model of a machine assembly is created in order to determine the deformation due to change in temperature and inertia effects. The aim of this article is to include the deformation determined by Finite element analysis in the dimension chain thereby controlling clearances in the mechanical assembly.

NEURAL NETWORK-BASED COST-TOLERANCE FUNCTIONS

Neural networks have received a lot of attention in many research and application areas. One of the major benefits

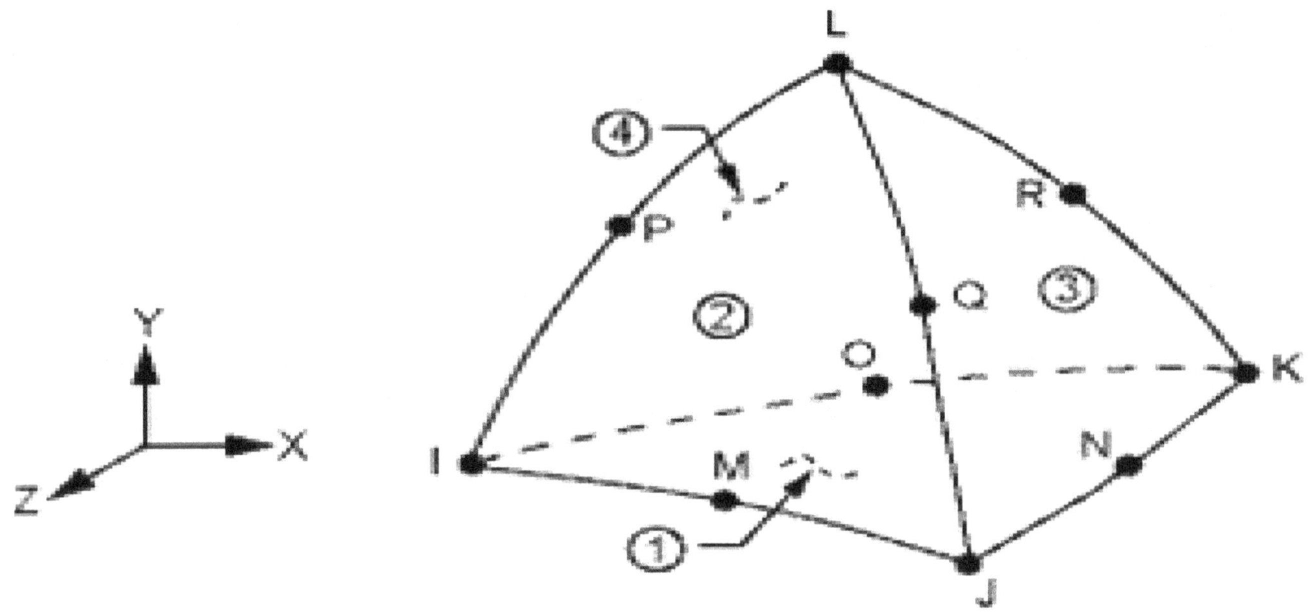

Figure 1. Element geometry.

of neural networks is the adaptive ability of their generalization of data from the real world. Exploiting this advantage, many researchers apply neural networks for nonlinear regression analysis and have reported positive experimental results in their applications (Stern, 1996). Recently, neural networks have received a great deal of attention in manufacturing areas. Zhang and Huang (1995) presented an extensive review of neural network applications in manufacturing. Neural networks are defined by Rumelhart and McClelland (1989) as `massively parallel interconnected networks of simple (usually adaptive) elements and their hierarchical organizations which are intended to interact with objects of the real world in the same way as biological nervous systems do.

The approach towards constructing the cost-tolerance relationships is based on a supervised back-propagation (BP) neural network. Among several well-known supervised neural networks, the BP model is the most extensively used and can provide good solutions for much industrial application (Lippmann, 1987). A BP network is a feed-forward network with one or more layers of nodes between the input and output nodes. An imperative item of the BP network is the iterative method, that propagates the error terms required to adopt weights back from nodes in the output layer to nodes in lower layers. The training of a BP network involves three stages: 1) the feed forward of the input training pattern, 2) the calculation and BP of the associated error, and 3) the adjustment of the weights. After the network reaches a satisfactory level of performance, it will learn the relationships between input and output patterns and its weights can be used to recognize new input patterns.

Figure 1 depicts a BP network with one hidden layer. The hidden nodes of the hidden layer perform an important role in creating internal representation. The following nomenclatures are used for describing the BP learning rule:

net_{pi} = net input to processing unit i in pattern p (a pattern corresponding to a vector of
factors),
w_{ij} = connection weight between processing unit I and processing unit j,
a_{pi} = activation value of processing unit i in pattern p,
δ_{pi} = the effect of a change on the output of unit I in pattern p,
g_{pi} = target value of processing unit i,
ε = learning rate.

The net inputs and the activation values of the middle processing nodes are calculated as follows:

$$net_{pi} = \sum_{j} w_{ij} a_{pj}, \quad (1)$$

$$a_{pi} = \frac{1}{1 + \exp\left(net_{pi}\right)} \quad (2)$$

The net input is the weighed sum of activation values of the connected input units plus a bias value. Initially, the connection weights are assigned randomly and are varied continuously. The activation values are in turn used to calculate the net inputs and the activation values of the output processing units using the same Equations (1) and (2). Once the activation values of the output units are calculated, we compare the target value with activation value of each output unit. The discrepancy is propagated using

$$\delta_{pi} = \left(g_{pi} - a_{pi}\right) f_i'\left(net_{pi}\right) \tag{3}$$

For the hidden processing units in which the target values are unknown, instead of Equation (3), the following equation is used to calculate the discrepancy. It takes the form

$$\delta_{pi} = f_i'\left(net_{pi}\right) \sum_k \delta_{pk} w_{ki}. \tag{4}$$

From the results of Equations (3) and (4), the weights between processing units are adjusted using

$$\Delta w_{ij} = \varepsilon \delta_{pi} a_{pj}. \tag{5}$$

In this paper, neural network is used to generate a Cost-tolerance model, the inputs being parameters and tolerance levels.

FINITE ELEMENT ANALYSIS

The finite element method (FEM) is a numerical technique for finding approximate solutions of partial differential equations (PDE) as well as of integral equations. The solution approach is based either on eliminating the differential equation completely (steady state problems), or rendering the PDE into an approximating system of ordinary differential equations, which are then numerically integrated using standard techniques.

The Finite Element Method is a good choice for solving partial differential equations over complicated domains (like cars and oil pipelines), when the domain changes (as during a solid state reaction with a moving boundary), when the desired precision varies over the entire domain, or when the solution lacks smoothness.

FEM allows detailed visualization of structures bend or twist, and indicates the distribution of stresses and displacements. FEM software provides a wide range of simulation options for controlling the complexity of both modeling and analysis of a system. Similarly, the desired level of accuracy required and associated computational time requirements can be managed simultaneously to address most engineering applications. FEM allows entire designs to be constructed, refined, and optimized before the design is manufactured.

In this paper, we have chosen the finite element analysis method to find the deformation in the machine assembly due to change in temperature and inertia effects .This analysis is done by using the ANSYS11.0 software. The analysis of the machine assembly is carried out by commercial FEM code ANSYS 11.0 with SOLID 98 element. SOLID98 is a 10-node tetrahedral element with quadratic displacement behavior and is well suited to model irregular meshes (such as produced from various CAD/CAM systems). When it is used in structural and piezoelectric analyses, SOLID98 has large deflection and stress stiffening capabilities. The element is defined by ten nodes with up to six degrees of freedom at each node.

ELITIST NON-DOMINATED SORTING GENETIC ALGORITHM (NSGA-II)

Kalyanmoy Deb proposed the NSGA-II algorithm (2002). Essentially, NSGA-II differs from non-dominated sorting Genetic Algorithm (NSGA) implementation in a number of ways. Firstly, NSGA-II uses an elite-preserving mechanism, thereby assuring preservation of previously found good solutions. Secondly, NSGA-II uses a fast non-dominated sorting procedure. Thirdly, NSGA-II does not require any tunable parameter, thereby making the algorithm independent of the user. Initially, a random parent population P_o is created. The population is sorted based on the non-domination. A special book-keeping procedure is used in order to reduce the computational complexity to $O(MN^2)$. Each solution is assigned a fitness equal to its non-dominated level (1 is the best level). Thus, minimization of fitness is assumed. Binary tournament selection, recombination, and mutation operators are used to create a child population Q_o of size N, thereafter; the algorithm that follows is used in every generation.

$R_t = P_t U Q_t$

F=fast-non-dominated-sort(R_t)
$P_{t+1} = \phi$ and i=1
Until $|P_{t+1}| + |F_i| \leq N$
$P_{t+1} = P_{t+1} U F_i$

crowding-distance-assignment(F_i)
i=i+1
Sort($F_i \propto_n$)
$P_{t+1} = P_{t+1} U P_{t+1}[1:(N - |P_{t+1}|)]$
Q_{t+1} = make-new-pop(P_{t+1})
t=t+1

First, a combined population $R_t = P_t U Q_t$ is formed. This

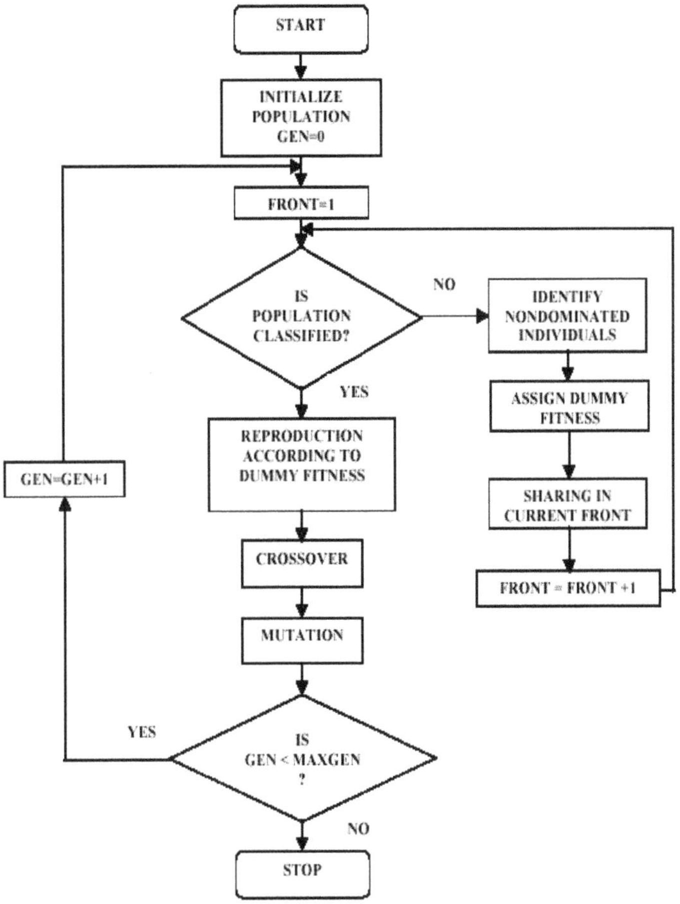

Figure 2. An iteration procedure of the NSGA-II algorithm.

allows parent solutions to be compared with the child population, thereby ensuring elitism. The population R_t is of size 2N. Then, the population R_t is sorted according to non-domination and non-dominated fronts F1, F2, and so on are found. The algorithm is illustrated in the following:

The new parent population P_{t+1} is formed by adding solutions from the first front F_1 and continuing to other fronts successively till the size exceeds N. Individuals of each front are used to calculate the crowding distance – the distance between the neighboring solutions. Thereafter, the solutions of the last accepted front are sorted according to a crowded comparison criterion and a total of N points are picked. Since the diversity among the solutions is important, the crowded comparison criterion uses a relation α_n as follows: solution i is better than solution j in relation α_n if ($i_{rank} < j_{rank}$) or (($i_{rank} = j_{rank}$) and ($i_{distance} > j_{distance}$)). That is, between two solutions with differing non-domination ranks, the preference is the point with the lower rank.

Otherwise, if both the points belong to the same front then the preference is the point, which is located in a region with smaller number of points (or with larger crowded distance). This way solutions from less dense regions in the search space are given importance in deciding which solutions to choose from R_t. This constructs the population P_{t+1}. This population of size N is now used for selection, crossover and mutation to create a new population Q_{t+1} of size N. A binary tournament selection operator is used but the selection criterion is now based on the crowded comparison operator α_n. The aforestated procedure is continued for a specified number of generations. It is clear from the earlier description that NSGA-II uses (i) a faster non-dominated sorting approach, (ii) an elitist strategy, and no niching parameter. It has been proved that the aforementioned procedure has $O(MN^2)$ computational complexity. The outline of the proposed optimization strategy is shown in Figure 2.

THE APPLICATION EXAMPLE

Assembly is the process by which various parts and subassemblies are brought together to form a complete assembly or product which is designed to fulfill a certain

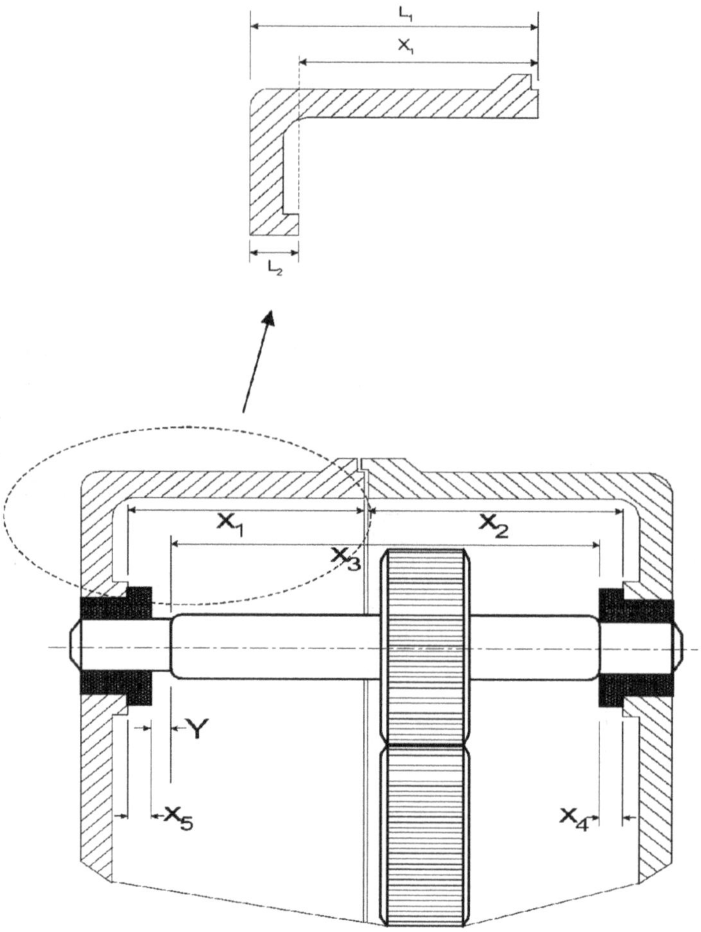

Figure 3. The gear box assembly.

mechanical function. A proper allocation and analysis of tolerance among the assembly components is important that the functionality and quality of the designs are met. Figure 3 is a classic Bjorke gearbox assembly (Bjorke, 1992). The gearbox assembly is the application example for the proposed tolerance design. The gearbox assembly consists of components X_1, X_2, X_3, X_4 and X_5. The assembly function that describes the quality value is:

$$Y = -X_3 - X_4 + X_2 + X_1 - X_5 \qquad (6)$$

The associated component dimensions and tolerances, U_1, U_2, U_3, U_4, U_5; t_1, t_2, t_3, t_4 and $t_{5,}$ $_{must}$ be determined so that the gap Y, between the bushing and hub fall within the functionality limits, T ± S, where T is 0.900 mm and S is 0.200 mm. Table 1 shows process capability limits for each component. The associated low, middle and high levels for input factors U_i and t_i are decided as shown in Table 2 and 3. Table 4 shows the parameter level for each component. The tolerance design involves following stages. Initially a neural network model of cost-tolerance function is developed based on the experimental results (Chase.K.W et al, 1990). The response variable in this study is Total cost which is sum of manufacturing cost and quality losses and it is expressed as

$$TC_i = \sum_{j=1}^{q} k_j \left[(U_{ij} - T_j)^2 + \sigma_{ij}^2 \right] + \sum_{k=1}^{m} C_M(t_{ik}), \qquad (7)$$

Where m is the total number of components from q assembly dimensions in a finished product, K_j the cost coefficient of the jth resultant dimension for quadratic loss function, U_{ij} the jth resultant dimension from the ith experimental results, σ_{ij} the jth resultant variance of statistical data from the ith experimental results, T_j the design nominal value for the jth assembly dimension, t_{ik} the tolerance established in the ith experiment for the kth component, and $C_M(t_{ik})$ the manufacturing cost for the tolerance t_{ik}.

Then neural network model of cost-tolerance function is developed as follows. The 2/3rd of experimental results drawn randomly are used to train the neural network. Before applying the neural network for modeling, the

Table 1. Process capability limits for each component.

Component I	Lower limit (mm)	Upper limit (mm)
t_1	0.014	0.042
t_2	0.018	0.052
t_3	0.024	0.072
t_4	0.009	0.027
t_5	0.010	0.030

Table 2. Feasible design space for each component.

Component I	Lower limit (mm)	Upper limit (mm)
X_1	15.9879	16.0121
X_2	17.9850	18.0150
X_3	28.9792	29.0208
X_4	1.7922	1.8078
X_5	2.2913	2.3087

Table 3. Tolerance and cost for each component.

Component I	Lower level $(mm)	Middle level $ (mm)	Upper limit $ (mm)
t_1	733.7(0.014)	579.8(0.028)	517.5 (0.042)
t_2	674.8(0.018)	541.7(0.035)	497.4 (0.052)
t_3	1385.8(0.024)	975.0(0.048)	899.3 (0.072)
t_4	541.2(0.009)	436.5(0.018)	403.6 (0.027)
t_5	522.8(0.010)	425.6(0.020)	398.7 (0.030)

Table 4. Parameter levels U_i for each component.

Component I	Lower level (mm)	Middle level (mm)	Upper level (mm)
X_1	15.9879	16.0000	16.0121
X_2	17.9850	18.0000	18.0150
X_3	28.9792	29.0000	29.0208
X_4	1.7922	1.80000	1.8078
X_5	2.2913	2.3000	2.3087

architecture of the network has been decided; that is the number of hidden layers and the number of neurons in each layer. As there are 10 inputs and 1 output, the number of neurons in the input and output layer has to be set to 10 and 1 respectively. Also, the back propagation architecture with one hidden layer is enough for majority of the applications. Hence only one hidden layer has been adopted. A procedure was employed to optimize the number of neurons in the hidden layer. Accordingly, an experimental approach was adopted, which involves testing the trained neural networks against the remaining 1/3rd of experimental results. Experimental and predicted outputs for different number of neurons have been compared. The regression statistics for different architecture are determined. The training function used in this research is Gradient descent with momentum back-propagation. The transfer function used in this research is tan-sigmoid and gradient. Descent w/momentum weight/bias learning function has been used. The learning rate = 0.7, momentum = 0.65 and training epochs = 2000. The weights (and biases) are randomly initialized between -0.5 and 0.5.Once the neural network gets trained, it can provide the result for any arbitrary value of input data set. Thus the neural

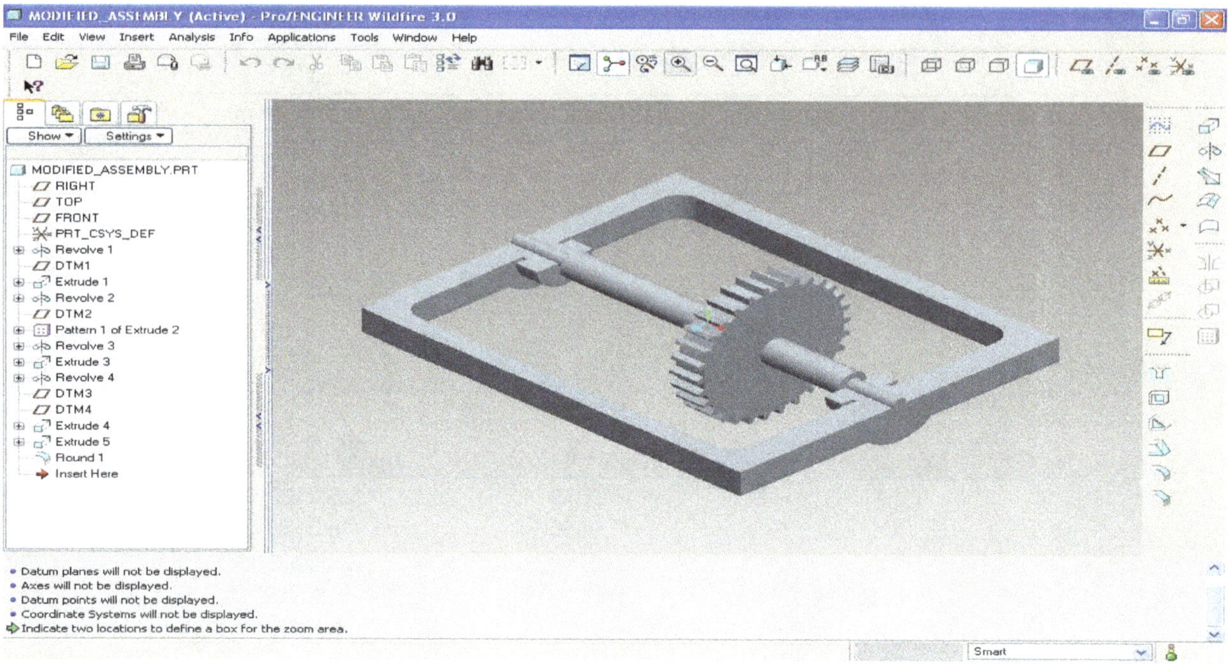

Figure 4. The Pro/E model.

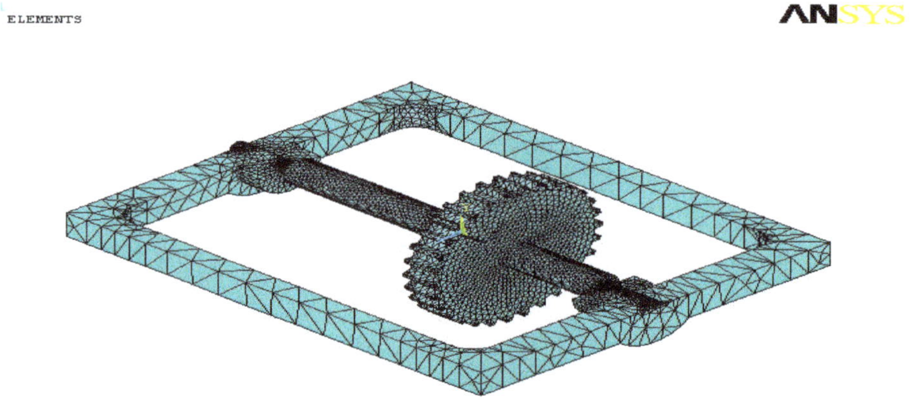

Figure 5. The meshed model.

network model for the above problem is developed as per the approach discussed previously. Once the neural network model of the cost-tolerance function is developed, than Finite element analysis of the assembly is done.

The major constraints in the presented design are variation of thermal environment both within and among various application categories and inertia effects. Hence the design which withstands temperature variation and inertia must be considered in the present case. If the temperature is 25°C when the gearbox is assembled and then varies between 10°C and 40°C during application; if the self weight of the shaft is considered and inertia effect due to angular velocity of the shaft is considered, then the deformation is determined using Finite Element Analysis. First, a 3D model of the gearbox assembly is created using Pro/E wildfire 3.0 software (Figure 4). Then it is converted into a file type (.sat) suitable for importing the same in the Ansys software version 11.0. Once the model is imported, material properties for the three components, shaft, bushing and casing is given. Then the model is meshed with SOLID98, which is a 10-node tetrahedral element with quadratic displacement behavior and it is well suited to model irregular meshes (such as produced from various CAD/CAM systems). The meshed model has 71223 elements and 107,726 nodes as shown (Figure 5). Then the loads and constraints are applied as shown (Figure 6). In order to account for inertia effects

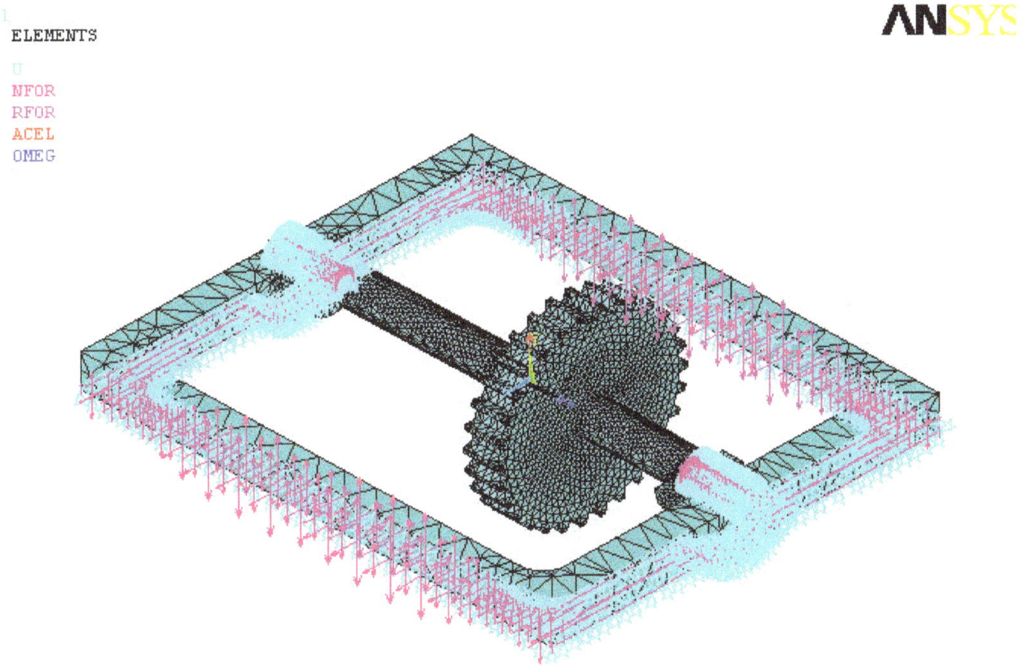

Figure 6. Applied loads and constraints.

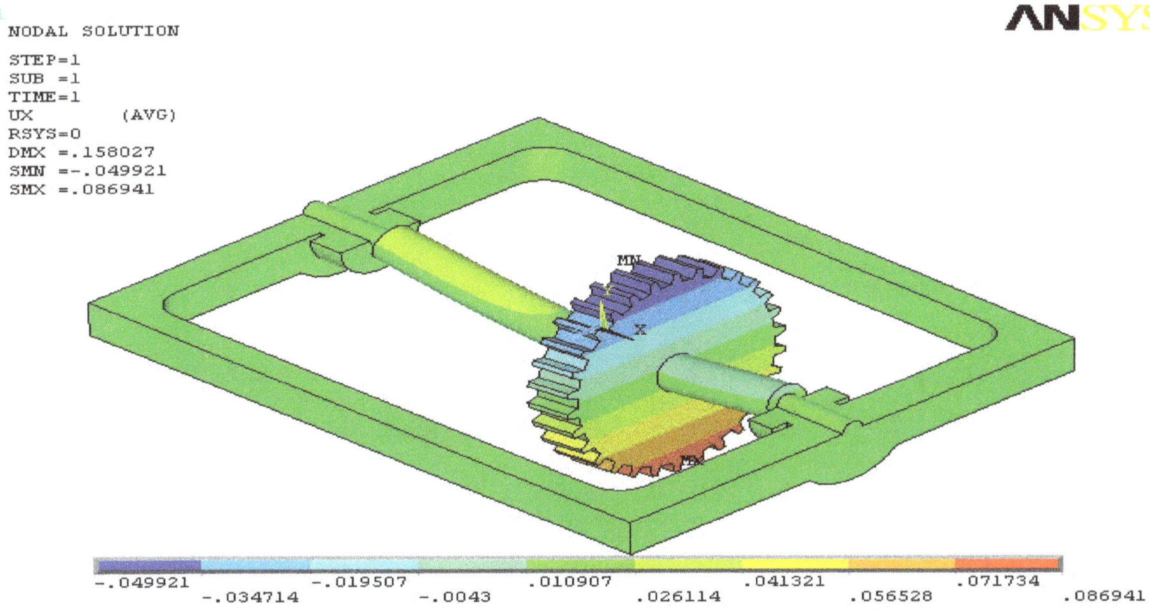

Figure 7. Deformation plot for 10℃.

like gravity, angular velocity, etc., appropriate values for g (9.81 m/s^2) and ω (rad/sec) are given. Then the deformation is calculated for three levels of temperature within the operating range (that is 10℃, 25℃ and 40℃). Figure 7 shows the deformation pattern for 10℃. The variation of deformation along the length of the shaft is determined for all the three values of temperature (Figure 8 to 10). The deformation due to thermal and inertia effect for various temperatures is plotted (Figure 11). Table 5 has deformation values for different levels of temperature. It is observed that the deformation has a linear relationship with the temperature and the same is

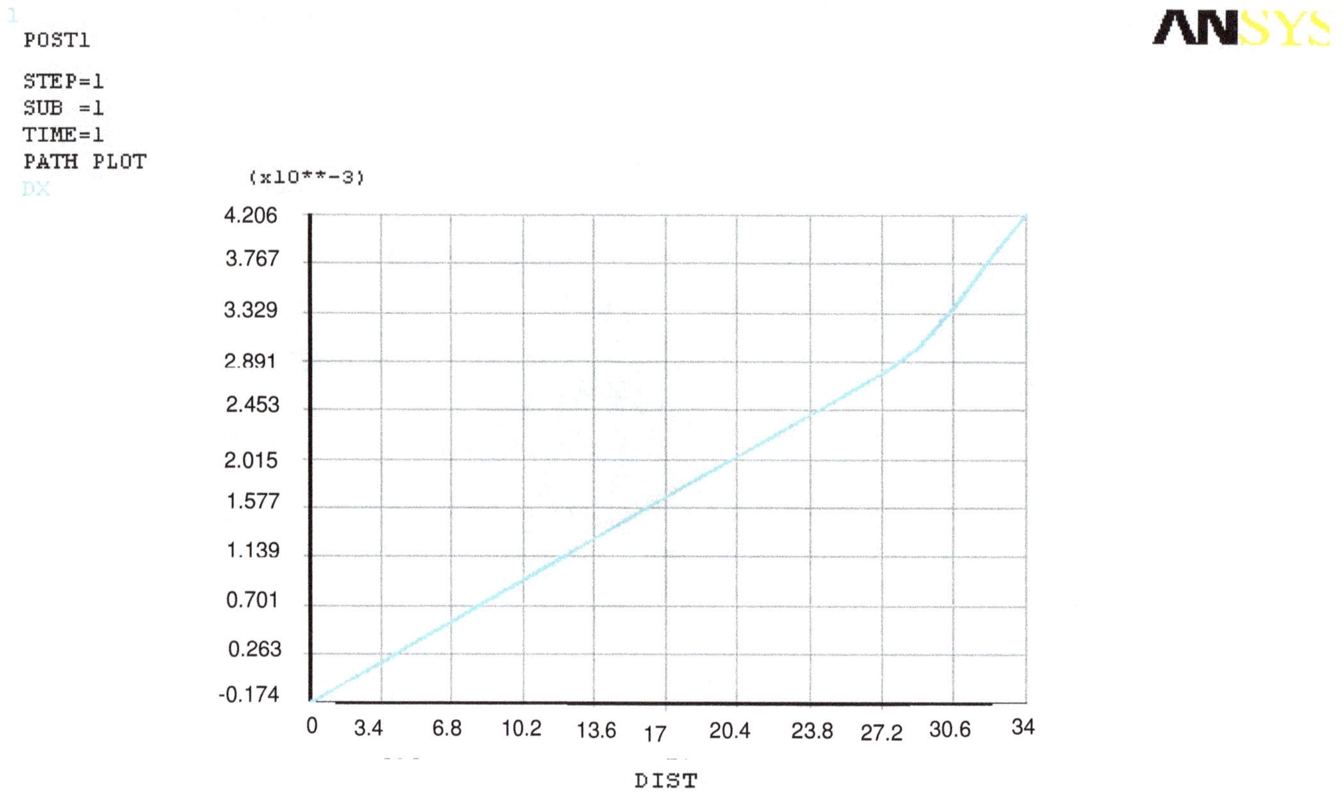

Figure 8. Deformation vs length plot (10℃).

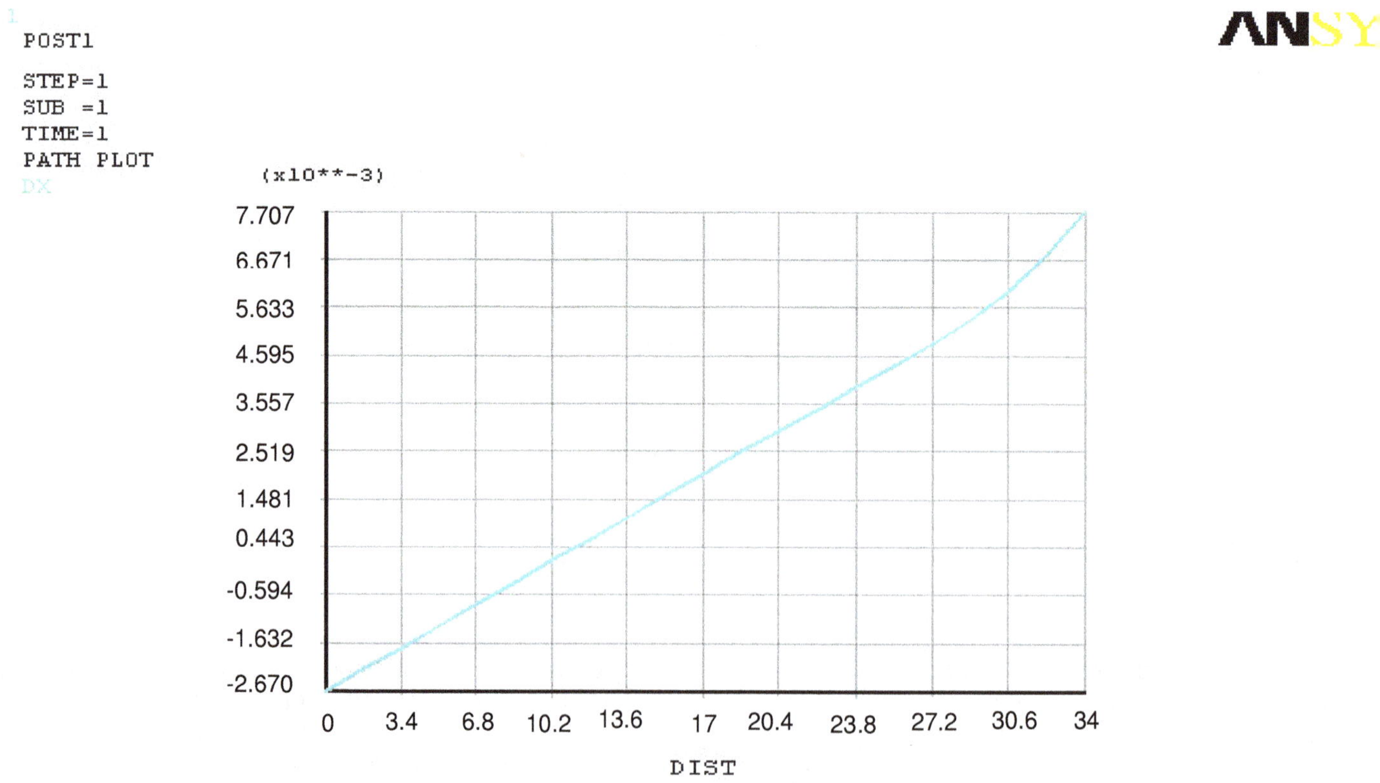

Figure 9. Deformation vs length plot (25℃).

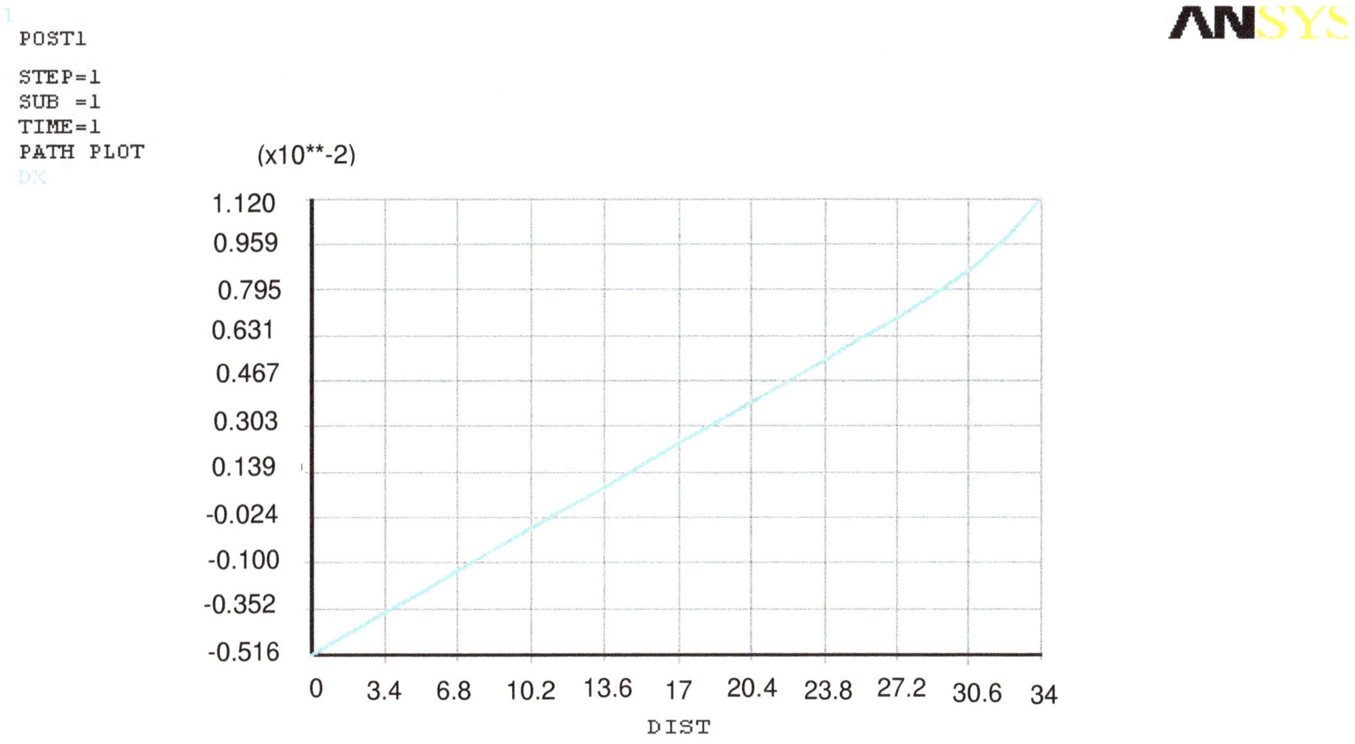

Figure 10. Deformation vs length plot (40℃).

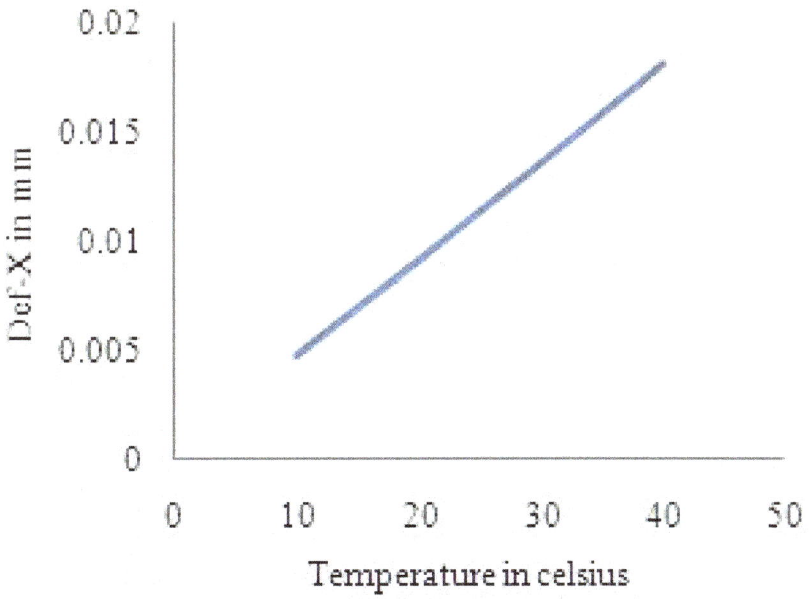

Figure 11. Deformation vs Temperature.

determined.

Once the deformation is determined and linear relationship between the deformation and the process variable (temperature) is determined, then the optimal values of the component dimensions and tolerances are determined by using NSGA II. Table 6 shows the NSGA II specific data. The solution of the gearbox assembly case can be found by solving the following mathematical

Table 5. Deformation for various temperatures.

S/No	Temperature (°C)	Deformation (mm)
1	10	0.4791E-2
2	25	1.1395E-2
3	40	1.18115E-2

Table 6. Optimal values U_i and t_i for TC at various temperatures T.

Temperature	10°C	25°C	40°C
Optimal values	TC	TC	TC
t_1	0.0276	0.0278	0.0282
t_2	0.0243	0.0245	0.0254
t_3	0.0256	0.0257	0.0264
t_4	0.0215	0.0218	0.0219
t_5	0.0295	0.0297	0.0298
X_1	15.9882	16.0000	16.0121
X_2	17.9923	18.0000	18.0150
X_3	28.9792	29.0000	29.0208
X_4	1.7922	1.80000	1.8078
X_5	2.3086	2.3000	2.3087
Optimal response value	$ 2535.52	$2531.25	$2528.12

Table 7. The NSGA II specific data.

Variable type	Real variable
Population size	100
Cross over probability	0.7
Real parameter mutation probability	0.2155
Real parameter SBX parameter	10
Real parameter mutation parameter	100
Total no of generation	100

models:

$$\begin{aligned}&\text{Minimize} TC = F(t_1, t_2 \ldots t_5, U_1, U_2 \ldots\ldots U_5)\\&\text{subjectedto:}\\&U_1 + U_2 - U_3 - U_4 - U_5 \le 0.900\\&\sqrt{t_1^2 + t_2^2 + t_3^2 + t_4^2 + t_5^2} \le 0.200 + \delta\\&\text{where} \delta = f\ (\text{temperature})\end{aligned} \tag{8}$$

A clearance of 0.9 mm has to be maintained, between the bushing and hub should be 0.900 mm. Another functional constraint is the constraint equation developed using statistical tolerance design method, where δ is the deformation determined using FEA. The deformation δ (Figure 11) is found to have a linear relationship with the variable temperature while the inertia effects like gravity and angular velocity are constant. The Problem (7) is solved by the proposed NSGA II discussed previously. The outline of the proposed optimization strategy is shown in Figure 2. The optimization strategy is explained as follows. Initially, the cost-tolerance function is established by the neural network model. Once the neural network based cost – tolerance function is established, and then optimization of the problem (Equation 7) is carried out using NSGA II. The optimization program determines the set of tolerance and component dimensions with minimum cost. The least cost is found for all the three values of temperature (Table 7).

DISCUSSION

Once a customer starts using a product, the quality of that product can vary for many reasons. Temperature impact and gravity effects have been found to be one of the reasons for variation in quality of the product. Product specifications must be perfectly fulfilled in order to manufacture a quality product. Furthermore, products may consist of a few components which are associated with various materials, grades, and tolerances. The cost involved for each component varies due to selected materials or assigned tolerance values. Hence, product design that considers the thermal impact and inertia effects is required to design those products successfully. The customer's perception of the quality of a design is closely related to the sensitivity of the design to environmental impact, which is the temperature and inertia effect. Design engineers must minimize the effects of the previously discussed factors on performance of the product; that is, a robust design regulating thermal impact is required. There are three ways to minimize quality variation caused by thermal impact:

1) Eliminate the reality of the thermal impact and inertia effects.
2) Design a product with a feature and parameter which can eliminate the deformation.
3) Have a robust design that enables to reduce deformation due to thermal and inertia effects.

It can be very costly, inconvenient, and inefficient to realize the first and second ways of eliminating temperature impacts, because some thermal impacts cannot be controlled and others are too expensive or difficult to control. A product or process is said to be robust when it is insensitive to the effects of sources of variation, even when the sources have not been eliminated. This leads to the third way which is a robust design - a process that results in a product performance which is minimally affected by temperature impact. Robust design focuses on minimizing variation or creating a system less sensitive to variation, making it possible to decrease cost, because expensive means for controlling quality are no longer necessary. Hence, this third method of eliminating temperature impact should be attempted before the first and second ways are tried. The objective of this study is to develop a robust product or process design that functions as intended under a wide range of temperatures for the duration of the design stage.

Conclusion

Using neural network, FEA and NSGA II, a statistical optimization of parameter and tolerance determination for assembly under various temperature and inertia effects have been developed. With the presented approach, critical component parameters and tolerances can be identified,and optimal component parameter and tolerance values can be determined. The component parameter and tolerance values found are the most robust to withstand temperature variation during the products application. Benefits also include low failure related costs and high product reliability. These benefits make it possible to create high-quality and cost effective parameter and tolerance design at the earliest stages of product development.

REFERENCES

Bejan A, Tsatsaronis G, Moran M (1996) Thermal Design and Optimization, John Wiley.

Evans DH (1974). "The State of Art: Part 1 - Background", J. Qual. Technol., 6: 188-195.

Hu J, Camelio J (2006). Modeling and control of compliant assembly systems. CIRPAnn. Manuf. Technol. 55-1,19-22.

Jeang A (2001), "Combined optimization of parameter and tolerance design for quality and cost", Int. J. Production Res., to appear in 2001.

Jeang A, Chang CL (2001). "Concurrent optimization of parameter and tolerance design via computer simulation and statistical method", Int. J. Adv. Manuf. Technol., to appear in 2001.

Kacker RN (1985). "Off-line quality control, parameter design, and Taguchi method", J. Qual. Technol., 17: 176-209.

Kalyanmoy D (2002) Optimization for engineering design. (Prentice-Hall), New York.

Lippmann RP (1987). An introduction to computing with neural nets, IEEE ASSP Magazine 4: 4-22.

Manarvi IA, Juster NP (2004). Framework of an integrated tolerance synthesis model and using FE simulation as a virtual tool for tolerance allocation in assembly design, J. Mater. process Technol., 150: 182 - 193.

Nair VN (1992). "Taguchi's parameter design: a panel discussion", Technometrics, 34: 127-161.

Nigam SD, Turner JU (1995). "Review of statistical approaches to tolerance analysis", Computer-Aided Design, 27(1): 6-15.

Padke MS (1989). Quality Engineering Using Robust Design, Prentice-Hall. Englewood Cliffs, New Jersey.

Pierre L, Teissandier D, Nadeau JP (2009). Integration of thermo mechanical strains into tolerancing analysis, Int. J. Interact. Des. Manuf., 3(4): 247-263.10.1007/s 12008-009-0058-8.

Rumelhart DE, McClelland JL (1989). Parallel Distributed Processing Explorations in the Microstructure of Cognition: Vol. I, MIT Press, Cambridge MA, USA.

Sergent J, Krum A (1998). Thermal Management Handbook, McGraw-Hill.

Soderberg R, Lindkvist L, Dahlstorm S (2006). Computer aided robust analysis for compliant assemblies. J. Eng. Des.17: 411-428.

Stern HS (1996). Neural networks in applied statistics (with discussion), Technometrics, 38: 205-220.

Stewart ML, Chase KW (2005). Variation simulation of fixture assembly for compliant structures using piece wise- linear analysis. Am. Soc. Mech. Eng. 16-1, 591-600.

Welch WJ, Yu TK, Kang SM, Sacks J (1990). "Computer experiments for quality control by parameter design", J. Qual. Technol, 22: 15-22.

Xie K, Wells L, Camelio JA, Youn BD (2007). Variation propagation analysis of compliant assemblies considering contact interaction. J. Manuf. Sci. Eng., Trans, ASME 129-5, 934-942, 2007.

Zhang HC, Huang H (1995). Applications of neural networks in manufacturing: A-state-of-the-art survey, Int. J. Production Res. 33(3): 705-728.

Mixed convection flow of second grade fluid with variable heat flux past a vertically stretching flat surface

Mushtaq M.[1], Ashraf M.[1]*, Asghar S.[1,2] and Hossain M. A.[3]

[1]Department of Mathematics, COMSATS Institute of Information Technology, Islamabad-Pakistan.
[2]Department of Mathematics, King Abdul Aziz University, Jeddah, Saudi Arabia.
[3]Department of Mathematics, University of Dhaka, Dhaka, Bangladesh.

In the present investigation, a numerical study of the flow and heat transfer analysis of viscoelastic second grade fluid due to heated, continuous stretching of a vertical sheet has been carried out. The stretching velocity is assumed to vary linearly with the distance measured from the leading edge. The surface heat flux is assumed to be varied in power of distance measured from the leading edge. The governing differential equations are transformed by introducing proper non-similarity variables and solved numerically using two different methods, namely, the local non-similarity method with second level of truncation and the implicit finite difference method for values of ξ ($=Gr_x/Re_x^2$) ranging from 0 to 10. The comparisons of the results obtained by the aforementioned methods are found in excellent agreement. Effects of the viscoelastic parameter, Λ (Deborah number) on the skin-friction and the heat transfer coefficients have been shown graphically for the fluid with Prandtl number equal to 0.7, 7.03 and 15.0.

Key words: Mixed convection, second grade fluid, stretching flat surface, variable surface heat flux.

INTRODUCTION

Recently, Mushtaq et al. (2007) explored the effects of thermal buoyancy on flow of a viscoelastic second grade fluid past a vertical, continuous stretching sheet of which the velocity and temperature distributions are assumed to vary according to a power-law form. In the present investigation we have analyzed the mixed convection flow of the same fluid along a continuously stretched vertically placed heated surface subject to non-uniform surface heat flux.

In actual practice, the flow of viscous incompressible fluid over a continuous material moving through a quiescent fluid is induced by the movement of the solid material and by thermal buoyancy. Therefore, these two mechanisms, surface motion and buoyancy force, will determine the momentum and thermal transport processes. The thermal buoyancy force arising due to the heating or cooling of a continuously moving surface may alter significantly the flow and thermal fields and thereby the heat transfer behavior in the manufacturing process. The literature survey shows that, a continuously moving surface through a quiescent medium has many engineering applications; such as hot rolling, wire drawing, spinning of laments, metal extrusion, crystal growing, continuous casting, glass fiber production, and paper production (Altan et al., 1979; Fisher, 1976; Tadmor and Klein, 1970). On the other hand, Karwe and Jaluria (1988, 1991) showed that the thermal buoyancy effects are more prominent when the plate moves vertically, that is, aligned with gravity, than when it is horizontal. In their analysis, they treated the mixed convection flow of aforementioned fluid over a continuous plate moving at a uniform speed, which may have applications in material processes, such as hot rolling, extrusion, and drawing. The numerical solution for the boundary layer flow was first presented by Sakiadis (1961). Later on, Magyari et al. (2001) and Magyari and

*Corresponding author. E-mail: mashraf682003@yahoo.com.

Keller (2000) reported numerical solution for rapidly decreasing velocity using self-similar method and the analytical solution for permeable surface moving with a decreasing velocity. Chen and Strobel (1980) dealt with the problem of combined forced and free convection flow of viscous incompressible fluid in boundary layers adjacent to a continuous horizontal sheet maintained at a constant temperature and moving with a constant velocity. Ingham (1986) investigated the existence of the solutions for the free convection boundary-layer flow of viscous fluid near a continuously moving vertical plate with temperature inversely proportional to the distance up the plate. Ali and Al-Yousef (1998) have investigated the problem of laminar mixed convection flow adjacent to a uniformly moving permeable vertical plate. Also, an analysis of mixed convection heat transfer from a vertical, continuously stretching sheet has been presented later on by Chen (1998). All the previous investigations were confined to the case of Newtonian fluid only.

In recent years non-Newtonian fluids, such as, Walter's fluid (Walters, 1962) or the viscoelastic second grade fluid, have become increasingly important from the point of industrial applications; for example, in certain polymer processing, one deals with flow of a non-Newtonian fluid over a stretching surface; since, these fluids show viscoelastic behavior, meaning that very short part of the history of the deformation gradient has an effect on the stress. In an incompressible fluid of differential type, apart from a constitutively indeterminate pressure, the stress is just a function of its velocity gradient and some number of its higher-time derivatives. These fluids do not exhibit the phenomenon of stress relaxation which means that with the instantaneous cessation of all local motion, the stress becomes pure pressure. Issues concerning the status of second grade fluid were discussed by Dunn and Rajagopal (1995).

Steady flow of a viscoelastic second-order fluid past a stretching sheet was investigated by Rajagopal et al. (1984). Bhattacharyya et al. (1998) studied the temperature distribution in the steady boundary layer flow of a second-order fluid past a stretching surface. Later, Chen et al. (1990) studied the flow and heat transfer in the boundary layer of a viscoelastic second grade fluid over a stretching surface subject to either constant temperature or uniform heat flux.

In the present investigation, we have analyzed the mixed convection flow of viscoelastic second grade, fluid along a heated and continuously moving surface subject to non-uniform surface heat flux. We further assume that the surface velocity is proportional to x and the wall heat flux is proportional to x^m, where x measures the distance from the leading edge of the stretched surface. The dimensionless boundary layer equations that govern the flow and heat transfer are transformed to local non-similarity equations using suitable transformations which are then solved numerically applying (*i*) the local non-similarity method as well as (ii) the implicit finite difference method. The numerical results thus obtained are presented in terms of local skin-friction and local Nusselt number for different values of the physical parameters, such as, the viscoelastic parameter, λ (also known as the Deborah number) and for Prandtl number, Pr, choosing the values of ξ within the range of 0 to 10. Conjugate effects of the viscoelastic parameter λ as well as the local mixed convection parameter ξ on the velocity and temperature field have also been shown graphically.

FORMULATION OF THE PROBLEM

We consider a steady two-dimensional mixed convection flow of a viscoelastic second grade fluid along a vertically stretching flat surface subject to variable heat-flux proportional to x^m, that is, $q_w(x)=q_0(x/L)^m$, where x measures the distance from the leading edge of the plate and m is a real number. The flow configuration is shown in Figure 1, in which y measures the distance from the surface in the direction normal to x. The ambient temperature is assumed to be, T_∞, constant. Here u and v are the x- and y-components of the fluid velocity and T being the temperature of the fluid in the boundary layer region. Finally, the stretching velocity is assumed to be a linear function of x, that is, $U_w(x)=U_0(x/L)$. Under the usual Boussinesq approximation, the governing dimensionless boundary layer equations for the conservation of mass, momentum and energy are given in Equations 1 to 3:

$$\frac{\partial U}{\partial X}+\frac{\partial V}{\partial Y}=0 \qquad (1)$$

$$U\frac{\partial U}{\partial X}+V\frac{\partial U}{\partial Y}=\frac{\partial^2 U}{\partial Y^2}+\lambda\left[\frac{\partial}{\partial X}\left(U\frac{\partial^2 U}{\partial Y^2}\right)+\frac{\partial U}{\partial Y}\frac{\partial^2 V}{\partial Y^2}+V\frac{\partial^3 U}{\partial Y^3}\right]+Ri\Theta \qquad (2)$$

$$U\frac{\partial \Theta}{\partial X}+V\frac{\partial \Theta}{\partial Y}=\frac{1}{\Pr}\frac{\partial^2 \Theta}{\partial Y^2} \qquad (3)$$

$$U=X, \quad V=0, \quad \frac{\partial \Theta}{\partial Y}=-X^m \text{ at } Y=0 \qquad (4)$$

$$U\to 0, \qquad \Theta\to 0 \text{ as } Y\to\infty \qquad (5)$$

The aforementioned dimensionless equations are being obtained by introducing the following dependent and independent dimensionless quantities;

$$X=\frac{x}{L}, \quad Y=\frac{y}{L}\mathrm{Re}_L^{1/2}, \quad Ri=\frac{Gr_L}{\mathrm{Re}_L^{5/2}}$$

$$U(X,Y)=\frac{u}{U_0}, \quad V(X,Y)=\frac{L}{\nu\,\mathrm{Re}_L^{1/2}}v, \quad \Theta(X,Y)=\frac{k\,\mathrm{Re}_L^{1/2}}{Lq_0}(T-T_\infty) \qquad (6)$$

$$U(x)=U_0(x/L), \; q_w(x)=q_0(x/L)^m$$

Where, u, v, T are the dependent variables and Θ is dimensionless temperature at the surface of the plate. The dimensional U_0 and q_0 are prescribed constants, m is the exponent

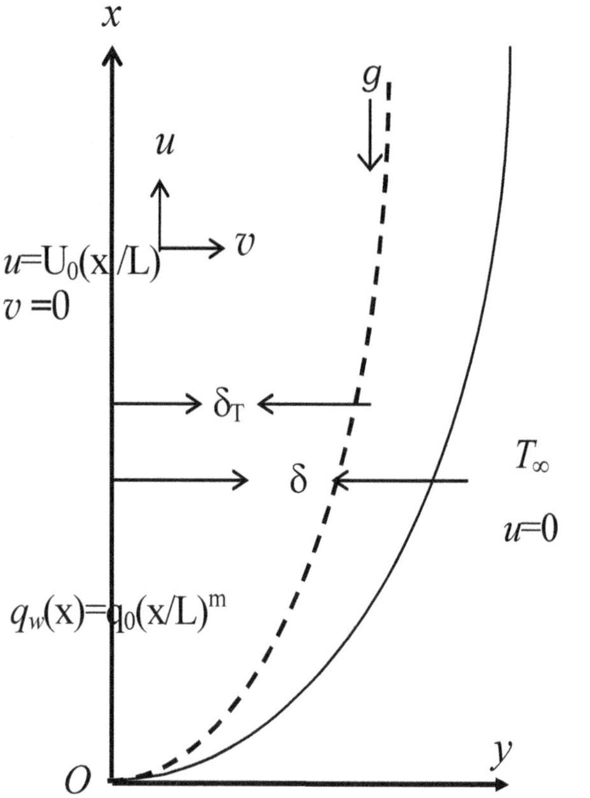

Figure 1. The flow configuration and the coordinate system.

and T_∞ is the temperature of the ambient fluid. Further, Ri is the Richardson number together with Equation 7 as the Reynolds number and the modified Grashof number for the heat-flux case. Further, in Equation 2, $\lambda = KRe_L/L^2$ denotes the viscoelastic parameter (or Deborah number):

$$\mathrm{Re}_L = U_0 L/\nu, \quad Gr_L = g\beta q_0 L^4/k\nu^2 \tag{7}$$

It can be seen that Equations 1 to 3 admit self-similar solution and hence we may introduce the following free group of transformations:

$$\eta = Y, \quad \xi = RiX^{m-1}, \quad \psi = X f(\eta,\xi), \quad \Theta = X^m \theta(\eta,\xi) \tag{8}$$

Introducing the new dimensionless variables given in Equation 8, Equations 2 and 3 can be written in the following dimensionless form:

$$f''' + ff'' - f'^2 + \lambda\left(2ff''' - (f'')^2 - ff^{iv}\right) + \xi\theta = (m-1)\xi\left[f'\frac{\partial f'}{\partial\xi} - f''\frac{\partial f}{\partial\xi} - \lambda\left(f'\frac{\partial f'''}{\partial\xi} + f'''\frac{\partial f'}{\partial\xi} - f''\frac{\partial f''}{\partial\xi} - f^{iv}\frac{\partial f}{\partial\xi}\right)\right] \tag{9}$$

$$\frac{1}{\Pr}\theta'' + f\theta' - m\theta f' = (m-1)\xi\left(f'\frac{\partial\theta}{\partial\xi} - \theta'\frac{\partial f}{\partial\xi}\right) \tag{10}$$

With the aforementioned transformation corresponding boundary conditions, Equations 4 and 5 will turn into:

$$f(0,\xi) = 0, \quad f'(0,\xi) = 1, \quad f'(\infty,\xi) = 0 \tag{11}$$

$$\theta'(0,\xi) = -1, \quad \theta(\infty,\xi) = 0 \tag{12}$$

Where, a prime denotes differentiation with respect to η.

In the case of local mixed convection parameter $\xi = 0$ and surface heat flux exponent $m = 2$, Equations 9 to 12 will take the following form:

$$f''' + ff'' - f'^2 + \lambda\left(2ff''' - (f'')^2 - ff^{iv}\right) = 0 \tag{13}$$

$$\frac{1}{\Pr}\theta'' + f\theta' - 2\theta f' = 0 \tag{14}$$

$$f(0,\xi) = 0, \quad f'(0,\xi) = 1, \quad f'(\infty,\xi) = 0 \tag{15}$$

$$\theta'(0,\xi) = -1, \quad \theta(\infty,\xi) = 0 \tag{16}$$

Equations 13 to 16 have become well-known equations that govern the flow of second grade viscoelastic fluid and heat transfer in the boundary layer region along a stretching sheet with power law surface heat flux. A similar problem had been investigated by Liu (2005).

Following the methodology used by Liu (2005), the analytical solutions of Equations 13 to 16 are obtained as shown:

$$f(\eta) = \frac{1}{r}\left(1 - e^{-r\eta}\right) \tag{17}$$

$$\theta(\eta) = \frac{1}{r}e^{-rs\eta}M\left(s-2, s+1, -se^{-r\eta}\right)\left[sM(s-2, s+1, -s) - s(s-2)M(s-1, s+2, -s)/(s+1)\right]^{-1} \tag{18}$$

Where:

$$M(a,b,z) = 1 + \sum_{n=1}^{\infty}\left((a)_n/(b)_n\right)z^n/n! \tag{19}$$

is the Kummer's function, and:

$$(a)_n = a(a+1)(a+2)\ldots(a+n-1) \tag{20}$$

$$r = \frac{1}{\sqrt{1+\lambda}}, \quad s = \Pr(1+\lambda) \tag{21}$$

The skin-friction coefficient, $f''(0)$, and the heat transfer coefficient, $1/\theta(0)$, are found to be:

$$r=\frac{1}{\sqrt{1+\lambda}},\quad s=\Pr(1+\lambda) \tag{22}$$

$$1/\theta(0)=\left[sM(s-2,s+1,-s)-s(s-2)M(s-1.s+2,-s)/(s+1)\right]/\left[\sqrt{1+\lambda}M(s-2,s+1,-s)\right] \tag{23}$$

The numerical values of $1/\theta(0)$ obtained using Equation 23 and also obtained by the implicit finite difference method (discussed subsequently) at selected values of viscoelastic parameter λ and Prandtl number *Pr* are entered in Table 2. From this table, it can be seen that the present numerical values are in excellent agreement with that of Liu (2005); since, the maximum difference between these two results is less than 0.5%.

SOLUTION METHODOLOGIES

As a first attempt, we employ the local non-similarity method of Sparrow and Yu (1971) in finding the solutions of Equations 9 to 12 by treating the local variable ξ as the non-similar parameter. Solutions for all ξ are then obtained by the implicit finite difference method together with Keller-box elimination technique (Keller, 1978) as well. Shot description of the aforementioned methods will be provided subsequently.

Local non-similarity method

With the exception of a few specially proposed boundary conditions (Sparrow and Gregg, 1958; Semenov, 1984), the vertical free convection boundary layer problems are normally locally non-similar. The local similar method reduces the coupled partial differential equations into a set of nonlinear ordinary differential equations to be solved numerically. However, the error introduced by this technique cannot be easily estimated. Sparrow et al. (1970) introduced the local non-similar method to improve this concept. However, like the local non-similar method, this technique is locally autonomous. Solutions at any specified stream wise station can be obtained without first obtaining upstream solutions. Keller and Yang (1972) employed a Görtler-type series to study the free convection boundary-layer flow along a non-isothermal vertical plate assuming that the wall temperature can be represented by a power series in the stream wise coordinate. Later, Kao et al. (1977) proposed the method of strained coordinates for the computation of the wall heat transfer parameter for a plate with an arbitrary prescribed surface temperature. In this method, the coordinate along the plate was transformed by using an integral function of the specified wall temperature so that the problem can be solved with any specified surface conditions. The non-similar solution can then be obtained for the local similarity solution by determining an approximate wedge parameter in such a way that the local similarity results give a value that would be obtained if one considers the local non-similarity solution method. Following this technique, the determination of the approximate wedge parameter involves an estimation and iterative procedure. Further, Yang et al. (1982) proposed an alternative method to evaluate the surface heat transfer rate and the wall shear stress for free convection boundary-layer flow considered by Kao et al. (1977) using a Merk-type series solution (Merk, 1959). The governing coupled partial differential equations in Kao et al. (1977) were transformed into a sequence of coupled ordinary differential equations which were then solved numerically by a fourth-order Runge-Kutta scheme with an incorporated least-squares convergence criteria for the zeroth-order solution and with the Newton-Raphson iteration scheme for higher-order solutions.

This section is concerned with the local non similarity method initiated by Sparrow and Yu (1971) and Minkowycz and Sparrow (1978) which has later been applied by many investigators such as Mushtaq el al. (2007), Hossain and Takhar (1996), Hossain et al. (1994) and Chen (1988). Formulation of the system of equations for the present local non similarity model, with reference to the present problem, will be demonstrated subsequently.

First level of truncation

At the first level of truncation, the terms of the form $\xi\partial(\ )/\partial\xi$ are deleted from the right hand side of Equations 9 and 10 (Sparrow and Yu, 1978) to yield the following system of equations:

$$f'''+ff''-f'^2+\lambda\left(2ff'''-(f'')^2-ff^{iv}\right)+\xi\theta=0 \tag{24}$$

$$\frac{1}{\Pr}\theta''+f\theta'-m\theta f'=0 \tag{25}$$

$$f(0,\xi)=0,\ f'(0,\xi)=1,\ f'(\infty,\xi)=0 \tag{26}$$

$$\theta'(0,\xi)=-1,\ \theta(\infty,\xi)=0 \tag{27}$$

It can be seen that Equations 24 and 25 can be regarded as a system of ordinary differential equations for the functions f and θ with ξ as a parameter for a given Prandtl number Pr and a viscoelastic parameter λ.

Second level of truncation

To find the higher level of truncation, we introduce the following functions:

$$g(\eta,\xi)=\frac{\partial f}{\partial\xi},\quad \phi(\eta,\xi)=\frac{\partial\theta}{\partial\xi} \tag{28}$$

For the second level, the governing Equations 9 and 10 are retained in full as:

$$f'''+ff''-f'^2+\lambda\left(2ff'''-(f'')^2-ff^{iv}\right)+\xi\theta \tag{29}$$

$$=(m-1)\xi\left[f'g'-f''g-\lambda\left(f'g'''+f'''g'-f''g''-f^{iv}g\right)\right] \tag{30}$$

$$\frac{1}{\Pr}\theta''+f\theta'-m\theta f'=(m-1)\xi\left(f'\phi-\theta' g\right) \tag{31}$$

$$\theta'(0,\xi)=-1,\quad \theta(\infty,\xi)=0 \tag{32}$$

The equations and boundary condition for g and ϕ can be devised by taking derivatives of Equations 29 to 32 with respect to ξ. This leads to:

$$g'''+fg''+mf''g-(m+1)f'g'+\lambda\left[(m+1)f'g'''+(m+1)f'''g'-(m+1)f''g''\right.$$
$$\left.-mf^{iv}g-fg^{iv}\right]+\theta+\xi\phi=(m-1)\xi\left[g'^2-gg''-\lambda\left(2g'g'''-g''^2-gg^{iv}\right)\right] \tag{33}$$

$$\frac{1}{\Pr}\phi''+f\phi'+mg\theta'-mg'\theta-(2m-1)f'\phi=(m-1)\xi\left[g'\phi-\phi'g\right] \tag{34}$$

$$g(0,\xi)=0,\quad g'(0,\xi)=0,\quad g'(\infty,\xi)=0 \tag{35}$$

$$\phi'(0,\xi)=0,\quad \phi(\infty,\xi)=0 \tag{36}$$

At this level of truncation, $\partial g/\partial\xi$ and $\partial\phi/\partial\xi$ and their derivatives with respect to η are neglected. This was first considered by Chen (2000) to obtain more accurate results. Considering the present scheme, solutions for f, g, θ and ϕ are obtained using the Natscheim-Swigert iteration technique. The required missing values of $f''(0,\xi)$ and $\theta(0,\xi)$ are obtained to measure the values of local skin-friction coefficient, $C_f Re_x^{1/2}/2$, and local heat transfer coefficient, $NuRe_x^{-1/2}$. The numerical values obtained for these quantities are depicted in Table 1 and compared with the solution obtained by the implicit finite difference method discussed subsequently and an excellent agreement is achieved.

Implicit finite difference solutions

The governing Equations 9 and 10 are revisited and the numerical solution is initiated using Keller Box scheme. We recast these equations into a set of simultaneous equations by introducing the variables U, V, W and P:

$$f'=F,\quad F'=G,\quad G'=H,\quad \theta'=P \tag{37}$$

$$G'+fG-F^2+\lambda\left[2FH-G^2-fH'\right]+\xi\theta$$
$$=(m-1)\xi\left[F\frac{\partial F}{\partial\xi}-G\frac{\partial f}{\partial\xi}-\lambda\left(F\frac{\partial H}{\partial\xi}+H\frac{\partial F}{\partial\xi}-G\frac{\partial G}{\partial\xi}-H'\frac{\partial f}{\partial\xi}\right)\right] \tag{38}$$

$$\frac{1}{\Pr}P'+fP-m\theta F=(m-1)\xi\left(F\frac{\partial\theta}{\partial\xi}-P\frac{\partial f}{\partial\xi}\right) \tag{39}$$

$$f(0,\xi)=0,\quad F(0,\xi)=1,\quad P(0,\xi)=-1 \tag{40}$$

$$F(\infty,\xi)=0,\quad \theta(\infty,\xi)=0 \tag{41}$$

We now place a net on the (η,ξ) plane defined by:

$$\eta_0=0,\ \eta_i=\eta_{i-1}+h_i,\qquad i=1,2,3,\dots,M$$

$$\xi_0=0,\ \xi_j=\xi_{j-1}+k_j,\qquad j=1,2,3,\dots,N$$

Which are then expressed in finite difference form by approximating the functions and their derivatives in terms of the central differences in both coordinate directions. Denoting the mesh points in the (η,ξ) plane by η_i and ξ_j, where i = 1, 2, 3, …, M and j = 1, 2, 3, …, N, central difference approximations are made such that the equations involving ξ explicitly, are centred at $(\eta_{i-1/2}, \xi_{j-1/2})$ and the remainder at $(\eta_{i-1/2}, \xi_j)$, where $\eta_{i-1/2} = (\eta_i + \eta_{i-1})/2$, etc. This results in a set of nonlinear difference equations for the unknowns at η_i in terms of their values at η_{i-1}. These equations are then linearised by the Newton's quasi-linearization technique and are solved using a block-tridiagonal algorithm, taking as the initial iteration of the converged solution, $\eta_i = \eta_{i-1}$. Now to initiate the process at $\eta = 0$, we first provide guess profiles for all five variables (arising the reduction to the first order form) and use the Keller box method to solve the governing ordinary differential equations. Having obtained the lower stagnation point solution it is possible to march step by step along the boundary layer. For a given value of η, the iterative procedure is stopped when the difference in computing the velocity and the temperature in the next iteration is less than 10^{-5}, that is, when $|\delta f^{\,i}|\le 10^{-5}$, where the superscript denotes the iteration number. The computations were not performed using a uniform grid in the ξ direction, but a non-uniform grid was used and defined by η_i = sinh((i-1)/p), with i = 1, 2, …,301 and p = 100.

RESULTS AND DISCUSSION

In the present investigation, two distinct methodologies were used to obtain the solution of the problem on mixed convection flow of viscoelastic second grade fluid along a non-isothermal stretched vertical surface. The physical parameters that control the flow and heat transfer are the Prandtl number, Pr, the viscoelastic parameter, λ, and the local Richardson number (also termed as the local mixed-convection parameter), ξ. Due to the presence of this local variable, ξ, the governing equations for the flow and heat transfer appeared as non-similarity equations. For values of, ξ, in the range of 0 to 10, the solutions are obtained by employing the local non similarity method and the implicit finite difference method. Once the functions f and θ and their derivatives are known, the important physical quantities, such as, the wall shear stress and the surface heat flux can be determined easily. The wall shear is expressed in terms of the local skin friction coefficient as:

$$C_{fx}=2\tau_w(x)/\rho(U_w(x))^2 \tag{42}$$

Table 1. Numerical values of $f'(0,\xi)$ and $1/\theta(0,\xi)$ obtained by local non-similarity method and implicit finite difference method against ξ when Pr = 0.70, λ = 1.0 and m = 2.0.

ξ	$f'(0,\xi)$		$1/\theta(0,\xi)$	
	LNS	FDM	LNS	FDM
0.00	-1.43788	-1.43764	1.02730	1.02730
0.25	-1.23718	-1.22929	1.06312	1.06263
0.50	-1.08240	-1.08563	1.08836	1.08674
0.75	-0.95084	-0.94250	1.10858	1.10619
1.00	-0.83465	-0.83850	1.12573	1.12274
1.25	-0.72974	-0.72161	1.14074	1.13727
1.50	-0.63357	-0.63629	1.15415	1.15029
1.75	-0.54446	-0.53821	1.16632	1.16212
2.00	-0.46119	-0.46150	1.17749	1.17299
2.50	-0.30882	-0.30727	1.19748	1.19247
3.00	-0.17144	-0.16924	1.21504	1.20961
4.00	0.07035	0.07249	1.24503	1.23893
5.00	0.28018	0.28207	1.27021	1.26361
6.00	0.46685	0.46834	1.29203	1.28507
7.00	0.63584	0.63673	1.31134	1.30408
8.00	0.79079	0.79096	1.32871	1.32116
9.00	0.93428	0.93366	1.34451	1.33668
10.00	1.06555	1.06544	1.35551	1.35081

And the local Nusselt number as:

$$N_{ux} = q_w(x)x / k(T_w - T_\infty) \tag{43}$$

Where:

$$\tau_w(x) = \mu\left(\frac{\partial u}{\partial y}\right)_{y=0} + \kappa\left(u\frac{\partial^2 u}{\partial x \partial y} + v\frac{\partial^2 u}{\partial y^2} + 2\frac{\partial u}{\partial x}\frac{\partial u}{\partial y}\right)_{y=0} \tag{44}$$

And:

$$(T - T_\infty)/(T_w - T_\infty) = \theta(\eta,\xi)/\theta(0,\xi) \tag{45}$$

Using Equations 42 and 43, we can express the local skin-friction, C_{fx} as:

$$\mathrm{Re}_x^{1/2}\, C_{fx} = 2(1+3\lambda)f''(0,\xi) \tag{46}$$

And the local Nusselt number as:

$$\mathrm{Re}_x^{-1/2}\, N_{ux} = 1/\theta(0,\xi). \tag{47}$$

The results are obtained in terms of local skin-friction and local Nusselt number against the mixed convection parameter ξ using the relations given in Equations 46 and 47.

Table 2 shows the comparison of analytical values of the local Nusselt number, Nu_x, using the methodology employed by Liu (2005) and those of present work obtained by the implicit finite difference method at selected values of λ and Pr when local mixed convection parameter ξ = 0 and surface heat flux exponent m = 2.

The numerical values of the local skin-friction coefficient, $f''(0, \xi)$, and local heat transfer coefficient, $1/\theta(0, \xi)$ for Pr = 0.70, λ = 1.0 and m = 2.0 obtained against local mixed convection parameter ξ in [0.0, 10.0] by the aforementioned methods are entered in Table 2. The results thus found are almost identical for all ξ in [0.0, 10.0]. Henceforth, the results discussed in the following paragraphs are due to implicit finite difference method only.

As a benchmark, the solution of Equations 13 to16 obtained by the present authors is compared with the one obtained by Chen et al. (1990) in Figure 2a and b as our special case when λ = 0 and m =1. It can be seen that these two results are in complete harmony with those presented in Chen et al. (1990).

Effect of physical parameters on skin-friction and Nusselt number

Numerical values of the local skin-friction coefficient as well as local heat transfer coefficient are depicted, respectively, in Figure 3a and b against ξ, while values of Pr equal to 0.70, 7.03 and 15.0 and that of λ = 1 and m =

Table 2. Comparison of analytical (Liu et al., 2005) and numerical values (present work) at selected values of λ and *Pr* when $\xi = 0$ and $m = 2$.

λ	*Pr*	**1 / θ(0)**	
		Liu (2005)	**Present**
	0.70	1.06932	1.06933
0.0	7.03	3.98056	3.98047
	15.00	5.93201	5.93169
	0.70	1.15164	1.15126
1.0	7.03	4.05531	3.89170
	15.00	6.00567	5.83579
	0.70	1.18707	1.18264
2.0	7.03	4.08805	3.60160
	15.00	6.03805	5.65077

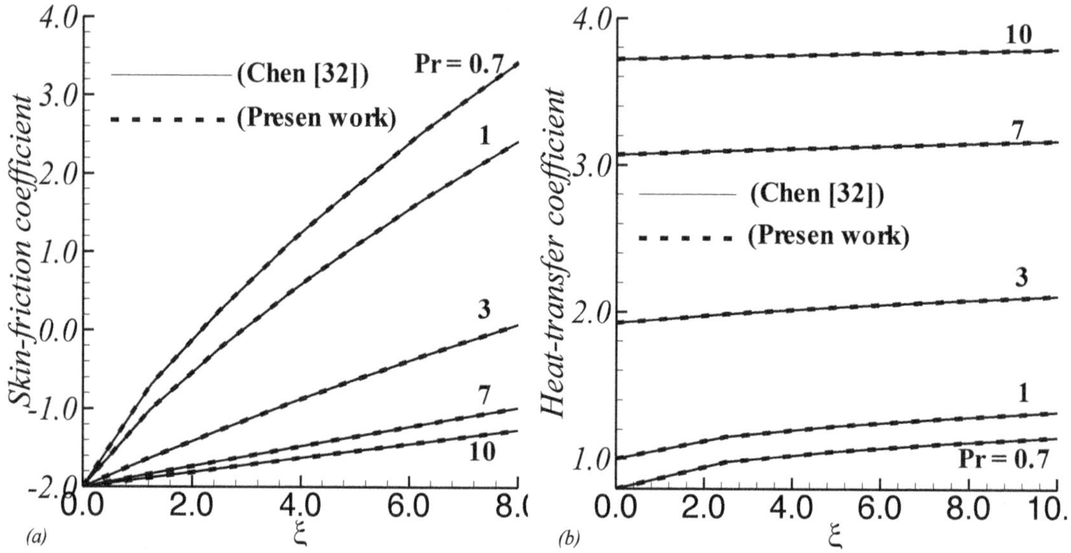

Figure 2. Numerical values of (a) $Re_x^{1/2}Cf_x/2$, local skin-friction coefficient and (b) $Re_x^{-1/2}Nu_x$, local heat transfer coefficient against ξ at selected values of *Pr* while $\lambda = 0$ and $m = 1$.

2.0. It may be observed that, for the given value of *Pr*, owing to increase in the value of local mixed convection parameter ξ, there is increase in both the local skin-friction $Re_x^{1/2}Cf_x/2$ and the local heat transfer coefficient, $Re_x^{-1/2}Nu_x$, which is expected, since in the downstream region the flow is dominated by the buoyancy force rather then the stretching rate of the plate. We further observe that increase in the value of $Re_x^{1/2}Cf_x/2$ is faster than that of $Re_x^{-1/2}Nu_x$ at a given value of *Pr*. From the same figure it may be noticed that, for an increase in *Pr*, the value of the local skin-friction coefficient, $Re_x^{1/2}Cf_x/2$, decreases where as there is increase in the value of the local heat transfer coefficient, $Re_x^{-1/2}Nu_x$. Now, we discuss the effect of the Deborah number (λ) on the local skin-friction coefficient, $Re_x^{1/2}Cf_x/2$ and the local heat transfer coefficient, $Re_x^{-1/2}Nu_x$. Numerical values of the local skin-friction coefficient, $Re_x^{1/2}Cf_x/2$ and the local heat transfer coefficient, $Re_x^{-1/2}Nu_x$ are depicted, respectively, in Figure 4a and b for values of λ equal to 0.0, 1.0 and 2.0 while the fluid's *Pr* is 7.03 and surface heat flux exponent *m* is 2.0 for $\xi \in [0, 10]$. From these figure one may observe that very near the leading edge there is decrease in the values of the local skin-friction coefficient and local heat transfer coefficient, owing to increase in λ. On the other hand, in the downstream-regime, values of both these physical quantities increase with the increase in λ.

This is expected, since in the downstream region the affect of the elastic property of the fluid get suppressed

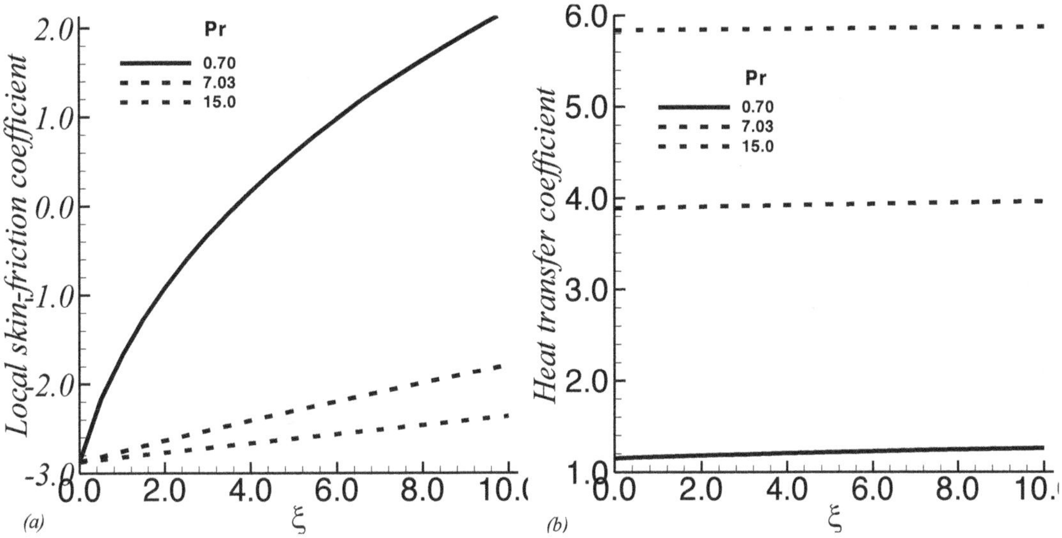

Figure 3. Variation of (a) local skin-friction coefficient and (b) local heat transfer coefficient with ξ at selected values of Prandtl number *Pr* when viscoelastic parameter λ =1 and surface heat flux exponent *m* =2.

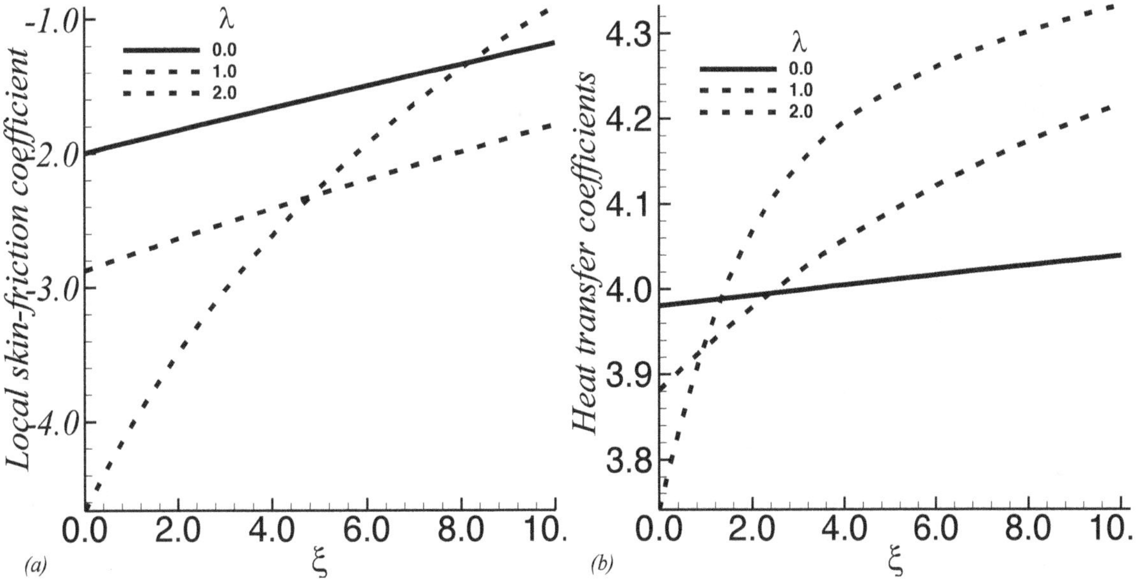

Figure 4. Variation of (a) local skin-friction coefficient and (b) local heat transfer coefficient with ξ at selected values of viscoelastic parameter λ when Prandtl number *Pr* = 7.03 and surface heat flux exponent *m*=2.

by the buoyancy force whereas it is strong near the leading edge.

Figure 5a and b illustrate the effect of varying surface heat flux exponent, *m*, on the local skin-friction coefficient, $Re_x^{1/2}Cf_x/2$, and the local heat transfer coefficient, $Re_x^{-1/2}Nu_x$, against local mixed convection parameter $\xi \in [0, 10]$ for λ = 1 and the *Pr* = 7.03. It can be noticed from Figure 5a that owing to increase in the surface heat flux exponent, *m*, the local skin-friction coefficient increases and the rate of increase first decreases and then increases, away from the leading edge. It may be observed from Figure 5b that, owing to increase in the surface heat flux exponent, *m*, the local heat transfer coefficient increases near the leading edge. From these figure, it can be noticed that both the local skin-friction coefficient, $Re_x^{1/2}Cf_x/2$, and the local heat transfer coefficient, $Re_x^{-1/2}Nu_x$ increases as moving away from the leading edge but increase in local skin-friction coefficient is higher than that of the local heat transfer coefficient.

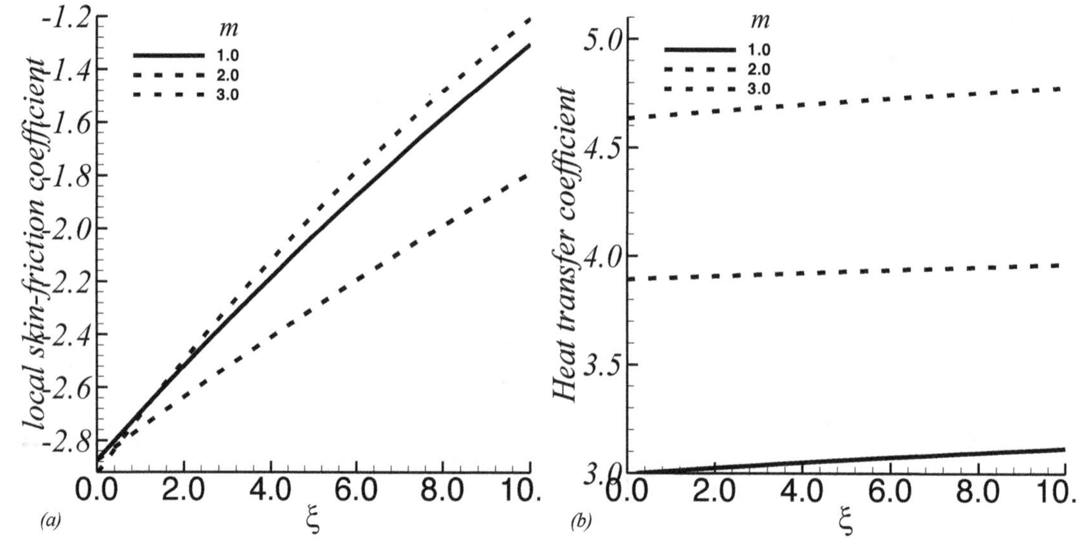

Figure 5. Variation of (a) local skin-friction coefficient and (b) local heat transfer coefficient with ξ at selected values of surface heat flux exponent m when Prandtl number Pr = 7.03 and viscoelastic parameter λ =1.

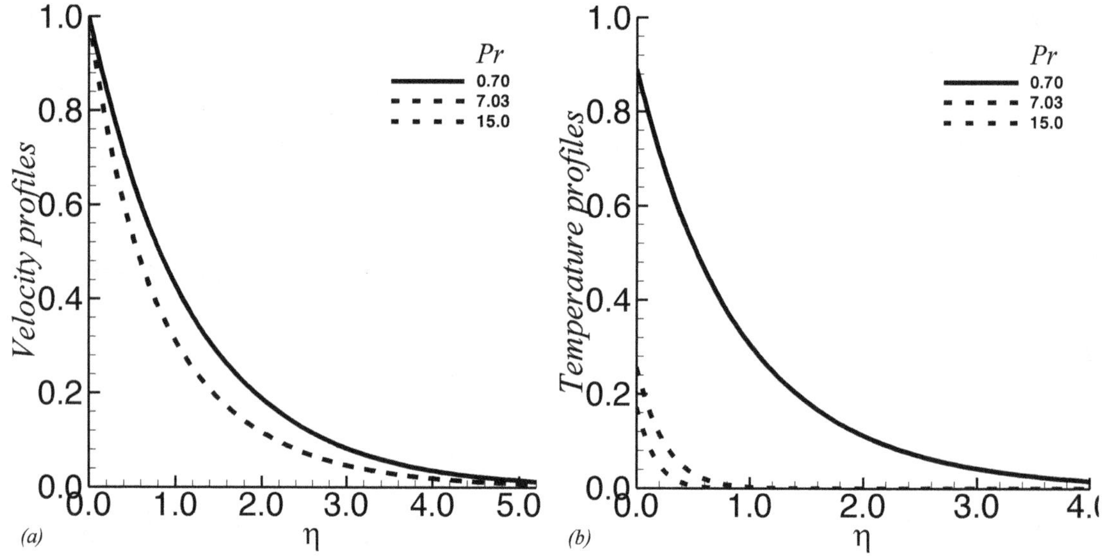

Figure 6. Variation of (a) velocity profiles and (b) temperature profiles against η for selected values of the Prandtl number Pr when viscoelastic parameter λ = 1.0, surface heat flux exponent m = 2 and local mixed convection parameter ξ = 1.

Effect of physical parameters on velocity and temperature profiles

Dimensionless velocity and temperature profiles are shown in Figure 6a and b, respectively, against η for Pr = 0.70, 7.03 and 15.0 while λ = 1, ξ = 1 and m = 2. It is depicted from Figure 6a that as λ increases from 0.70 to 15.0, both the velocity profiles and the momentum boundary layer thickness decreases. It is particular to note that the velocity profiles become insensitive to Prandtl number beyond Pr = 7.03. Also it can be seen from Figure 6b that Pr has least effect on the temperature profiles as Pr goes higher than 7.03. Owing to increase in Pr, both the temperature profiles and the thermal boundary layer thickness decrease.

The effect of increasing value of λ on velocity and temperature profiles for local mixed convection parameter ξ =1, Pr = 7.03 and m = 2 are presented in Figure 7a and b, respectively. From these Figure, one can observe that as the value of λ increases, there is decrease in the

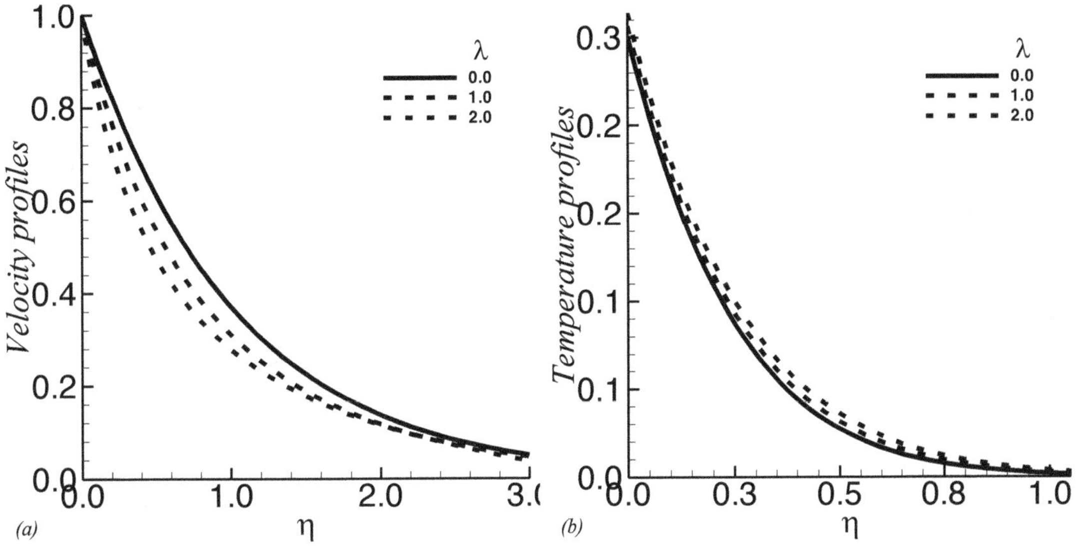

Figure 7. Variation of (a) velocity profiles and (b) temperature profiles against η for selected values of the viscoelastic parameter λ when Prandtl number Pr = 7.03, surface heat flux exponent m = 2 and local mixed convection parameter ξ = 1.

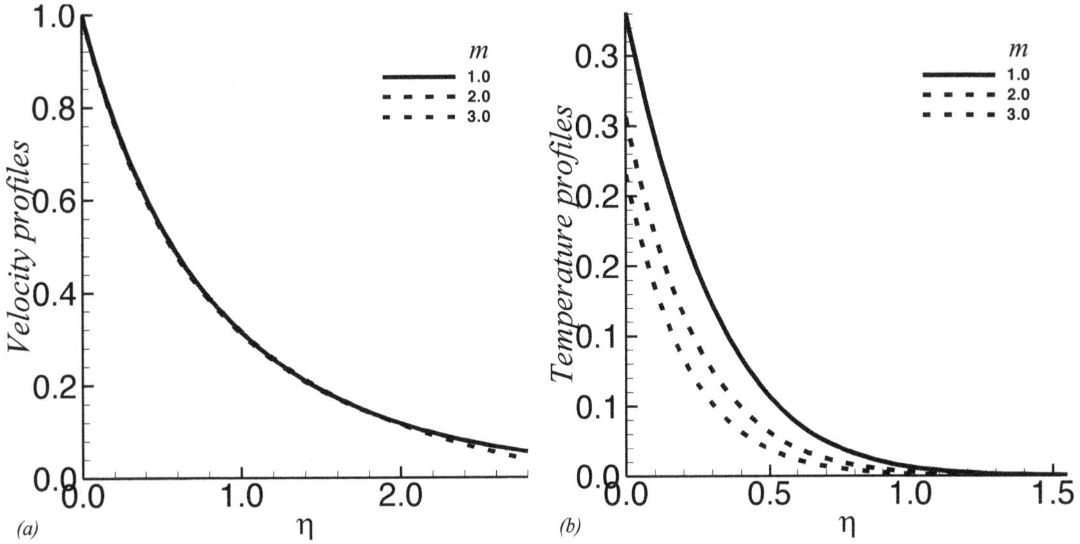

Figure 8. Variation of (a) velocity profiles and (b) temperature profiles against η for selected values of surface heat flux exponent m when Prandtl number Pr = 7.03, viscoelastic parameter λ = 1 and local mixed convection parameter ξ = 1.

velocity profiles and increase in the temperature profiles in the boundary layer regions and that leads to decrease and increase in the momentum and thermal boundary layer thickness, respectively.

The velocity profiles seems to be have no effect due to increase in *m,* while Pr = 7.03, λ and ξ = 1 but away from the surface it has effect of decreasing velocity as well as decreasing momentum boundary layer thickness (Figure 8a). From Figure 8b, it is observed that both the temperature profiles and thermal boundary layer thickness decrease owing to increase in *m*.

It is known that when ξ = 0, the velocity profile corresponds to pure forced convection flow of the second grade fluid along a stretched surface. But with the increase in ξ, buoyancy force becomes stronger and hence the velocity profile of the fluid increases in the thick region near the surface of the plate. It can further be observed that buoyancy effects tend to disappear away

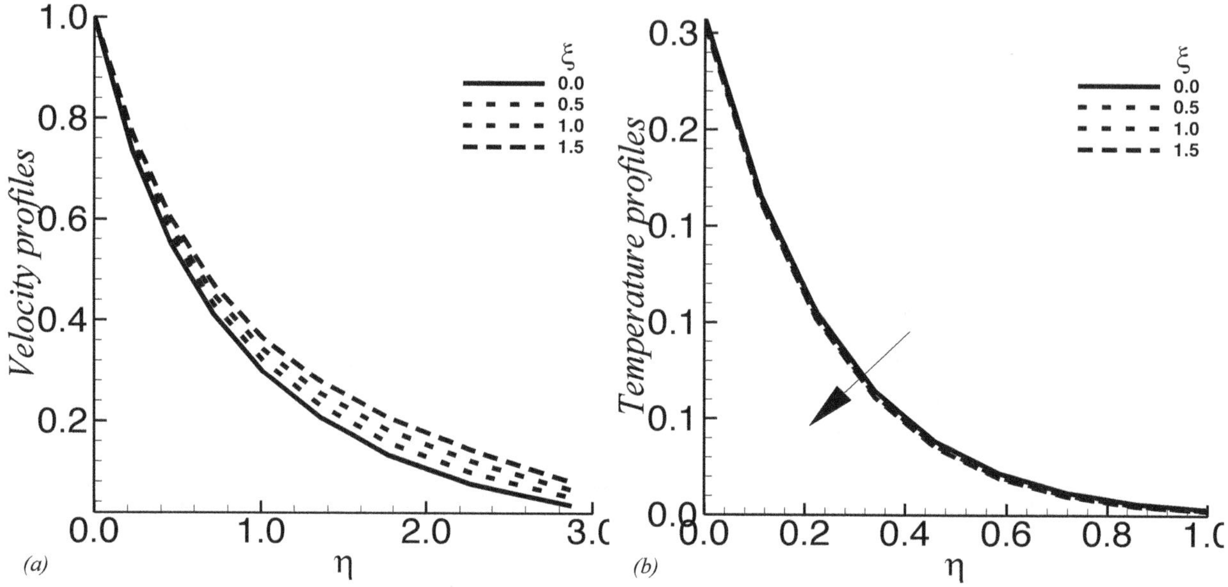

Figure 9. Variation of (a) velocity profiles and (b) temperature profiles against η for selected values of local mixed convection parameter ξ when Prandtl number Pr = 7.03, viscoelastic parameter λ = 1 and surface heat flux exponent m = 2.

from the surface of the plate (Figure 9a). From Figure 9b, we further see that for Pr = 7.03, λ and m = 2, owing to increase in the mixed convection parameter (local Richardson number), ξ , temperature profiles decreases whereas the thermal boundary layer thickness seems to be remained the same.

Conclusions

In the present analysis, we have investigated the mixed convection flow of viscoelastic second grade fluid past a heated and continuously stretched vertical flat plate, numerically. Solutions of the governing equations for momentum and energy for the mixed convection regime have been obtained using two different methods. Here, the numerical results have been provided in terms of the local skin-friction coefficient, local heat transfer coefficient, and velocity and temperature profiles.

From the present investigation, the following conclusions may be drawn:

1. In the mixed convection regime the values of the local skin-friction coefficient and local heat transfer coefficient increase with the increase in ξ.
2. The values of both the skin-friction coefficient and the heat transfer coefficients decreases near the leading edge owing to increase of λ, and this behavior is reverse in the upstream regime.
3. Increase in the value of λ, leads to decrease the momentum boundary layer thickness and increase in the thermal boundary layer thickness.
4. Owing to increase in m, both the temperature and thermal boundary layer thickness decrease whereas no significant change in velocity profiles.
5. Both the momentum and thermal boundary layer thickness are least sensitive to ξ.

List of symbols

C_{fx}= local skin friction coefficient
g = gravitational acceleration (m/sec^2)
Gr_x= local Grashof number
K = kinematic elasticity (= κ/ρ)
Nu_x= local Nusselt number
P = fluid pressure (Pa)
Pr = Prandtl number
q_w= heat transfer per unit area at the surface
Re_x= local Reynolds number
Ri = Richardson number
T = temperature (°C)
T_w= surface temperature(°C)
T_∞= ambient temperature(°C)
U_w= velocity of the moving surface (m/s)
u = velocity in x-direction (m/s)
v = velocity in y-direction (m/s)
U = dimensionless velocity in x-direction
V = dimensionless velocity in y-direction
x, y = Cartesian coordinates (m)
X, Y = dimensionless Cartesian coordinates

Greek symbols

α = thermal diffusivity of the ambient fluid ($k/\rho c_p$)

β = coefficient of thermal expansion of fluid (K^{-1})
δ = momentum boundary layer thickness (m)
δ_T = thermal boundary layer thickness (m)
η = similarity variable (m)
θ = dimensionless temperature
κ = second grade parameter
λ = viscoelastic parameter (Deborah number)
μ = effective dynamic viscosity (Pa/s)
ν = effective kinematic viscosity (μ/ρ)
ξ = local mixed convection parameter
ρ = fluid density at reference temperature (T_0)
τ_w = shear-stress at the surface
ψ = stream function (m^2/s)

REFERENCES

Ali M, Al-Yousef F (1998). Laminar mixed convection from a continuously moving vertical surface with suction or injection. Heat Mass Transf., 33: 301-306.

Altan T, Oh S, Gegel H (1979). Metal forming fundamentals and applications, American Society of Metals, Metals Park, OH (Book).

Bhattacharyya S, Pal A, Gupta AS (1998). Heat transfer in the flow of a viscoelastic fluid over a stretching surface. Heat Mass Transf., 34: 41-45.

Chen CH (1998). Laminar mixed convection adjacent to vertical, continuously stretching sheets. Heat Mass Transf., 33: 471-476.

Chen CH (2000). Mixed convection cooling of a heated, continuously stretching surface. Heat Mass Transf., 36: 79-86.

Chen KC, Char MI, Cleaver JW (1990). Temperature field in non-Newtonian flow over a stretching plate. J. Math. Anal. Appl., 151: 301-307.

Chen TS (1988). Parabolic systems: Local nonsimilarity method in W. J. Minkowycz et al. (eds.), Handbook of numerical heat transfer, Wiley, New York.

Chen TS, Strobel FA (1980). Buoyancy effects in boundary layer adjacent to a continuous, moving horizontal flat plate. ASME J. Heat Trans., 102: 170-172.

Dunn JE, Rajagopal KR (1995). Fluids of differential type: Critical review and thermodynamic analysis. Int. J. Eng. Sci., 33: 689-729.

Fisher EG (1976). Extrusion of plastics, Wiley, New York,

Hossain MA, Banu N, Nakayama A (1994). Non-darcy forced convection boundary layer flow over a wedge embedded in a saturated porous medium, 26: 399-414.

Hossain MA, Takhar HS (1996). Radiation effect on mixed convection along a vertical plate with uniform temperature. J. Heat Mass Transf., 31: 243-248.

Ingham DB (1986). Singular and non-unique solutions of the boundary-layer equations for the flow due to free convection near a continuously moving vertical plate. Appl. Math. Phys., (ZAMP) 37: 559-572.

Kao TT, Domoto GA, Elrod HG (1977). Free convection along a nonisothermal vertical flat plate. Trans. ASME J. Heat Transf., 99: 72-78.

Karwe MV, Jaluria Y (1988). Fluid flow and mixed convection transport from a moving plate in rolling and extrusion processes. ASME J. Heat Transf., 110: 655–661.

Karwe MV, Jaluria Y (1991). Numerical simulation of thermal transport associated with a continuously moving flat sheet in materials processing. ASME J. Heat Transf., 113: 612-619.

Keller HB (1978). Numerical methods in boundary layer theory. Annu. Rev. Fluid Mech., 10: 417-433.

Keller M, Yang KT (1972). A Görtler-type series for laminar free convection along a non-isothermal vertical flat plate. Quart. J. Mech. Appl. Math., 25: 447-457.

Liu IC (2005). Flow and heat transfer of an electrically conducting fluid of second grade in a porous medium over a stretching sheet subject to a transverse magnetic field. Int. J. Non-Linear Mech., 40: 465-474.

Magyari E, Ali ME, Keller B (2001). Heat and mass transfer characteristics of the self-similar boundary-layer flows induced by continuous surface stretched with rapidly decreasing velocities. Heat Mass Transf., 38: 65-74.

Magyari E, Keller B (2000). Exact solutions for self-similar boundary-layer flows induced by permeable stretching walls. Eur. J. Mech. B: Fluids, 19: 109-122.

Merk HJ (1959). Rapid calculation for boundary-layer transfer using wedge solutions and asymptotic expansions. J. Fluid Mech., 5: 460-480.

Minkowycz WJ, Sparrow EM (1978). Numerical solution scheme for local nonsimilarity boundary layer analysis. Numer. Heat Transf., 1: 69-85.

Mushtaq M, Asghar S, Hossain MA (2007). Mixed convection flow of second grade fluid along a vertical stretching flat surface with variable surface temperature. Heat Mass Transf., (in press).

Rajagopal KR, Na TY, Gupta AS (1984). Flow of a visco-elastic fluid over a stretching sheet. Rheol. Acta., 23: 213-215.

Sakiadis BC (1961). Boundary layer behavior on continuous solid surfaces: I. boundary-layer equations for two-dimensional and axisymmetric flow. A.I.Ch.E. J., 7(1): 26-28.

Semenov VI (1984). Similar problems of steady-state laminar free convection on a vertical plate. Heat Transf. Sov. Res., 16: 69-85.

Sparrow EM, Gregg JL (1958). Similar solutions for free convection from nonisothermal vertical plate. Trans. ASME J. Heat Transf., 80: 379-384.

Sparrow EM, Quack H, Boerner CJ (1970). Local non-similar boundary layer solutions. AIAA J., 8: 1936-1942.

Sparrow EM, Yu HS (1971). Local non-similarity thermal boundary layer solutions. Trans. ASME J. Heat Transf., 93: 328-334.

Tadmor Z, Klein I (1970). Engineering principles of plasticating extrusion: Polymer Science and Engineering Series, Van Nostrand Reinhold, New York.

Walters K (1962). Non-Newtonian effects in some elastico-viscous fluids whose behaviour at small rates of shear is characterised by a general equation of state. Q. J. Mech. Appl. Math., 15: 63-76.

Yang J, Jeng DR, DeWitt KJ (1982). Laminar free convection from a vertical plate with nonuniform surface conditions. Numer. Heat Transf., 5: 165-184.

Investigation of cutting temperatures distribution in machining heat treated medium carbon steel on a lathe

A. G. F. Alabi[1], T. K. Ajiboye[1]* and H.D. Olusegun[2]

[1]Department of Mechanical Engineering, Faculty of Engineering, University of Ilorin, Kwara State, Nigeria.
[2]Federal University of Technology, Minna, Niger State, Nigeria.

This study was carried out on medium carbon steel subjected to various form of heat treatment operations by assessing the temperature distribution during the machining process. The model of oblique band heat source, moving in the direction of cutting in an infinite medium with an appropriate image heat source was used in this investigation. The model analysis was carried out separately on the chip, the tool and the work material to numerically determine the temperature distribution during machining process using finite element method with nodal grids. Johnson-Cook model was used to determine the work materials flow stress upon which the material properties were determined. Stress/strain tests were conducted on the specimens to determine the materials' constant parameters which were used as the input parameters to the modeling equation written in Visual Basic 6.0. The optimum shear angle of 20° was used for the machining process and the frictional characteristics were also determined. The temperature along the shear plane AB was determined with reference to coordinate axis within the tool, workpiece and the chip using the model. The results which compare favorably with other results from literatures, provide basis for the design of machining variables for optimum and quality machined products which can also be applied in the computer programming of NC machine for precision machining

Key words: machining, cutting tool, heat treatment, "as received", normalized, annealed, tempered, hardened, simulated, profile.

INTRODUCTION

The heat generated during machining or turning constitutes a major problem in the manufacturing industry. It is a source of cost expenditure and also constitute hazard to health especially when coolants are used. Heat generation during the machining process was so important that it was one of the foremost topics investigated. It had earlier been observed that heat generation was responsible for the wear of the cutting tool; hence, an empirical relationship between the cutting speed, tool temperature and tool life was developed (Taylor, 1934). As a result of these findings, a more heat resistant known as high speed steel (HSS) was developed for both high speed and very high temperature machining. An analytical investigations of the temperature generated in machining processes were also developed (Rosenthal, 1945; Hahn, 1951; Trigger et al., 1951; Loewen et al., 1954). Following these studies there had been many analytical, mechanical and numerical models which simulated the metal cutting processes. Most especially, numerical models are highly important in predicting chip formation, computing the cutting forces, analyzing strain and strain rate, determining the temperature distribution, and stresses on the cutting edge (Ozel et al., 2004).

The numerical models enable the prediction of the tool life as tool wear is accurately monitored. The models analyze temperature during machining, its effect as it generates flow stress, strain and strain rate, and how it causes surface asperity on the workpiece. The asperity determines the quality of the workpiece surface finish.

*Corresponding author. E-mail: engrtkajiboye@yahoo.com.

Another important parameter considered by the models is the effect of friction. Friction between the workpiece, the cutting tool and the chip plays a significant role in the simulation of the cutting forces, stresses and temperature distribution (Karpat et al., 2005). The model of metal cutting which uses the finite element concept for calculating the temperature fields as proposed by Tay et al. (1990) was applied in this study. The investigation was carried out on the medium carbon steel specimens subjected to various form of heat treatment processes of hardening, normalizing, tempering and annealing in the temperature range of 350 to 850°C. The materials' properties determined from the stress strain curve was used as input parameters to an Oxley type constitutive model which was developed and programmed in Visual Basic to determine the machining temperature distribution of the workpiece, tool and chip.

MATERIALS PREPARATION

Test specimens' preparation

The material used for this study is a medium carbon steel with carbon content of 0.30% carbon as determined by X-ray diffraction technique. Standard heat treatment procedures were adopted, Totten (2007) to heat treat the medium carbon steel. Five different samples were prepared for each test. The first five specimens were not heat-treated and were kept as "untreated sample". The next batches of five samples were place in a furnace at a temperature of 850°C for a period of 3 h for the microstructure of the specimens to completely transform. They were removed from the furnace and immediately quenched to maintain the hardened structure. The batch to be tempered after heated to 850°C were later heated to about 350°C and then allowed to cool in still air, so that the samples became softer. The batch to be normalized were heated to 850°C and then allowed to cool outside the furnace in still air. A homogenous structure of ferrite and pearlite were formed. The annealed specimens were heated in the furnace at 850°C for 3 h. The furnace was turned off and the specimen allowed to cool in the furnace while they were contained inside a cast iron box with a mixture of cast iron borings, charcoal, lime and fine sand to disallow decarburization of the specimen.

THEORETICAL ANALYSIS

The finite difference method is one of the most popular tools when considering the analysis of temperature distribution in machining processes. Consider a classical orthogonal cutting process for the continuous two dimensional chip formation case as shown in Figure 1. The shearing is assumed to occur on the shear plane with friction at the rake face acting on the tool-chip interface over the contact length, *l*. It is assumed that the total workdone in machining is converted to heat, Q. The heat generating regions are also assumed to be at the shear zone and rake face. The application of the cutting tool at the point of action is equivalent to the application of a source of energy. The heat losses at the side surface of the chip, and at the tool flank and rake face other than at the tool-chip interface are considered to be negligible. The temperatures at the back face of the tool wedge and through a cross-section of the chip, some distance from the cutting region are assumed to be constant. It is assumed that the tool and chip achieve steady state and quasi-steady state conditions, respectively.

There are two important zones that give rise to the total heat, Q_T, generated during machining. They are the shear zones which generate heat given as Q_s, and the heat produced on the rake face, Q_f. It is assumed that Q_f is uniformly distributed over the contact area and that no frictional heat is transferred to the workpiece (Cerett et al., 1996).

The final temperature distribution of the chip and tool system is obtained by using finite difference analysis with nodal grids and boundary conditions (Ng et al., 1999). From Figures 2 and 3, the total heat generated per unit time, Q_T is given in;

$$Q_T = Q_s + Q_f \tag{1}$$

$$Q_s = F_s V_s = \frac{(\tau w t_u V Cos\, \lambda_n)}{\{Sin\, \phi\, Cos\, (\phi - \alpha)\}} \tag{2}$$

$$Q_f = FV_c = \frac{(\tau w t_u V Sin\, \lambda_n)}{\{Cos\, (\phi + \lambda_n - \alpha) Cos\, (\phi - \alpha)\}} \tag{3}$$

$$Q_T = F_p V = \frac{\{\tau w t_u V Cos\, (\lambda_n - \alpha)\}}{Sin\, \phi\, Cos\, (\phi + \lambda_n - \alpha)} \tag{4}$$

The rise in the average bulk temperature of the chip due to shearing (T_s) can be found from one of the convectional shear zone thermal analysis given in

$$T_s = \frac{((1 - B_1) Q_s)}{\rho c w t_u V} \quad \text{(Kapat et al., 2005)} \tag{5}$$

And

$$B_1 = \frac{1}{\left[1 + 0.753 \left(\frac{R_T}{\varepsilon} \right)^{\frac{1}{2}} \right]} \tag{6}$$

where Q_T, Q_s, and Q_f are as defined in the preceding equations; F_s is shear force, F_c is cutting force, F is frictional resistance of the tool acting on the chip, V_s is the velocity of the chip relative to the workpiece, V_c is velocity of chip relative to the tool, τ is shear stress, w is the width of cut, V is the resultant cutting speed, λ_n is friction angle, ϕ is shear angle and α is normal rake angle, B is the proportion of Q_s entering the workpiece, R_T is the thermal number, ε is the plastic strain of the material of the workpiece.

$$R_T = \frac{t_u V}{\alpha} \text{ = thermal number.} \tag{7}$$

$$\varepsilon = Cot\, \phi + \tan(\phi - \alpha) \tag{8}$$

When the chip is considered as a body in quasi-static thermal equilibrium with a moving heat source and the tool as a body in steady state thermal equilibrium with a stationary heat source, it is possible to find the final temperature in the chip and tool. By using the Cartesian co-ordinate system and applying it to the Figure 5, the generated heat conduction equations for the chip and tool are given as follows:

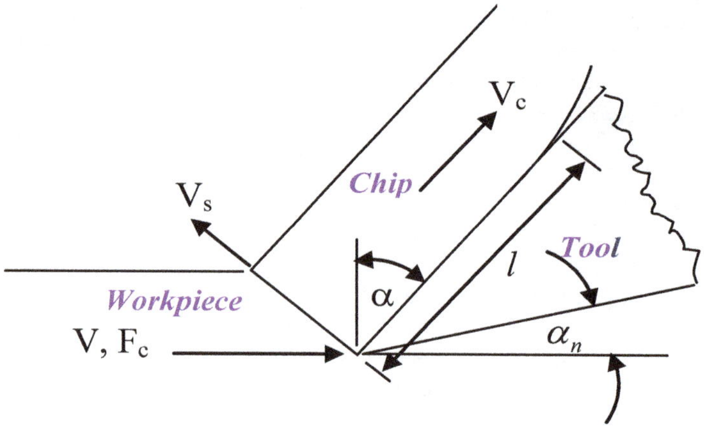

Figure 1. Continuous chip formation in orthogonal cutting process.

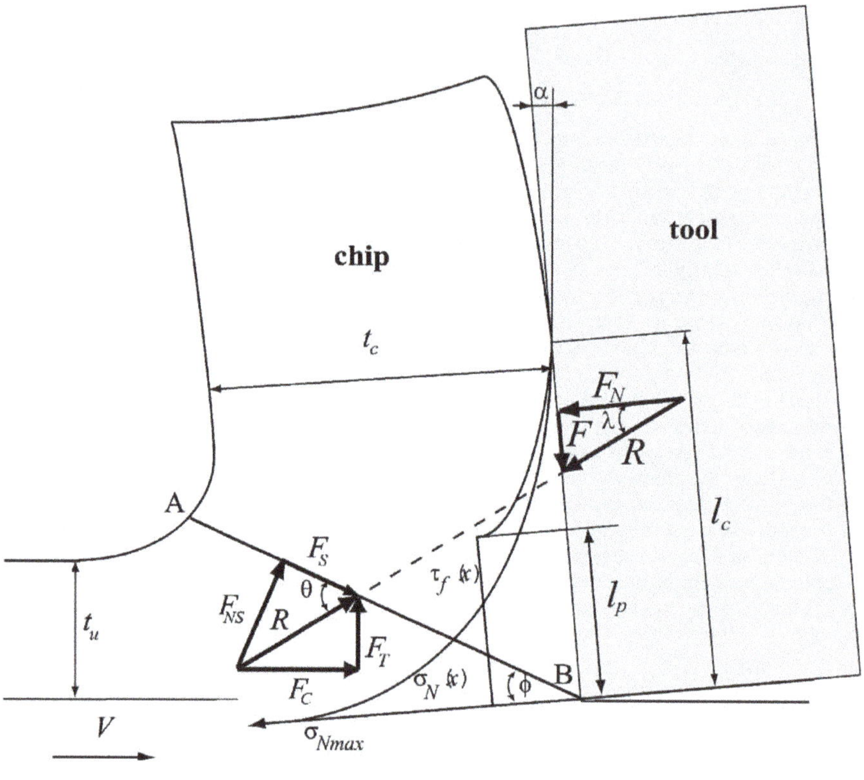

Figure 2. Forces acting on the shear plane and the tool with resultant stress distribution on the tool rake face (Kapat et al., 2005).

$$dvk\left[\frac{\partial^2 T}{\partial x^2}+\frac{\partial^2 T}{\partial y^2}+\frac{\partial^2 T}{\partial z^2}\right]+q = dv\rho c\frac{\partial T}{\partial t_u} = dv\rho cv\frac{\partial T}{c\partial x} \quad (9)$$

$$dvk\left[\frac{\partial^2 T}{\partial x^2}+\frac{\partial^2 T}{\partial y^2}+\frac{\partial^2 T}{\partial z^2}\right]+q = dv\rho c\frac{\partial T}{\partial t_u} = 0 \quad (10)$$

where k is thermal conductivity, c is specific heat capacity, and T is the average bulk temperature rise of the chip due to shearing. In finite difference form, Equations 9 and 10 become, For the chip,

$$A + q = \delta v\rho c\left[\frac{T(x+\delta x, y, z) - T(x-\delta x, y, z)}{2\delta x}\right] \quad (11)$$

For the tool;

$$A + q = 0 \quad (12)$$

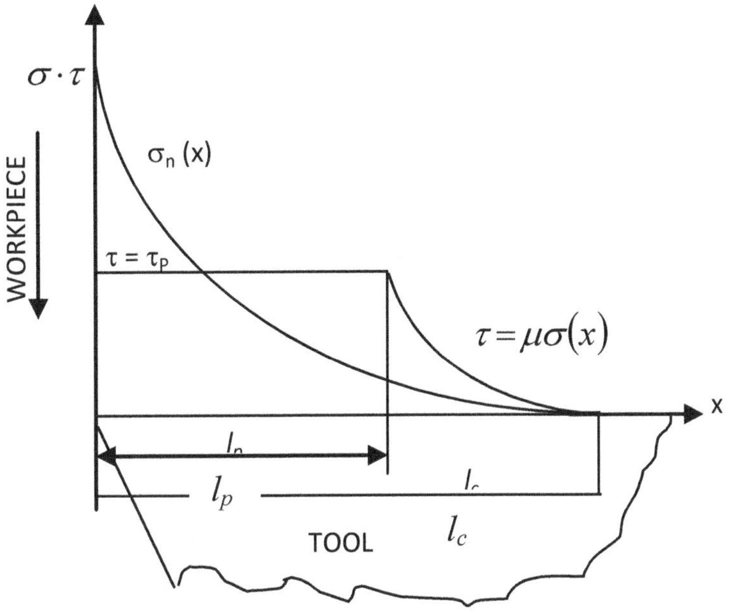

Figure 3. Normal and frictional stress distributions on the tool rake face.

where;

$$A = \delta x \delta y \delta z \; k\left[\frac{T(x+\delta x, y, z)+T(x-\delta x, y, z)-2T(x, y, z)}{\delta x^2} + \frac{T(x, y+\delta y, z) + T(x, y-\delta y, z)-2T(x, y, z)}{\delta x^2} + \frac{T(x, y, z+\delta z) + T(x, y, z-\delta z) - 2T(x, y, z)}{\delta x^2}\right] \quad (13)$$

If both the chip and tool are considered, the cutting edge and common rake face plane are used to select the x- and y axes as shown on Figure 4a. The heat generated at the nodes inside the chip, Figure 5b, for example CDEFG is assumed to be zero since heat will only be transferred by friction and shearing. By the concept of mirror imagining, at positions C and D, (Figure 4a and b), we have; At C,

$$T(x, y+\delta y, z) = T(x, y-\delta y, z) \quad (14)$$

At D,

$$T(x+\delta x, y, z) = T(x-\delta x, y, z) \quad (15)$$

The values of Equations 14 and 15 are used in Equation 13. The tool is represented by Figure 5. For internal nodes like J, q = 0. At point on the boundary such as K and L, the principles of mirror imaging is again applied.

From Figure 5 the situations differ for the nodes at the edge and those inside the grid. To develop the finite difference equations from first principles the following steps were followed. At M, the finite difference equation is,

$$A_1 = 0 \quad (16)$$

$$A_1 = k\frac{(1+e)(1+b)\delta x\, \delta y\, \delta z}{4} = \left[2\left(\frac{eT\theta(x+\delta x, y, z)+T(x-e\delta x, y, z)-(1+e)T(x, y, z)}{e(1+e)\delta x^2}\right) + 2\left(\frac{T(x, y+d\delta y, z)+dT(x, y-\delta y, z)-(1+d)T(x, y, z)}{d(1+d)\delta y^2}\right) + \frac{T(x, y, z+\delta z)+T(x, y, z-\delta z)-2T(x, y, z)}{\delta z^2}\right] \quad (17)$$

and at N;

$$A_2 = 0 \quad (18)$$

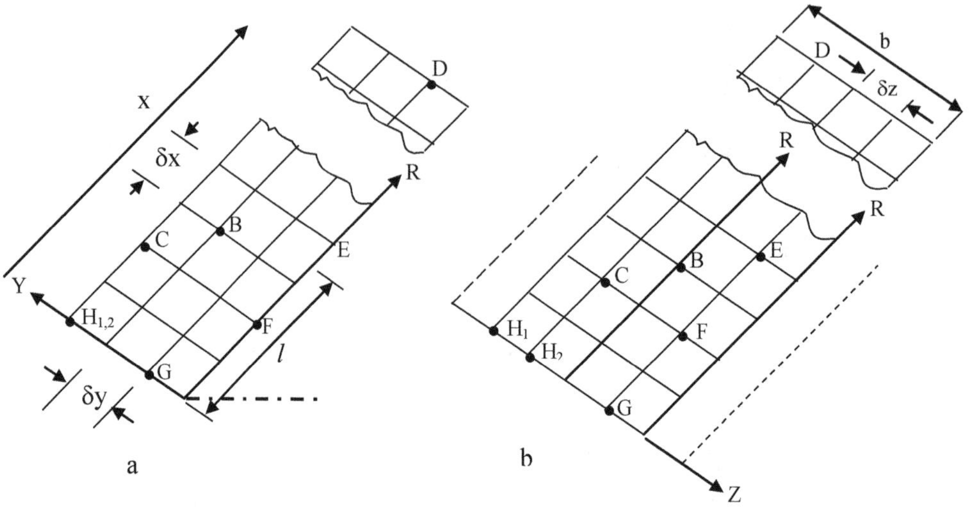

Figure 4. Nodal points inside chips.

where;

$$A_2 = \delta v k\left[\frac{T(x,y,z+\delta z)+T(x,y,z-\delta z)-2T(x,y,z)}{\delta z^2} + \frac{2}{d\delta y\, e\,\delta x}\left\{T_1 - T(x,y-d\delta y,z) - x^2\,T(x,y,z) - \frac{(e+g)(f\,\delta y - g\delta x\,x)\delta x\,T(x,y-d\delta y,z)}{g\delta x^2(g+e) - f\delta y^2(f+d)}\right\}\right] \quad (19)$$

where,

$$T_1 = \theta(x - T\delta x,\ y - d\delta y,\ z)$$

$x = \tan\alpha$ and d, e, f, g are fractions of regular nodal spacings on Figure 5. It is important to note that the Cartesian finite grid is ideal for the chips, but the analysis is more complete, when considering the wedge tool because of the extra equations needed for the flank face. For the flank face, the equation for a one dimensional heat flow is given as,

$$\frac{\partial^2 T}{\partial y^2} = \frac{R_T}{t_u}\frac{\partial T}{\partial x} \quad (20)$$

When only temperature rise at the heat source is considered the temperature rise at point I on Figure 6 caused by the entire moving band heat source is given as,

$$T_I = \frac{q_{pl}}{2\pi\lambda}\int_{-l}^{+l} e^{-VR\,Cos\left(\frac{\beta-\phi}{2a}\right)} K_o\left(\frac{V(X-l_i)}{2a}\right) dl_i \quad (21)$$

where T_l is temperature rise at point *l*, q_{pl} is heat liberation intensity, V is resultant cutting speed, R is the distance between the moving line heat source and the point I, β is angle between R and X-axis, a is thermal diffusivity, ϕ is shearing angle, K_o is zero order Bessel function which determines the location of the nodal points, *li* is the instantaneous point of interest on the shear plane *AB*, and λ is the thermal conductivity.

Modeling of temperature distribution in the chip

The general heat flow Equation (20) can now be modeled for the primary shear zone and tool-chip interface. According to the general solution of Equation 21, the temperature rise at any point on the chip due to primary heat source can be found by Equation 21, with the heat liberation intensity, q_{pl}, now replaced by shear heat intensity, q_{shear}, as a result of cutting and with the coordinate system given in Figure 6.

$$T_{shear}(X,Z) = \frac{q_{shear}}{2\pi\lambda_c}\int_{l_i=0}^{L_{AB}} e^{-(U)\frac{V_c}{2a_c}} \times \left\{K_o\left[\frac{V_c}{2a_c}\sqrt{(U)^2 + (U')^2}\right]\right\}dl_i + \frac{q_{shear}}{2\pi\lambda_c}\int_{l_i=0}^{L_{AB}} e^{-(U)\frac{V_c}{2a_c}}\left\{K_o\left[\frac{V_c}{2a_c}\sqrt{(U)^2 + (2t_{ch} - U')^2}\right]\right\}dl_i \quad (22)$$

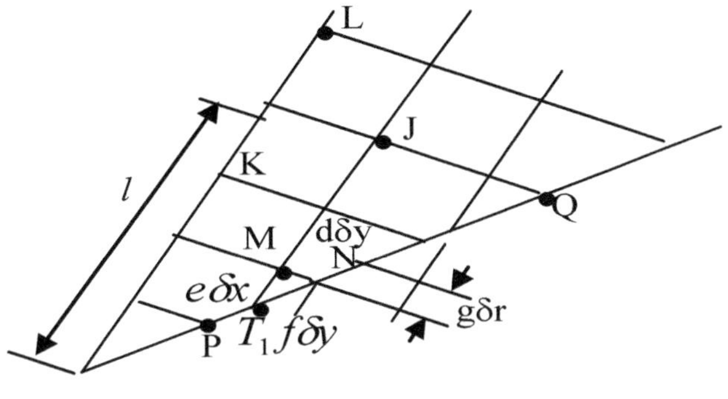

Figure 5. Nodal points in the tool.

where

$$U = X - l_c + l_i Sin(\phi - \alpha)$$

$$U' = Z - l_i Cos(\phi - \alpha), \text{ and}$$

$$X_i = l - l_i Sin(\phi - \alpha),\ Z_i = l_i Cos(\phi - \alpha),\ l = {t_{ch}}/{Cos(\phi - \alpha)}$$

l_{AB} is the length of shear zone/plane, l_c is tool-chip contact length, and K_o, ϕ, α, V_c, a, λ_n are as defined previously.

The effect of secondary heat source along the tool-chip interface on the chip side modeled as a non-uniform moving band heat source and the temperature rise at any point in the chip due to this heat source can be calculated with Equation 23.

The tool rake face and the upper surface of the chip are considered to be adiabatic. Since an imaginary heat source is added to the tool rake face to make this surface adiabatic, this image heat source will also have double heat intensity.

$$T_{chip-friction} = \frac{1}{\pi \lambda_c} \int_{l_i=0}^{l_c} B_i(l_i) q_f(l_i) e^{-(X-l_i)\frac{V_c}{2a_c}} \times \left[K_o\left(R_i \frac{V_c}{2a_c}\right) + K_o\left(R_i' \frac{V_c}{2a_c}\right)\right] dl_i \quad (23)$$

where;

$R_i = \sqrt{(X-x)^2 + Z^2}$, and $R_i' = \sqrt{(X-x)^2 + (2t_{ch} - Z)^2}$, which are the distances from heat sources respectively.

Non-uniform heat partition for the chip side and non-uniform heat intensity are denoted by $B_i(l_i)$ and $q_{pl}(l_i)$ respectively. As a result, the temperature rise at any point in the chip can be found by using Equation 24 with the addition of room temperature T_o.

$$T_{chip} = T_{shear} + T_{chip-friction} + T_o \quad (24)$$

The heat intensity of the primary and secondary heat sources in this study is considered uniform, and can be computed as given in Equation 25;

$$q_s = \frac{F_s V_s}{l w} \quad (25)$$

where F_s is shear force on AB, q_s is shear heat intensity, and V_s, *l*, w are as defined previously.

and;

$$q_f = \frac{F V_c}{l_c w} \quad (26)$$

where F is cutting force, V_c is chip velocity, q_f is frictional heat intensity, l_c and w are as defined before.

Modeling of temperature rise in the tool

The tool side of the secondary heat source is modeled as a non-uniform stationary rectangular heat source. In the flank surface of the tool, it is assumed that the heat lost is zero. Also, in the lower surface of the chip, there is going to be a heat generation and for effective matching of the temperature, the heating effect of the primary shear zone on the tool rake face needs to be included as given in Equation 27.

$$T_{tool-chip}(X', Y', Z') = \frac{1}{2\pi \lambda_t} \int_{y_i=-w/2}^{w/2} \int_{x_i=0}^{l_c} (1 - B_i(x_i')) \times q_{pl}(x_i') \left(\frac{1}{R_i} + \frac{1}{R_i'}\right) dx_i dy_i \quad (27)$$

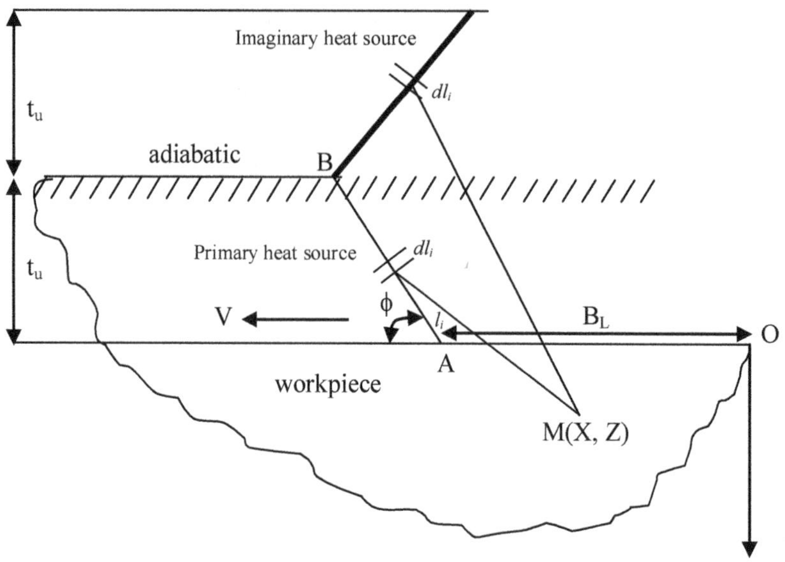

Figure 6. Thermal modeling of primary heat source on the workpiece side, adapted from Boothroyd (1963).

where;

$R_i = \sqrt{(X'-x_i')^2 + (Y'-y')^2 + Z'^2}$ $R_i' = \sqrt{(X'-2l_c + x')^2 + (Y'-y')^2 + Z'^2}$

λ_t = thermal conductivity of the tool material.

The tool side of the rubbing heat source can be modeled as in Figure 7, where in the tool rake face; it is assumed that the heat loss is zero. According to the coordinate system given in Figure 7, the temperature rise at any point on the tool due to the rubbing heat source can be written as:

$$T_{M\,tool-rubbing}(X'',Y'',Z'') = \frac{1}{2\pi\lambda_t}\int_0^{B_L}\int_{-w/2}^{w/2}\left[1-B_2(X'')\right]\times q_{rub}(X'')\left(\frac{1}{R_i}+\frac{1}{R_i'}\right)dy''\,dx'' \tag{28}$$

where;

$R_i = \sqrt{(X''-x'')^2 + (Y''-y'')^2 + Z''^2}$, $R_i' = \sqrt{(2B_L - X'' - x'')^2 + (Y''-y'')^2 + Z''^2}$,
B_L = length of blunt end of the tool, and for a pointed tool, the B_L = 0.

The heat intensity of the rubbing heat source is modeled as nonlinear by multiplying the cutting velocity and shear stress distribution. The temperature rise at any point on the tool can then be expressed as;

$$T_{tool} = T_{tool-chip} + T_{tool-rubbingi} + T_o \tag{29}$$

In the case of a sharp tool, this rubbing heat generation will be zero; hence Equation 29 reduces to;

$$T_{tool} = T_{tool-chip} + T_o \tag{30}$$

Modeling of temperature rise on the workpiece

In order to model temperature rise on the workpiece, primary heat source is modeled again as an oblique moving band heat source which moves under the workpiece surface with cutting velocity (Guo et al., 2004). In the model shown in Figure 8, it also assumed that the heat loss at the uncut workpiece surface is zero. The origin of the coordinate system in this model is assumed to be the end of flank wear width. According to the given coordinate system, the temperature rise at any point on the workpiece is given as;

$$T_{workpiece-shear}(X''',Z''') = \frac{q_{shear}}{2\pi\lambda_c}\int_{l_i=0}^{L_{AB}} e^{\frac{(X'''-l_i Sin\phi - B_L)V}{2a_c}} \times\left\{K_o\left[\frac{V}{2a_c}\sqrt{(B_L + l_i Cos\phi - X''')^2 + (Z''' + l_i Sin\phi)^2}\right] + K_o\left[\frac{V}{2a_c}\sqrt{(B_L + l_i Cos\phi - X''')^2 + (2t_u + Z''' - l_i Sin\phi)^2}\right]\right\}dl_i \tag{31}$$

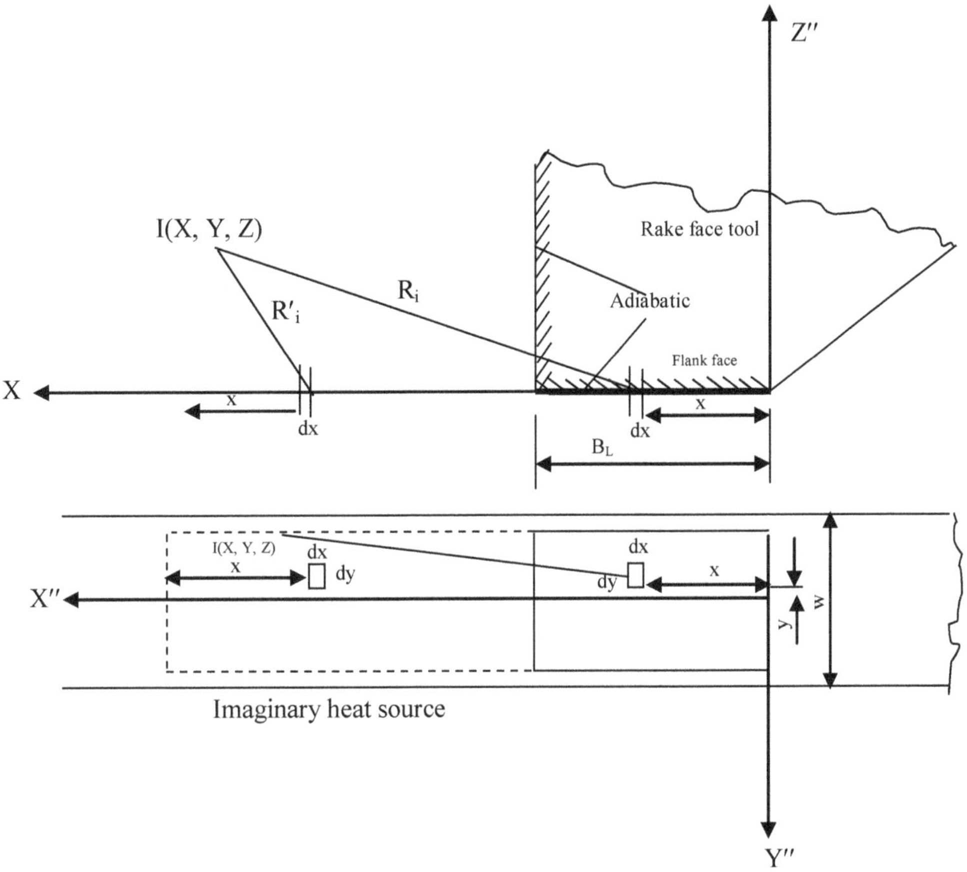

Figure 7. Thermal modeling of rubbing heat source on the tool rake and flank faces for a blunt tool modified (Murarka et al., 1981).

The rubbing heat source is modeled as a band heat source moving along the tool-workpiece surface with cutting velocity. The tool-workpiece interface is adiabatic and since the heat source does not move obliquely, the rubbing and its imaginary heat sources coincide as shown in Figure 8. The temperature rise on the workpiece was given as (Ceretti et al., 1996);

$$T_{wr}(X''', Z''') = \frac{1}{\pi \lambda_c} \int_0^{B_L} B_2(X''') q_{rub}(X''') e^{\frac{(X''' - x''')V}{2a_c}} \times \left[K_o \left(\frac{V}{2a_c} \sqrt{(X''' - x)^2 + Z'''^2} \right) \right] dx''' \quad (32)$$

Combining these, the temperature rise at any point in the workpiece is given as:

$$T_w(X''', Z''') = T_{workpieceshear} + T_{workpiece\,rubbing} + T_o \quad (33)$$

The heat partition ratios $B_1(x)$ *and* $B_2(x)$ indicate the energy going into the chip and workpiece respectively and can be estimated from the following empirical equations based on a compilation of experimental data (Trigger et al., 1951);

$$\left.\begin{aligned} B_1 &= 0.5 - 0.35\lg(R_T \tan\phi);\ 0.04 \le R_T \tan\phi \le 10.0 \\ B_1 &= 0.3 - 0.15\lg(R_T \tan\phi);\ R_T \tan\phi \ge 10.0 \end{aligned}\right\} \quad (34)$$

where; ϕ is shear angle, R_T is a non-dimensional thermal number,

$$R_T = \frac{\rho C_p V t_u}{K} \quad (35)$$

Where, ρ is density, C_p is specific heat, V, t_u, and K are as defined previously.

If B is the fraction of the shear plane heat conducted into the workpiece, that is, the heat partition ratio of the workpiece, then (1 – B) is the heat partition ratio in the chip and is expressed as B_2.

RESULTS AND DISCUSSION

The high rate of material flow during machining process makes it possible for the assumption of adiabatic heat transfer which makes it possible to directly compute the

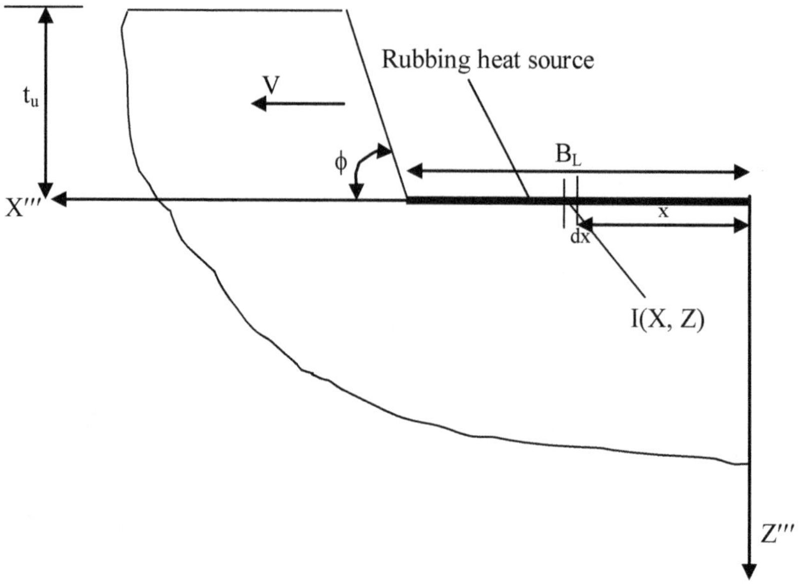

Figure 8. Thermal modeling of rubbing heat source on the workpiece side (Trigger et al., 1957).

Table 1. Tool materials and workpiece physical properties (Karpat et al., 2005).

Carbide tool properties	Thermal Conductivity [W/(m°C)]	46
	Heat Capacity (N/mm^2/°C)	2.7884
	Specific heat (J/kg/°C)	203
	Coefficient of thermal expansion[μm/(m°C)]	4.7 at room temp.
	Density (kg/m^3)	15000
High speed steel tool properties	Thermal conductivity [W/(m°C)]	47.8
	Heat capacity (N/mm^2/°C)	2.5325 + 0.002983T°C
	Specific heat (J/kg/°C)	960
	Coefficient of thermal expansion [μm/(m°C)]	8.45 at room Temp.
	Density [kg/m^3]	3399.5
Workpiece materials (medium carbon steel)	Thermal conductivity [W/(m°C)]	42.7
	Heat capacity (N/mm^2/°C)	3.5325+0.002983T°C
	Specific heat (J/kg/°C)	432.6
	Coefficient of thermal expansion [μm/(m°C)]	13.05 at room Temp.
	Density (kg/m^3)	7850

temperature distribution at any point of interest located within the shear cutting regime by the coordinates axis as an exact solution due to the plastic dissipation at that point. In the low thermal conductivity material like Ti6Al4V alloy, the highest temperatures occur inside the chip while the maximum temperatures are observed on the tool rake face in machining steels (Özel and Zeren, 2005). Using the developed model equations, the temperatures generation in the chip, the tool and the workpiece were determined. Based on the analysis, the temperature rise in the primary and secondary deformation zones are high and approach steady state very rapidly. As a result of the steadiness, the temperature distribution within the interface was determined as an exact value at any point of interest located within the cutting regime. The other materials' properties used with the modeling equation were determined from stress/strain tests carried out on the samples. The tool materials properties and the recommended machining speeds were given in Tables 1 and 2, respectively.

It was observed that the temperature generation was

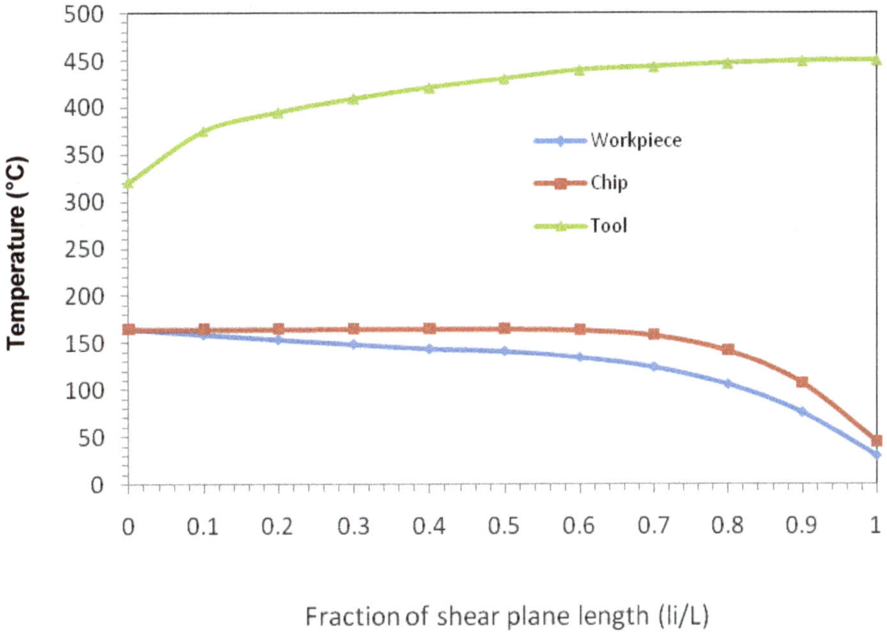

Figure 9. Temperature distribution along the shear plane.

Table 2. Materials with their recommended cutting speed (Sharma, 2003).

Material to be machined	Average cutting speed (m/min)
Cast Iron (average)	18 – 30
Cast Iron (hard)	12 – 18
Mild steel	21 – 25
Tool steel	12 – 15
Brass	45 – 60
Bronze	12 – 18

very high at the tool-chip interface, leaving majority of the heat with the tool immediately it separates from the tool. This was as a result of the heat generation at the secondary deformation zone and at tool-chip interface that was conducted to the cutting tool. The flank face of the tool with the workpiece, is of course of low heat generation, since the model was for a sharp tool, otherwise the rubbing effect will also lead to an intense heat generation in this zone. The model was also, able to generates the temperature along the shear plane, *AB* of length, *l*. The shear plane *AB* was expressed as a ratio of the instantaneous length l_i along the shear plane, *AB*. The temperature profile at any point of interest along the shear plane, *AB* expressed as a ratio l_i/l were generated until $l_i/l = 1$. The plot is shown in Figure 9.

The beauty of the model was that it gives a flexible approach to monitor the temperature profile along the shear plane, *AB* and to know the point of maximum or minimum temperature generation compared to the experimental which can only give the average results for a given machining condition.

The temperature distribution profile for the chip, tool and the workpiece shows that the tool has the highest temperature followed by the chip and the work piece. The highest temperature of the tool was as a result of is constant contact area during machining. The chip cools faster since the contact area is not constant. The work piece of course, has the least heat distribution, since shearing only occurs at the shear plane during the machining operation.

Conclusion

In this study, finite element modeling was utilized to simulate the temperature distribution for orthogonal cutting of medium carbon steel subjected to various form of heat treatment operations. Based on the materials' properties obtained from the stress/strain analysis, the annealed specimens gave a better machining condition, within the selected machining variables as compared to tempered, normalized, hardened, and the untreated. The modification of the grain structures during the heat treatment processes was observed to be responsible for these improved machining properties. The developed model gave a satisfactory result for the simulation of the chip formation and the development of the temperature distributions in the tool, chip and the workpiece. Very high and localized temperatures were observed for all the samples at the tool-chip interface due to a detailed friction model and the shearing action within the zone. The temperature profile along the shear plane *AB* was also analytically simulated. The temperature were however, expressed as a fraction of the instantaneous distance, l_i, located by the coordinate axis in the nodal grids structure for both the tool, chip and the workpiece to obtain the temperature profile. The temperature profile obtained indicated that the tool has a higher machining temperature when machining steel materials.

ACKNOWLEDGEMENTS

The authors acknowledge the staff of the mechanical workshop of the University of Ilorin and the University of Lagos, especially Mr. R. Raji and Mr. S. Arinde. We also acknowledge the assistance of the staff of the Instrument Laboratory, University of Ilorin.

Nomenclature

Q_T, Total heat generated per unit time (W); Q_s, heat generated along the shear zone (W); Q_f, heat generated on the rake face (W); NC, numerical control; F_s, shear force (N); F_c, cutting forces (N); F_T, thrust forces (N); F, frictional force (N); V_s, shear velocity (m/s); V_c, chip velocity (m/s); τ, shear stress (N/mm^2); w, width of cut (mm); V, resultant cutting speed (mm/s); λ_n, friction angle (°); ϕ, shear angle (°); α, normal rake angle (°); B, proportion of heat generated that enter the workpiece; R_T, thermal number; ε, plastic strain of material of workpiece; k, thermal conductivity (J/ms°C); c, specific heat capacity; T, average temperature rise of the chip due to shearing (°K); q_{pl}, heat liberation intensity of a moving plane heat source (J/m^2 s); T_l, temperature rise at point (°K); a, thermal diffusivity (m^2/s); K_o, zero order Bessel function which determines the location of the nodal points; X, the co-ordinate axis; l_i, the instantaneous point of interest on the shear plane; q_l, heat liberation intensity of a moving line heat source (J/m^2 s); λ, thermal conductivity (J/ms°C); R, distance between moving line heat source and the point M (mm); β, angle between R and X – axis (°); q_s, shear heat intensity (J/m^2 s); ρ, density (kg/m^3).

REFERENCES

Taylor GL, Quinney H (1934). "The latent energy remaining in a metal after cold working" Proceedings of the Royal Society of London, Series A. pp. 307 – 326.

Rosenthal D (1946). "The Theory of moving sources of heat and its application to metal treatments"' Transactions of ASME; 68: 849 – 866.

Hahn RS (1951). "On the temperature developed at the shear plane in the metal cutting process" Proceeding of 1st U. S. National congress of Applied Mechanics; pp. 661 – 666.

Karpat Y, Zeren E, Ozel T (2005). ""Workpiece material model based prediction for machining processes"; Trans. of NAMRI, 33: 413 – 420.

Tay OA, Stevenson MG, Davis DG (1990). "Using the finite element method to determine temperature distribution in orthogonal machining"; Proceedings of the Institute of Mechanical Engineers, London; 188: 627.

Ceretti E, Fallbohmer P, Wu WT, Altan T (1996). "Application of 2-D FEM to chip formation in Orthogonal cutting". J. Mater. Process. Technol., 59: 169 – 181.

Ng EG, Aspinwall DK, Brazil D, Monaghan J (1999). "Modeling of Temperature and Forces when Orthogonally Machining Hardened Steel", Int. J. Mach. Tools. Manufact., 39: 885 – 903.

Boothroyd G (1963). "Temperature in Orthogonal metal cutting", Proceeding of the Institution of Mechanical Engineers; 177(29): 789 – 810.

Murarka PD, Hindua S, Barrow G (1981). "Influence of Strain, Strain-rate and Temperature on the flow Stress in the primary deformation zone in metal cutting"; Int. J. Mach. Tool. Des. Res., 21(34): 207 – 216.

Guo YB, Yen DW (2004). "A FEM study on mechanisms of discontinuous chip formation in hard turning"; J. Mater. Process. Technol., pp. 155 – 156, 1350 – 1356.

Sharma PC (2003). "A Textbook of Production Engineering" S. Chand and Co. Ltd., New Delhi, India.

Totten GE (2007). Steel Heat Treatment Handbook. vol. 1. Metallurgy and Technologies, vol. 2, Equipment and Process Design, 2nd edition. CRC Press, Boca Raton.

Özel T, Zeren E (2005). Finite Element Modeling of Stresses induced by High Speed Machining with Round Edge Cutting Tools, Proceedings of ASME International Mechanical Engineering Congress and Exposition, Orlando, Florida; pp. 1-9.

11

Evaluation of thermodynamic properties of ammonia-water mixture up to 100 bar for power application systems

N. Shankar Ganesh* and T. Srinivas

Vellore Institute of Technology, Vellore-632014, India.

In Kalina power generation, as well as vapor absorption and refrigeration systems ammonia-water mixture has been used as working fluids. In this work, new MatLab code was developed to calculate the thermodynamic properties which will be used to simulate Kalina cycle. The progam developed in MatLab gives fast calculation of the thermodynamic properties. The correlations proposed by Ziegler and Trepp (1984), Patek and Klomfar (1995) and Soleimani (2007) were used to calculate the property diagrams in MatLab. The solved properties are bubble point temperature, dew point temperature, specific enthalpy, specific entropy, specific volume and exergy. A flowchart was developed to understand the computation of the properties. The property chart that is enthalpy-concentration, entropy-concentration, temperature-concentration and exergy-concentration charts have been prepared. The present work can be used to simulate the power generating systems to get the feasibility of the proposed ideas up to 100 bar. This work can be used to carry out the exergy analysis of Kalina power cycles.

Key words: Ammonia-water mixture, thermodynamic, power generation.

BACKGROUND

In ammonia-water mixture, ammonia has got low boiling point which makes it useful for utilizing the waste heat source and makes the possibility of boiling at low temperature. Ammonia-water mixture as non-azeotropic (for a non-azeotropic mixture, the temperature and composition continuously change during boiling) nature will have the tendency to boil and condense at a range of temperatures which possess a closer match between heat source and working fluid mixture. As ammonia have got a similar molecular weight as that of water, it makes it possible to utilize the standard steam turbine components.

For determining the thermodynamic properties of ammonia-water mixtures, various studies were published. Ziegler and Trepp (1984) described an equation for the thermodynamic properties of ammonia-water mixture in absorption units. In his work, the Gibbs excess energy equation was utilized for determining the specific enthalpy, specific entropy and specific volume. They developed the properties up to a pressure of 50 bar and temperature of 500 K. Barhoumi et al. (2004) presents modelling of the thermodynamic properties. Feng and Yogi (1999) combine the Gibbs free energy method for mixture properties and the bubble and dew point temperature equations for phase equilibrium were used. Patek and Klomfar (1995) give a fast calculation of thermodynamic properties. Senthil and Subbarao (2008) present fast calculation for determining enthalpy and entropy of the mixtures.

The main objectives of the present work are to combine correlations proposed by Ziegler et al. (1984) and carried out in MatLab, which avoids numerous procedure and

*Corresponding author. E-mail: nshankar_g@rediffmail.com.

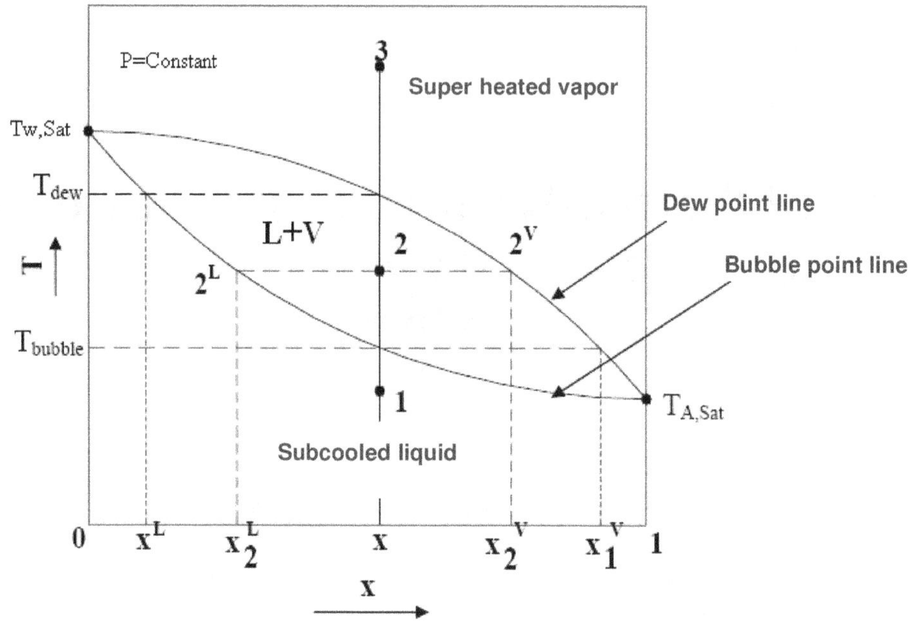

Figure 1. Equilibrium temperature-concentration curve for NH_3- H_2O at constant pressure.

time interval in obtaining the result. The presented work can be used for energy and exergy solutions to power generating systems. The exergy details and its concentration graph for the ammonia-water mixtures are not reported out in the literature, which is the gap identified and presented at various pressures.

THERMODYNAMIC EVALUATION OF NH_3-H_2O MIXTURE PROPERTIES

For ammonia-water mixture, to calculate the thermodynamic properties like specific enthalpy, specific entropy and specific volume, the need of bubble and dew point temperatures at various pressures and compositions are very essential and is the prior step. For estimating those temperatures, various correlations have been developed. The correlation developed by Patek and Klomfar (1995) is proposed in this work which avoids tedious iterations required by the complicated method fugacity coefficient of a component in a mixture and the correlation proposed by Ibrahim and Klein (1993).

Figure1 shows the details of bubble point and dew point temperature variations with ammonia concentration. The loci of all the bubble points are called the bubble point line and the loci of all the dew points are called the dew point line. The bubble point line is the saturated liquid line and the region between the bubble and dew point lines is the two phase region, where both liquid and vapor co-exist in equilibrium ("Vapor Absorption Refrigeration Systems Based on Ammonia-Water Pair", 2004).

Calculating bubble and dew point temperatures

The bubble point and dew point temperatures of the ammonia-water mixture are found from the correlations in Equations (1) and (2), developed by Patek and Klomfar (1995).

$$T_b(p,x) = T_o \sum_i a_i (1-x)^{m_i} \left[\ln\left(\frac{p_o}{p}\right) \right]^{n_i} \quad (1)$$

$$T_d(p,y) = T_o \sum_i a_i (1-y)^{m_i/4} \left[\ln\left(\frac{po}{p}\right) \right]^{n_i} \quad (2)$$

Figure 2 shows the bubble and dew point temperatures developed with the correlation by Patek and Klomfar (1995) up to pressure of 100 bar using MATLAB code. A flowchart was prepared to understand the mathematical calculations for properties.

Development of equations

The properties are derived from Gibbs free energy function from Ziegler and Trepp (1984).

Liquid phase

The Gibbs free energy for both liquid and gas phases were determined from Equations (3) and (4), developed by Ziegler and Trepp (1984) which is the summation of contributions of the pure components, the ideal free energy of mixing and the free excess energy.

$$g^l_{(T,p,x)} = \left\{ \begin{array}{l} (1-x)g^l{H_2O}_{(T,p)} + x\,g^l{NH_3}_{(T,p)} + \\ RT\left[(1-x)\log_e(1-x) + x\log_e x\right] + g_{E(T,p,x)} \end{array} \right\}$$

(3)

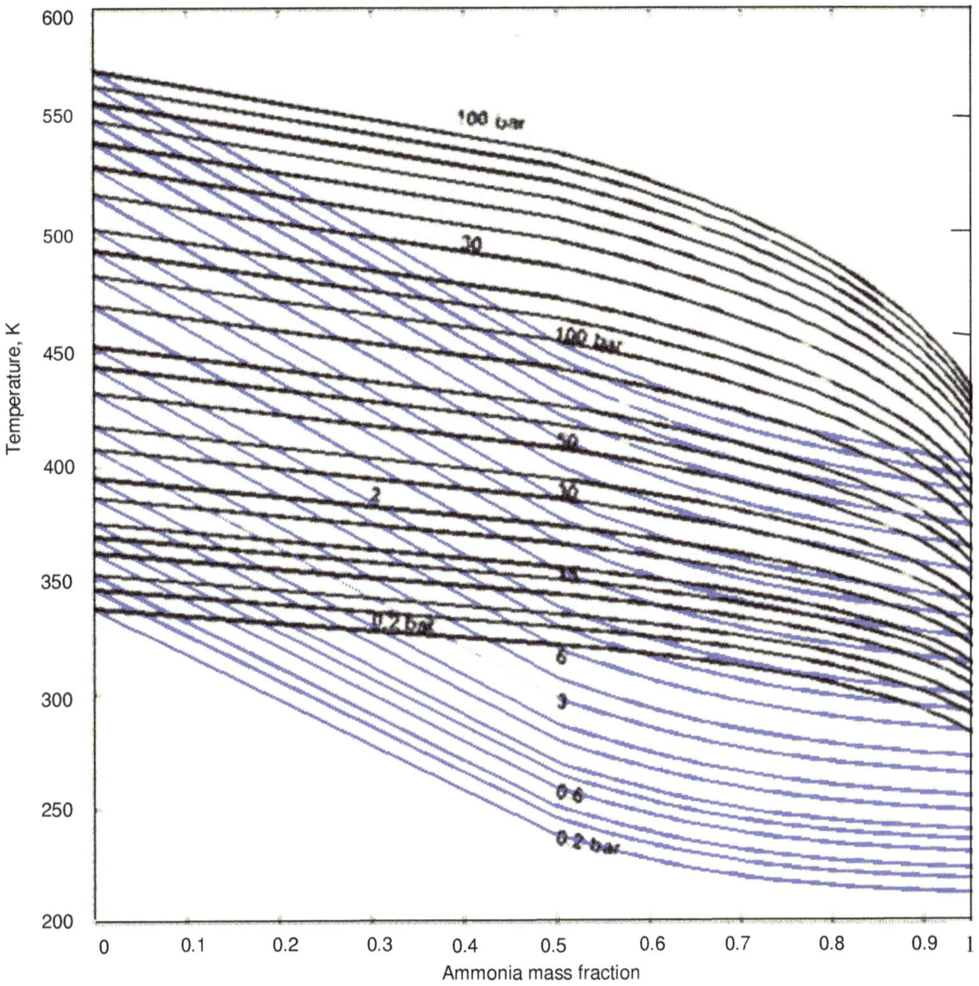

Figure 2. Bubble and dew point temperatures up to100 bar pressure.

Equation of state for pure component in liquid phase

The equation of state for pure components in liquid phase is given as follows:

$$g^l_{H_2O(T,p)} = \left[\begin{array}{l} h^l(T_o,p_o) - Ts^l(T_o,p_o) + \int c^l_p dT - T\int \frac{c^l_p}{T} dT + \\ \left(a_1 + a_3T + a_4T^2\right)(p - p_o) + a_2 \frac{\left(p^2 - p_o^2\right)}{2} \end{array} \right]_{H_2O} \quad (4)$$

Where,

$$c^l_{p(T,po)} = b_1 + b_2T + b_3T^2 \quad (5)$$

Similarly, the liquid heat capacity at constant pressure can be assumed to be second order in temperature according to Equation (5) (Ziegler and Trepp, 1984).On substituting (5) in Equation (4), the following Equations (6) and (7) were obtained for pure components in liquid phase.

$$g^l_{H_2O(Tp)} = \left\{ \begin{array}{l} \left[h^l(T_o,p_o) - Ts^l(T_o,p_o) + \left[\begin{array}{l} b_1(T - T_o) + \frac{b_2}{2}\left(T^2 - T_o^2\right) + \\ \frac{b_3}{3}\left(T^3 - T_o^3\right) \end{array} \right] \right] + \\ \left[\left(a_1 + a_3T + a_4T^2\right)(p - p_o) + a_2 \frac{\left(p^2 - p_o^2\right)}{2} \right] \end{array} \right\}_{H_2O} \quad (6)$$

$$g^l_{NH_3(Tp)} = \left\{ \begin{array}{l} h^l_{(T_o,p_o)} - Ts^g(T_o,p_o) + \left[b_1(T - T_o) + \frac{b_2}{T}\left(T^2 - T_o^2\right) + \frac{b_3}{3}\left(T^3 - T_o^3\right) \right] \\ + \left[\left(a_1 + a_3T + a_4T^2\right)(p - p_o) + a_2 \frac{\left(p^2 - p_o^2\right)}{2} \right] \end{array} \right\}_{NH_3} \quad (7)$$

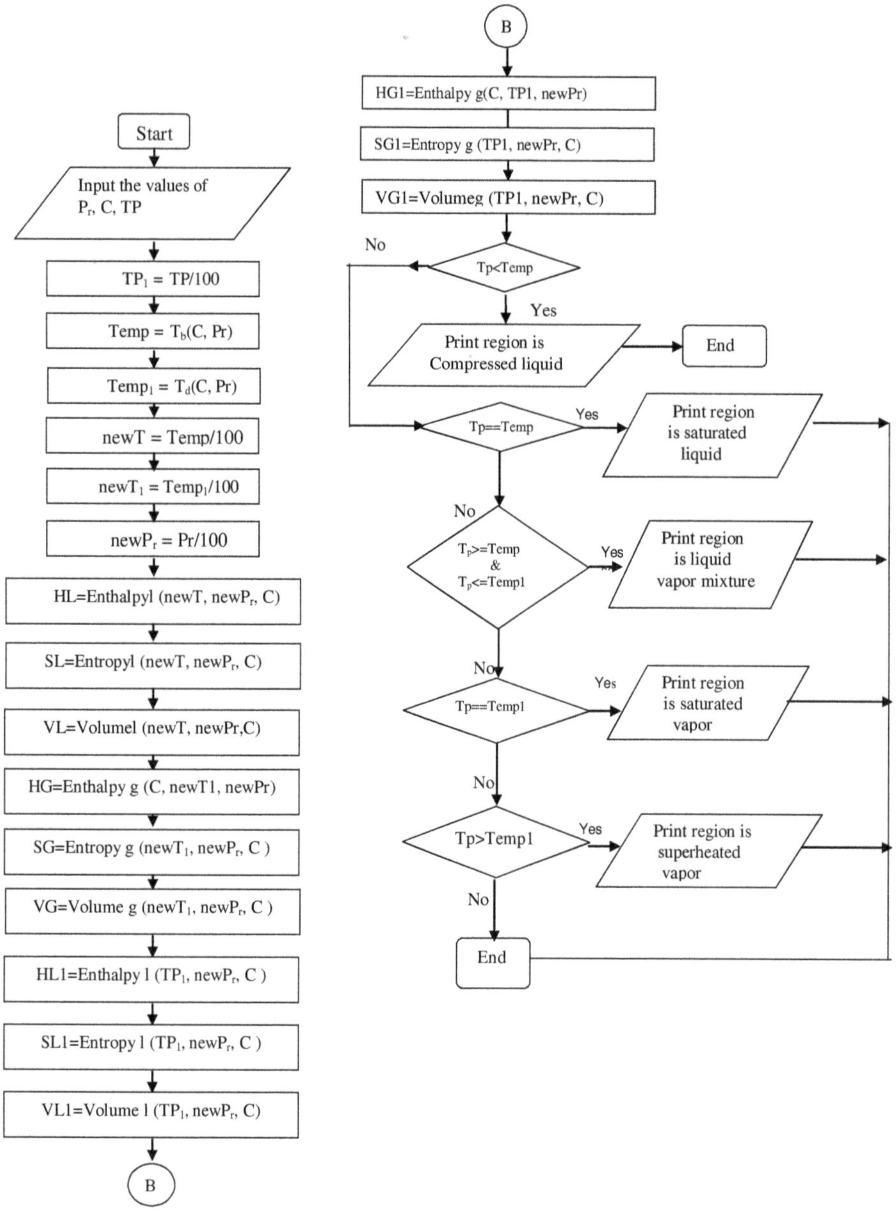

Figure 3. Flowchart to find thermodynamic properties of mixture.

Liquid mixture correlation

The Gibbs excess energy g_E for liquid mixtures is expressed as:

$$g_{E(T,p,x)} = \left\{ \begin{array}{l} \left(e_1 + e_2 p + (e_3 + e_4 p)T + \dfrac{e_5}{T} + \dfrac{e_6}{T^2} \right) + \\ (2x-1)\left(e_7 + e_8 p + (e_9 + e_{10} p)T + \dfrac{e_{11}}{T} + 3\dfrac{e_{12}}{T^2} \right) + \\ (2x-1)^2 \left(e_{13} + e_{14} p + \dfrac{e_{15}}{T} + \dfrac{e_{16}}{T^2} \right) \end{array} \right\} \quad (8)$$

In Figure 3, the procedure carried out in calculating the thermodynamic properties enthalpy, entropy and volume in both phases for the compressed region, saturated region, in between bubble and dew point region and superheated region were explained from the corresponding equations.

For a given pressure, concentration and temperature the bubble point and dew point temperatures were calculated and then the given temperature is compared with the bubble and dew temperatures and identifies the respective region as represented in the flowchart. The flowchart in Figure 3 shows the procedure for calculating the thermodynamic properties of the mixture. For a given property value, the corresponding regions can be identified from the flowchart.

Equation of state for pure component in gas phase

The general equation of state for pure component in gas phase is identified in the following equation.

$$g^{g}{}_{(T,P,Y)} = \begin{Bmatrix} (1-y)g^{g}H_2O_{(T,p)} + yg^{g}NH_{3(T,P)} + \\ RT[(1-y)\log_e(1-y) + y\log_e y] \end{Bmatrix}$$

(9)

For the gas phase, the Gibbs free energy equation is given below:

$$g^{g}{}_{H_2O}(T,p) = \begin{Bmatrix} h^{g}(T_o,p_o) - Ts^{g}(T_o,p_o) + \int C_p^{go}dT - T\int \frac{C_p^{go}}{Tdt} + RT\ln\frac{p}{p_o} + c_1(p-p_o) + \\ c_2\left(\frac{p}{T^3} - 4\frac{p_o}{T_o^3} + 3\frac{p_oT}{T_o^4}\right) + c_3\left(\frac{p}{T^{11}} - 12\frac{p_o}{T_o^{11}} + 11\frac{p_oT}{T_o^{12}}\right) + \\ c_3\left(\frac{p}{T^{11}} - 12\frac{p_o}{T_o^{11}} + 11\frac{p_oT}{T_o^{12}}\right) + \frac{c_4}{3}\left(\frac{p^3}{T^{11}} - 12\frac{p_o^3}{T_o^{11}} + 11\frac{p_o^3T}{T_o^{12}}\right) \end{Bmatrix}_{H_2O} \tag{10}$$

Where:

$$c_{p\,(T)}^{go} = d_1 + d_2T + d_3T^2 \tag{11}$$

Similarly on substituting (11) in Equation (10), the pure components in gaseous phase from ammonia and water were obtained in Equations (12) and (13).

$$g^{g}{}_{H_2O}(T,p) = \begin{Bmatrix} h^{g}(T_o,p_o) - Ts^{g}(T_o,p_o) + d_1(T_o - T) + \frac{d_2}{2}\left(T_o^2 - T^2\right) + \frac{d_3}{3}\left(T_o^3 - T^3\right) \\ + RT\log_e\frac{p}{p_o} + c_1(p - p_o) + c_2\left(\frac{p}{T^3} - 4\frac{p_o}{T_o^3} + 3\frac{p_oT}{T_o^4}\right) \\ + c_3\left(\frac{p}{T^{11}} - 12\frac{p_o}{T_o^{11}} + 11\frac{p_oT}{T_o^{12}}\right) + \frac{c_4}{3}\left(\frac{p^3}{T^{11}} - 12\frac{p_o^3}{T_o^{11}} + 11\frac{p_o^3T}{T_o^{12}}\right) \end{Bmatrix}_{H_2O} \tag{12}$$

$$g^{g}NH_{3(T,p)} = \begin{Bmatrix} h^{g}(T_o,p_o) - Ts^{g}(T_o,p_o) + d_1(T_o - T) + \\ \frac{d_2}{2}\left(T_o^2 - T^2\right) + \frac{d^3}{3}\left(T_o^3 - T^3\right) + \\ RT\log_e\frac{P}{P_o} + c_1(P - P_o) + \\ c_2\left(\frac{P}{T^3} - 4\frac{P^o}{T_o^3} + 3\frac{P_oT}{T_o^4}\right) \\ + c_3\left(\frac{P}{T^{11}} - 12\frac{P_o}{T_o^{11}} + 11\frac{P_oT}{T_o^{12}}\right) + \\ c_4\left(\frac{P^3}{T^{11}} - 12\frac{P_o^3}{T_o^{11}} + 11\frac{P_o^3T}{T_o^{12}}\right) \end{Bmatrix}_{NH_3} \tag{13}$$

Equation of state for pure components

Specific enthalpy at liquid and vapor phases

The molar enthalpy of the liquid phase and gaseous phase were specified and simplified in Equation (14) to (19). Equation (16) is derived in MATLAB to find the liquid enthalpy in the compressed and saturated regions. Equation (19) is derived in MATLAB to find the enthalpy in gaseous phase for the saturated and superheated regions.

$$h^{l} = -RT_BT^2\left(\frac{\partial\left(\frac{g^{l}(T,p,x)}{T}\right)}{\partial T}\right)_{p,x} \tag{14}$$

$$h^{l} = -RT_BT^2\begin{Bmatrix} (1-x)\frac{\partial}{\partial T}\left(\frac{g^{l}H_2O_{T,p}}{T}\right) + \\ x\frac{\partial}{\partial T}\left(\frac{g^{l}NH_{3T,p}}{T}\right) \\ + \frac{\partial}{\partial T}R[(1-x)\log_e(1-x) + x\log_e x] \\ + \frac{\partial}{\partial T}\left(\frac{{}^{g}E(T,p,x)}{T}\right) \end{Bmatrix}_{p,x} \tag{15}$$

$$h^{l} = \begin{Bmatrix} \left(\frac{R}{18}\times T_B\right)\times(1-x)\times\begin{bmatrix} h^{l}(To,po) + b_1(T-T_o) + \frac{b_2}{2}\left(T^2 - T_o^2\right) + \\ \frac{b_3}{3}\left(T^3 - T_o^3\right) + \left(a_1 - a_4T^2\right)(p - p_o) + \\ \frac{a_2}{2}\left(p^2 - p_o^2\right) \end{bmatrix}_{H_2O} + \\ \left(\frac{R}{17}\times T_B\right)\times(x)\times\begin{bmatrix} h^{l}_{(To,po)} + b_1(T-T_o) + \frac{b_2}{2}\left(T^2 - T_o^2\right) + \\ \frac{b_3}{3}\left(T^3 - T_o^3\right) + \left(a_1 - a_4T^2\right)(p - p_o) + \\ \frac{a_2}{2}\left(p^2 - p_o^2\right) \end{bmatrix}_{NH_3} + \\ \left[\frac{R}{(x\times 17 + (1-x)\times 18)}\times T_B\times(x)\times(1-x)\times\begin{pmatrix} \left(e_1 + e_2p + 2\frac{e_5}{T} + 3\frac{e_6}{T^2}\right) + \\ (2x-1)\left(e_7 + e_8p + 2\frac{e_{11}}{T} + 3\frac{e_{12}}{T^2}\right) + \\ (2x-1)^2\left(e_{13} + e_{14}p + 2\frac{e_{15}}{T} + 3\frac{e_{16}}{T^2}\right) \end{pmatrix}\right] \end{Bmatrix} \tag{16}$$

$$h^g = -RT_BT^2\left(\frac{\partial\left(\frac{g^g(T,p,y)}{T}\right)}{\partial T}\right)_{p,y}$$

(17)

$$h^g = -RT_BT^2\left\{\begin{array}{l}(1-y)\frac{\partial}{\partial T}\left(\frac{g^g{}_{H_2O\,T,p}}{T}\right)+y\frac{\partial}{\partial T}\left(\frac{g^g{}_{NH_3\,T,p}}{T}\right)+\\ \frac{\partial}{\partial T}R\left[(1-x)\log_e(1-x)+x\log_e x\right]\end{array}\right\}_{p,y}$$

(18)

The subscript l indicates liquid
g indicates gas
o indicates ideal gas state
$T_B = 100K$
$P_B = 10bar$

$$h^g = \left\{\begin{array}{l}-\left(\frac{R}{18}\times T_B\right)\times(1-y)\times\left[\begin{array}{l}h^g_{(To,po)}+d_1(T-T_o)+\frac{d_2}{2}\left(T^2-T_o^2\right)+\\ \frac{d_3}{3}\left(T^3-T_o^3\right)+c_1(p-p_o)+4c_2\left(\frac{p}{T^3}-\frac{p_o}{T_o^3}\right)+\\ 12c_3\left(\frac{p}{T^{11}}-\frac{p_o}{T_o^{11}}\right)+4c_4\left(\frac{p^3}{T^{11}}-\frac{p_o^3}{T_o^{11}}\right)\end{array}\right]_{H_2O}\\ -\left(\frac{R}{17}\times T_B\right)\times y\times\left[\begin{array}{l}h^g_{(To,po)}+d_1(T-T_o)+\frac{d_2}{2}\left(T^2-T_o^2\right)+\\ \frac{d_3}{3}\left(T^3-T_o^3\right)+c_1(p-p_o)+4c_2\left(\frac{p}{T^3}-\frac{p_o}{T_o^3}\right)+\\ 12c_3\left(\frac{p}{T^{11}}-\frac{p_o}{T_o^{11}}\right)+4c_4\left(\frac{p^3}{T^{11}}-\frac{p_o^3}{T_o^{11}}\right)\end{array}\right]_{NH_3}\end{array}\right\}$$

(19)

The coefficients used in Equations 15,17,20,23, 26 and 29 are given in (Table 1) and (Table 2).

Specific entropy at liquid and vapor phases

The molar entropy of the liquid and gaseous phases were specified and simplified in Equation (20) to (25). Equation (22) is derived in MATLAB to find the liquid entropy in the compressed and saturated regions. Equation (25) is derived in MATLAB to find the entropy in gaseous phase for the saturated and superheated regions.

$$s^l = -R\left(\frac{\partial g^l(T,p,x)}{\partial T}\right)_{(p,x)}$$

(20)

$$s^l = -R\left\{\frac{\partial}{\partial T}\left(\begin{array}{l}(1-x)g^l{}_{H_2O(T,p)}+xg^l{}_{NH_3(T,p)}\\ +RT\left[(1-x)\log_e(1-x)+x\log_e x\right]+g_{E(T,p,x)}\end{array}\right)\right\}_{(p,x)}$$

(21)

On reduction, the above equation becomes:

$$s^l = \left\{\begin{array}{l}\left[\frac{R}{18}\times(1-x)\left(\begin{array}{l}s^l_{(To,po)}+b_1\log\left(\frac{T}{T_o}\right)+b_2(T-T_o)\\ +\frac{b_3}{2}\left(T^2-T_o^2\right)+(-a_3-2a_4T)(p-p_o)\end{array}\right)\right]_{H_2O}+\\ \left[\frac{R}{17}\times x\left(\begin{array}{l}s^l_{(To,po)}+b_1\log\left(\frac{T}{T_o}\right)+b_2(T-T_o)\\ +\frac{b_3}{2}\left(T^2-T_o^2\right)+(-a_3-2a_4T)(p-p_o)\end{array}\right)\right]_{NH_3}-\\ \left[\frac{R}{x.17+(1-x).18}\times\frac{R}{x.17+(1-x).18}\times((1-x)\log_e(1-x)+x\log_e x)\right]+\\ \left[\frac{R}{x.17+(1-x).18}\times x\times(1-x)(se_1+se_2+se_3)\right]\end{array}\right\}$$

(22)

Where:

$$se_1 = -e_3 - e_4p + \frac{e_5}{T^2} + \frac{2e_6}{T^3}$$

$$se_2 = (2x-1)\left(-e_9 - e_{10}p + \frac{e_{11}}{T^2} + \frac{2e_{12}}{T^3}\right)$$

$$se_3 = (2x-1)^2\left(\frac{e_{15}}{T^2} + \frac{2e_{16}}{T^3}\right)$$

$$s^g = -R\left(\frac{\partial g^g(T,p,y)}{\partial T}\right)_{(p,y)}$$

(23)

$$s^g = -R\left\{\frac{\partial}{\partial T}\left(\begin{array}{l}(1-y)g^g{}_{H_2O(T,p)}+yg^g{}_{NH_3(T,p)}\\ +RT\left[(1-y)\log_e(1-y)+y\log_e y\right]+g_{E(T,p,y)}\end{array}\right)\right\}_{(p,y)}$$

(24)

$$s^g = \left\{\begin{array}{l}\left[-\frac{R}{18}\times(1-y)\left(\begin{array}{l}-s^g_{(To,po)}-d_1\log\left(\frac{T}{T_o}\right)+\frac{d_2}{2}(2T_o-2T)+\frac{d_3}{2}\left(T_o^2-T^2\right)+\frac{R}{18}\log\left(\frac{p}{po}\right)+\\ c_2\left(\frac{3p_o}{T_o^4}-\frac{3p}{T^4}\right)+c_3\left(\frac{11p_o}{T_o^{12}}-\frac{11p}{T^{12}}\right)+\frac{c_4}{3}\left(\frac{11p_o^3}{T_o^{12}}-\frac{11p^3}{T^{12}}\right)\end{array}\right)\right]_{H_2O}+\\ \left[-\frac{R}{17}\times(y)\left(\begin{array}{l}-s^g_{(To,po)}-d_1\log_e\left(\frac{T}{T_o}\right)+\frac{d_2}{2}(2T_o-2T)+\frac{d_3}{2}\left(T_o^2-T^2\right)+\frac{R}{18}\log_e\left(\frac{p}{po}\right)+\\ c_2\left(\frac{3p_o}{T_o^4}-\frac{3p}{T^4}\right)+c_3\left(\frac{11p_o}{T_o^{12}}-\frac{11p}{T^{12}}\right)+\frac{c_4}{3}\left(\frac{11p_o^3}{T_o^{12}}-\frac{11p^3}{T^{12}}\right)\end{array}\right)\right]_{NH_3}-\\ \left[-\left(\frac{R}{y.17+(1-y).18}\times\frac{R}{y.17+(1-y).18}\right)\times((1-y)\log_e(1-y)+y\log_e y)\right]\end{array}\right\}$$

(25)

Table 1. Coefficients for the equations for the pure components.

Coefficient	Ammonia	water
a_1	$3.971423.10^{-2}$	$2.748796.\ 10^{-2}$
a_2	$-1.790557.10^{-5}$	$-1.016665.10^{-5}$
a_3	$-1.308905.10^{-2}$	$-4.452025.10^{-3}$
a_4	$3.752836.10^{-3}$	$8.389246.10^{-4}$
b_1	$1.634519.10^{1}$	$1.214557.10^{1}$
b_2	-6.508119	-1.898065
b_3	1.448937	$2.911966.10^{-1}$
c_1	$-1.049377.10^{-2}$	$2.136131.10^{-2}$
c_2	-8.288224	$-3.169291.10^{1}$
c_3	$-6.647257.10^{2}$	$-4.634611.10^{4}$
c_4	$-3.045352.10^{3}$	0.0
d_1	3.673647	4.019170
d_2	$9.989629.10^{-2}$	$-5.175550.10^{-2}$
d_3	$3.617622.10^{-2}$	$1.951939.10^{-2}$
h^l	4.878573	21.821141
h^g	26.468879	60.965058
s^l	1.644773	5.733498
s^g	8.339026	13.453430
T_o	3.2252	5.0705
p_o	2.0000	3.0000

Table 2. Coefficients for the Gibbs excess energy function.

e_1	$-4.626129.10^{1}$
e_2	$2.060225.10^{-2}$
e_3	7.292369
e_4	$-1.032613.10^{-2}$
e_5	$8.074824.10^{1}$
e_6	$-8.461214.10^{1}$
e_7	$2.452882.10^{1}$
e_8	$9.598767.10^{-3}$
e_9	-1.475383
e_{10}	$-5.038107.10^{-3}$
e_{11}	$-9.640398.10^{1}$
e_{12}	$1.226973.10^{2}$
e_{13}	-7.582637
e_{14}	$6.012445.10^{-4}$
e_{15}	$5.487018.10^{1}$
e_{16}	$-7.667596.10^{1}$

Specific volume of liquid and vapor phases

The specific volume of the liquid and gaseous phases, were specified and simplified in Equation (26) to (31). Equation (28) is derived in MATLAB to find the liquid volume in the compressed and saturated regions. Equation (31) is derived in MATLAB to find the volume in gaseous phase for the saturated and superheated regions.

$$v^l = \frac{RT_B}{p_B}\left(\frac{\partial}{\partial p} g^l{}_{(T,p,x)}\right)_{(T,x)} \tag{26}$$

$$v^l = \frac{RT_B}{p_B}\left\{\begin{array}{l}(1-x)\left(\frac{\partial}{\partial P} g^l{}_{H_2O}\right)_{T,p,x} + (x)\left(\frac{\partial}{\partial P} g^l{}_{NH_3}\right)_{T,p,x} + \\ \frac{\partial}{\partial P} RT[(1-x)\log_e(1-x) + x\log_e x] + \frac{\partial}{\partial P} g_{E_{T,p,x}}\end{array}\right\}_{T,x} \tag{27}$$

In the same manner specific volumes were solved.

$$v^l = \left\{\begin{array}{l}\left(\frac{R}{18}\times\frac{T_B}{100p_B}\times(1-x)\times(a_1 + a_2p + a_3T + a_4T^2)\right)_{H_2O} + \\ \left(\frac{R}{17}\times\frac{T_B}{100p_B}\times(x)\times(a_1 + a_2p + a_3T + a_4T^2)\right)_{NH_3} + \\ \frac{R}{x.17+(1-x).18}\times\frac{T_B}{100p_B}\times x(1-x)\times(e_2 + e_4T + (2x-1)e_8 + e_{10}T + (2x-1)^2 e_{14})\end{array}\right\} \tag{28}$$

$$v^g = \frac{RT_B}{p_B}\left(\frac{\partial}{\partial p} g^g{}_{(T,p,y)}\right)_{(T,y)} \tag{29}$$

$$v^g = \left[\frac{RT_B}{p_B}\left\{\begin{array}{l}(1-y)\left(\frac{\partial}{\partial P} g^g{}_{H_2O}\right)_{T,p,y} + (y)\left(\frac{\partial}{\partial P} g^g{}_{NH_3}\right)_{T,p,y} + \\ \frac{\partial}{\partial P} RT[(1-y)\log_e(1-y) + y\log_e y]\end{array}\right\}_{T,y}\right] \tag{30}$$

$$v^g = \left\{\begin{array}{l}\left(\frac{R}{18}\times\frac{T_B}{100p_B}\times(1-y)\times\left(\frac{R}{18}\times\frac{T}{p}\right) + c_1 + \frac{c_2}{T^3} + \frac{c_3}{T^{11}} + \frac{c_4p^2}{T^{11}}\right)_{H_2O} + \\ \left(\frac{R}{17}\times\frac{T_B}{100p_B}\times y\times\left(\frac{R}{18}\times\frac{T}{p}\right) + c_1 + \frac{c_2}{T^3} + \frac{c_3}{T^{11}} + \frac{c_4p^2}{T^{11}}\right)_{NH_3}\end{array}\right\} \tag{31}$$

RESULTS AND DISCUSSION

In this study, from the simplified Equations 17 and 19 the liquid and vapor enthalpies were calculated and coded in MatLab. Similarly, Equations 22 and 25 were used to calculate the liquid and vapor entropies. The results generated using these equations were programmed in MatLab. With MatLab, the graphs were plotted and compared with the graphs from the Feng and Yogi (1999). The graphs obtained from this work, show a very close trend of comparison, Feng and Yogi (1999).

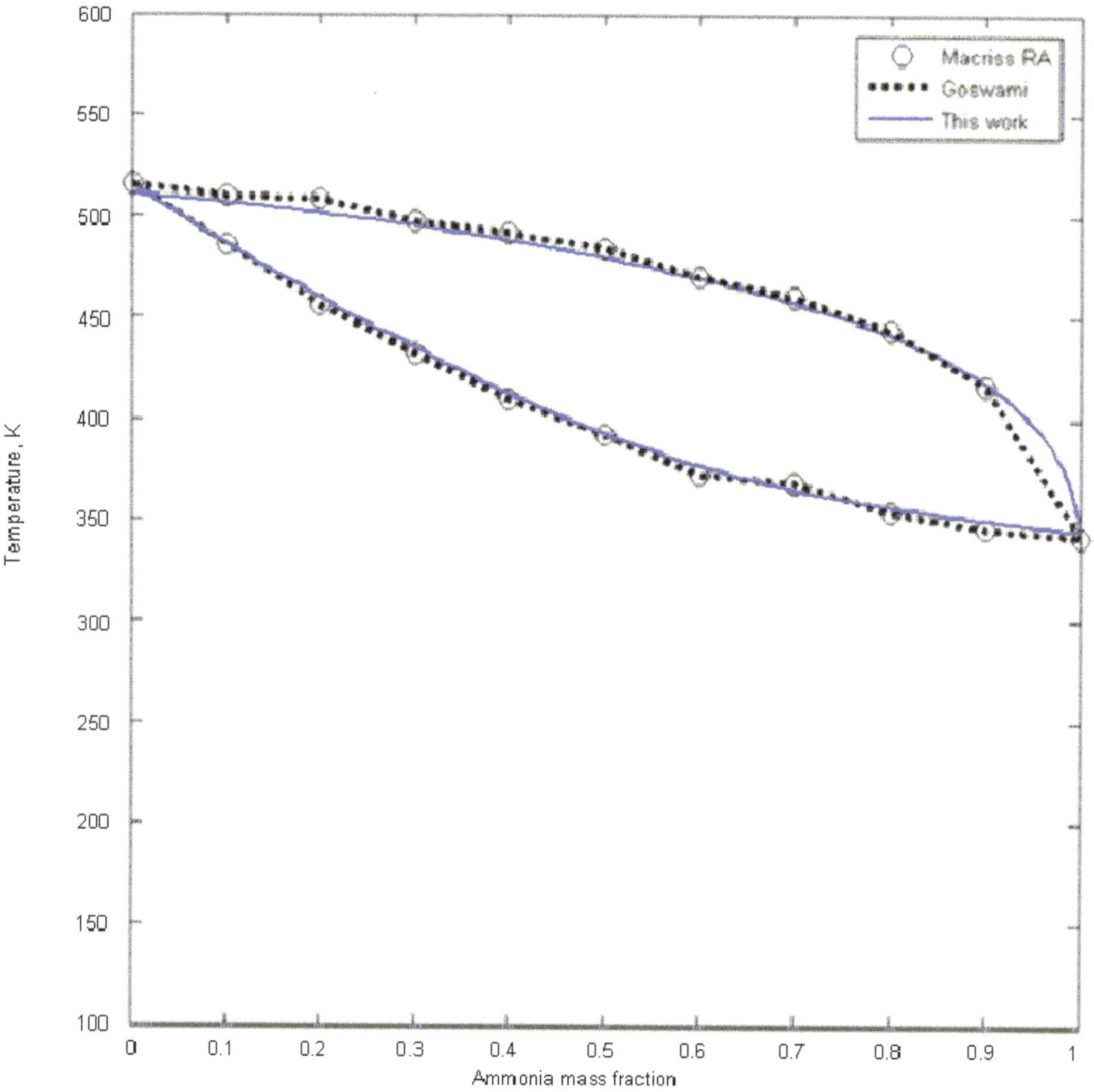

Figure 4. Bubble and dew point temperatures a 34.47 bar.

Figure 4 shows a plot between the temperature and ammonia mass fraction. Here, the data in the sense of the values of temperature at a particular concentration. The temperatures at a particular concentration obtained from Macriss and Goswami (1999) are very close with the produced result using MatLab. Figure 4 shows the bubble point temperature and dew point temperature curves at a specified pressure and for different concentrations. The bubble point temperature and dew point temperature values are identical at initial and final concentrations ensuring a closed curve. The differences between our computed values and the data are less than 0.5%.The simulated works were carried out in MATLAB, which shows a closer match with the literature. This work requires less calculation and can be utilized for the thermodynamic properties.

(Table 3) gives the property values at different regions. For a given pressure, temperature and concentration the bubble and dew point temperatures were calculated and the given temperature will be compared with those two temperatures and determine in which region the given temperature lies. If the given temperature is less than the bubble point temperature then region will be a compressed liquid region and for which the corresponding enthalpy, entropy and volumes were obtained using MATLAB. In calculating the dryness

Table 3. Thermodynamic properties value at different regions (p = 65 bar, x = 0.6, T = 125 °C, T_b = 138 °C and T_d = 228°C).

T °C	Condition	h_l kJ/kg	h_g kJ/kg	h kJ/kg	s_l kJ/kg-K	s_g kJ/kg-K	s kJ/kg-K	v_l m^3/kg	v_g m^3/kg	V m^3/kg
125	Compressed liquid	366.68	-	-	1.40	-	-	0.0015	-	-
138	Saturated liquid curve	435.95	-	-	1.57	-	-	0.0016	-	-
215	Two phase region-	-	-	1775.94	-	-	4.59	-	-	0.008
228	Saturated vapor curve	-	1978.23	-	-	5.04	-	-	0.011	-
250	Superheated	-	2049.47	-	-	5.18	-	-	0.012	-

fraction, the ammonia mole fraction of vapor phase is obtained from correlation by Soleimani (2007).

$$y(x,P) = 1- \exp[aP^b x + (c + d/P)\, x^2] \qquad (32)$$

The present results found a closer match with the existing results in the plots at a temperature less than 500°C and 100 bar. In finding the values of enthalpy, entropy and volume in between Tb and Td regions the dryness fraction is calculated from the equation developed in (19).

The liquid enthalpy and vapor enthalpy plots were shown in (Figure 5) and (Figure 6). From (Figure 5), the variation in the liquid enthalpy decreases first and then increases with increase in concentration at a specified pressure. The results obtained were validated and shows a closer match with the compared results. The differences are less than 3% for all the data. (Figure 6) shows the variation in the vapor enthalpy curve. The enthalpy value decreases continuously with the increase in the concentration. The enthalpy concentration with respect to the parameters pressure and concentration is shown in (Figure 7). The liquid enthalpy plot is obtained by considering the bubble point temperature and ammonia mole fraction of liquid phase. For plotting the auxiliary curve the liquid enthalpy is considered as a function of bubble point temperature and ammonia mole fraction of vapor phase. (Figure 7) of the present work has got similar curves at saturated liquid and vapor conditions as compared with existing graph by Ziegler and Trepp (1984). The values of enthalpies at any concentration and pressure from the result (22) is compared with the present values and got similar values. The ammonia mole fraction of vapor phase is obtained by correlation by Soleimani (2007). With the utilization of these correlations the result shows good agreement with the previous work. With the combination of correlations stated in the abstract, the present work was carried out using a new program code MatLab and shows the similar trends in all the graphs. Work is the properties from the combination of the three correlations and obtained in MatLab. (Figure 8) shows the entropy of saturated liquid at a specified temperature and various concentrations.

(Figure 8) shows the entropy of saturated liquid at a specified temperature and various concentrations. The entropy decreases and increases with the increase in concentration. The plot obtained is validated with the existing results and produces a good match. Whereas in the entropy of saturated vapor from (Figure 9) at a specified temperature, the plot decreases continuously with the increase in concentration. (Figure 10) shows the entropy concentration diagram for ammonia - water mixture at various pressures and concentrations. The gap on the left hand side between the liquid curves is less compared with the gap on the right side of the plot which can even be extended to 150 bar with the same correlations. The values obtained by this plot can be utilized for any thermodynamic cycle. Upon increasing the pressures vapor curve and auxiliary curve are embedded one over the other forming a close gap between each other.

Figures (11 and 12) shows the liquid volume and vapor volume which has been derived utilizing bubble point temperature. With both the plots at a specified pressure the volume decreases with the increase in the concentration. Exergy analysis is the maximum useful work obtained during an interaction of a system with equilibrium state. The total exergy of a system becomes a summation of physical exergy and chemical exergy.

$$E = E_{ch} + E_{ph} \qquad (33)$$

$$E_{ph} = (h-h_o)-T_o(s-s_o) \qquad (34)$$

$$E_{ch} = \left[\frac{x_i}{M_{NH_3}}\right] e^{o}_{ch,NH_3} + \left[\frac{(1-x_i)}{M_{H_2O}}\right] e^{o}_{ch,H_2O} \qquad (35)$$

Where e^o_{ch}, NH_3 and e^o_{ch}, H_2O are chemical exergies o ammonia and water. The standard chemical exergy of ammonia and water are taken from Ahrendts (1980).

The exergy concentration plot for ammonia-water mixture at various pressures is shown in (Figure 13). The liquid exergy curve decreases to certain concentration and approaches a near constant relation. The vapor exergy

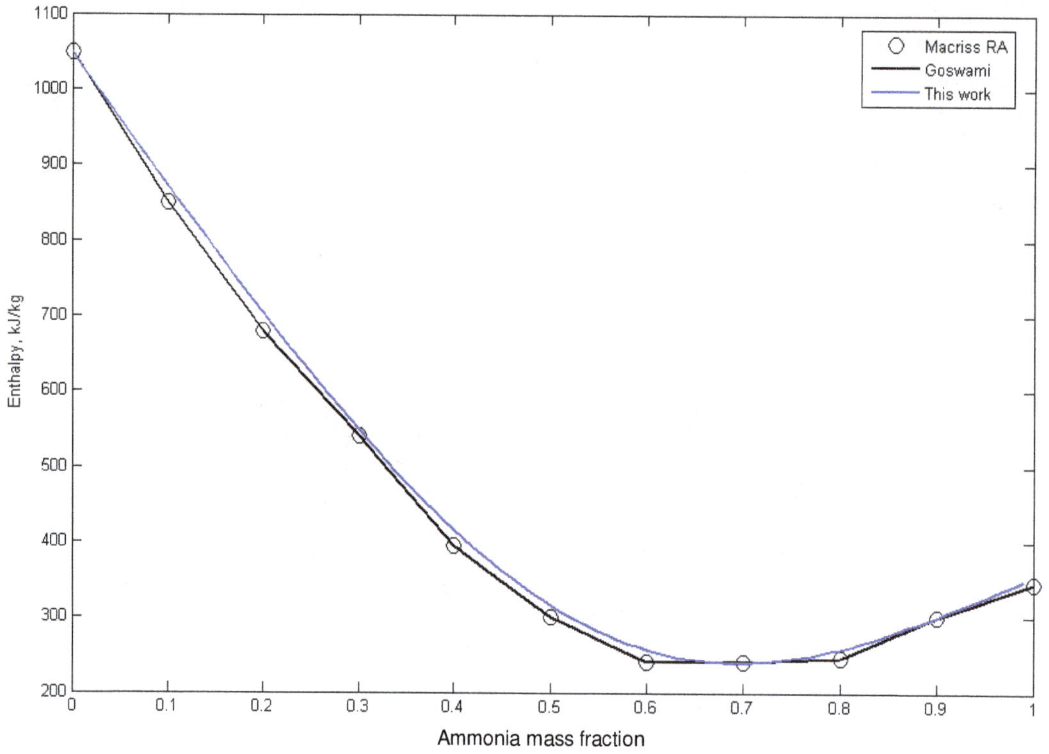

Figure 5. Enthalpy of saturated liquid at P=34.47 bar.

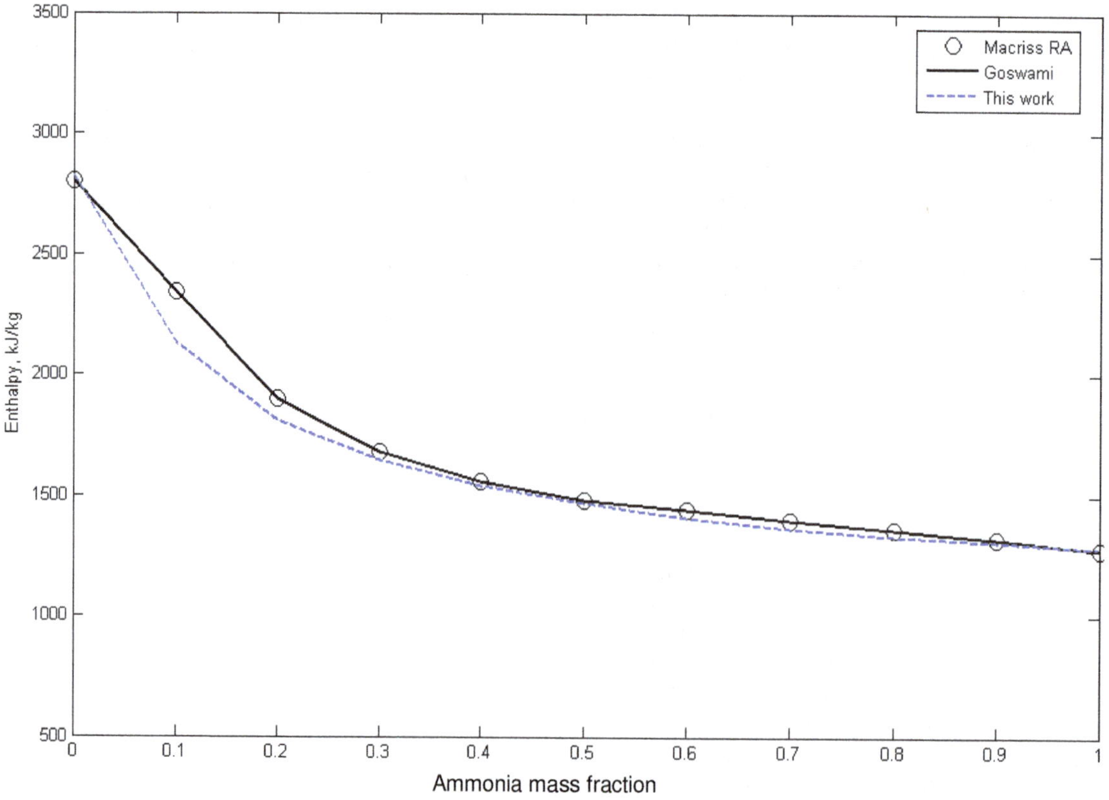

Figure 6. Enthalpy of saturated vapor at 34.47 bars.

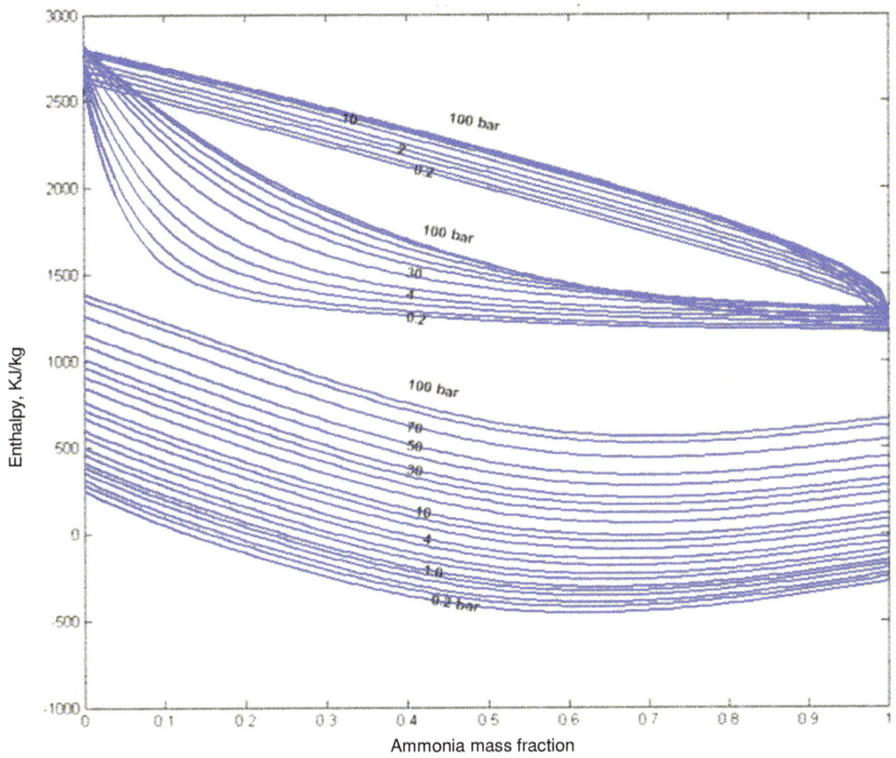

Figure 7. Ammonia-water enthalpy concentration diagram.

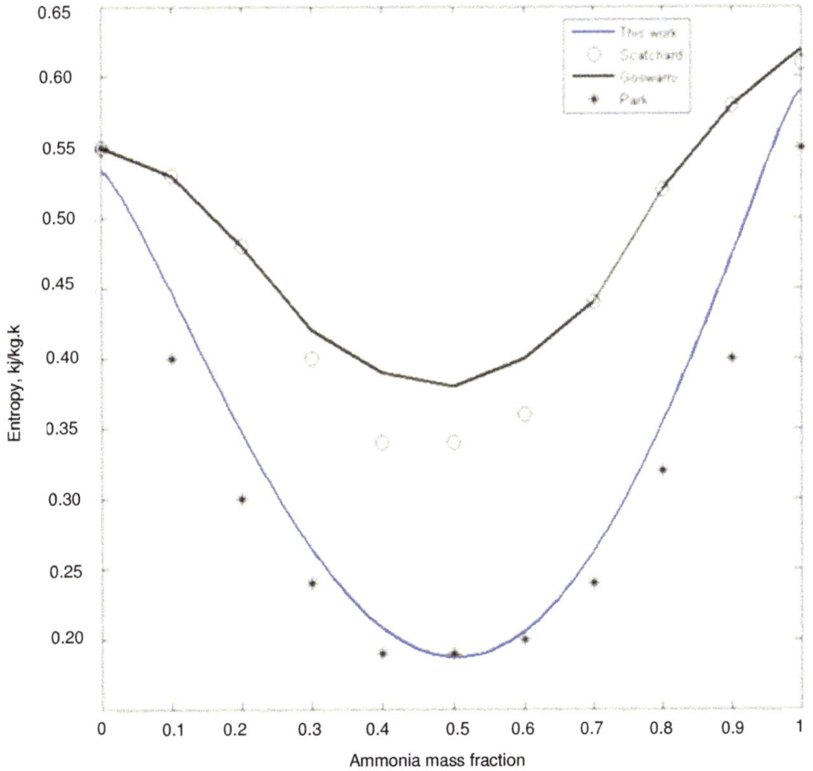

Figure 8. Entropy of saturated liquid at 37°C.

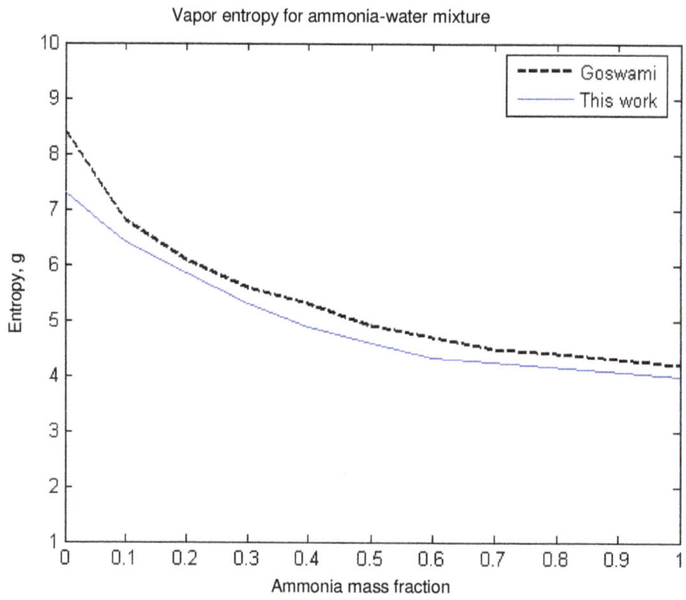

Figure 9. Entropy of saturated vapor at 37 °C.

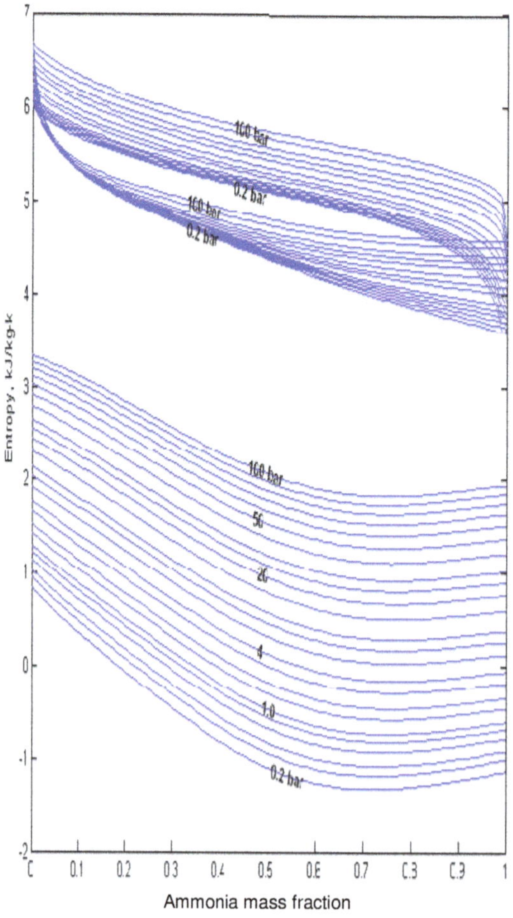

Figure 10. Entropy concentration diagram for ammonia-water mixture.

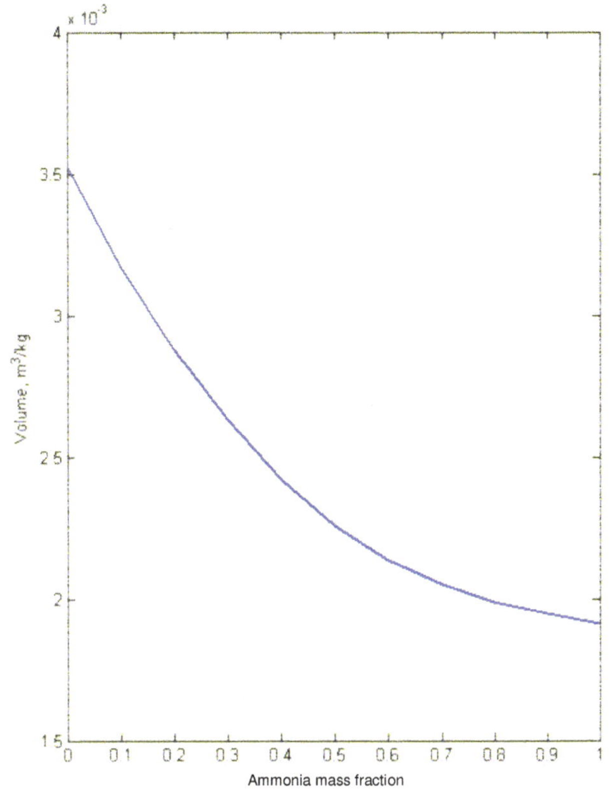

Figure 11. Volume of saturated liquid at P=34.47 bar.

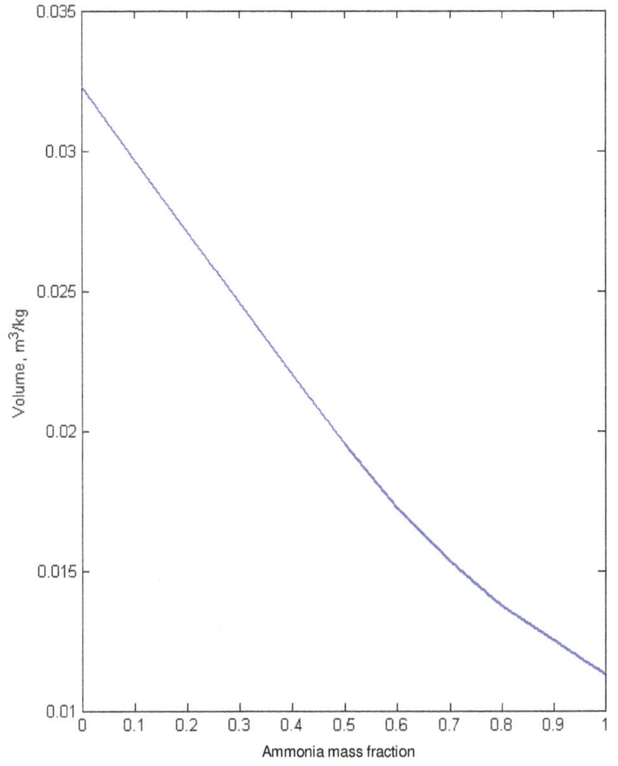

Figure 12. Volume of saturated vapor at P=34.47 bar.

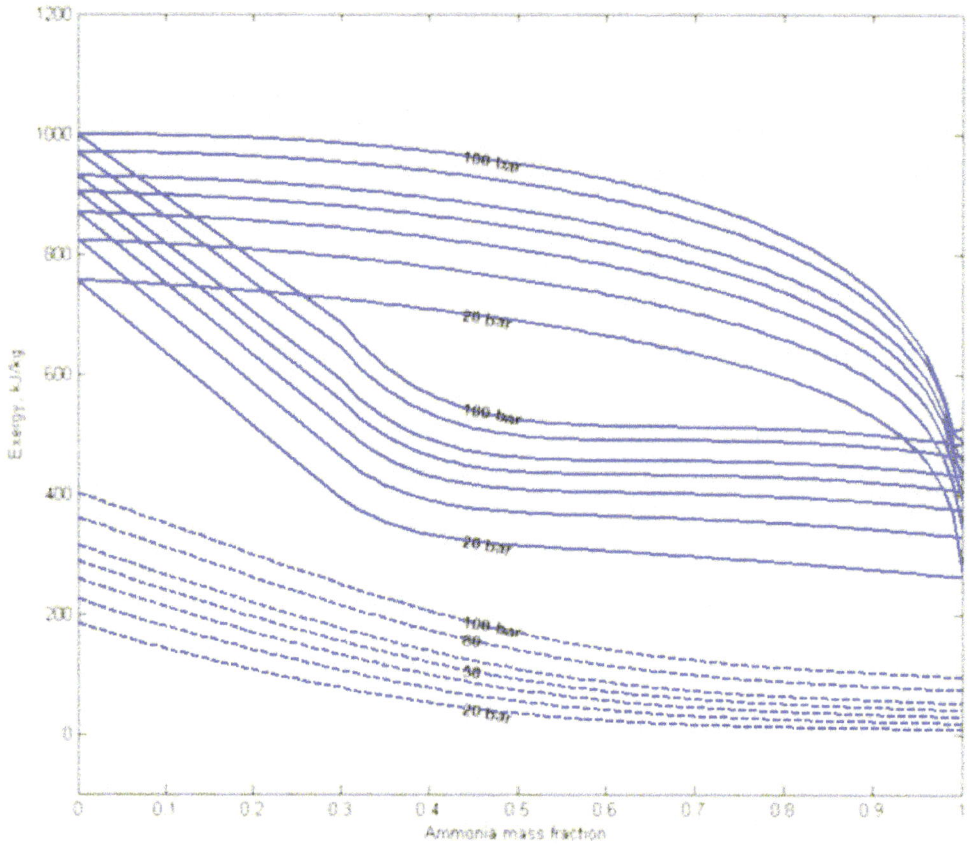

Figure 13. Exergy concentration diagram for ammonia-water mixture.

curve decreases continuously with the increase in concentration. The gap on the left hand side between the liquid curves is wider than the right hand side. The auxiliary lines are contracting at low ammonia fractions whereas the same lines are expanding at high ammonia fractions. The vapor exergy curve and auxiliary curves have identical values at initial and final concentrations which results in a closed loop. The space between the liquid exergy and the closed loop is reduced with the increase in pressures.

Conclusion

To develop thermodynamic properties of ammonia-water mixtures various correlations were analyzed. In this work three different correlations were utilized for developing the results. Bubble and dew point temperatures were obtained utilizing the correlation of Patek and Klomfar (1995), which reduces iterations, which is been utilized for finding the properties enthalpy, entropy and volume. The properties were derived using relations Ziegler and Trepp (1984). The mole fraction of ammonia in vapor phase was solved with the correlation by Soleimani (2007). With the utility of these correlations, the need of tedious iterations used in fugacity method was reduced. The results obtained in this work were validated by comparing with the published data and found closer matching. Here, the results of the graphs obtained from MatLab are compared with the existing results and proves the similar trends, which is the evidence and that is why it was mentioned, as found to have close match. The exergy for the ammonia-water system have been simulated with the help of the derived properties, to carry out the second law analysis to power systems.

Nomenclature: a_i, b_i, c_i, d_i, e_i, m_i, n_i, ρ, Coefficients; **h,** specific enthalpy, kJ/kg; **s,** specific entropy, kJ/kg-K; **v,** specific volume, m^3/kmol; **T,** temperature, K; p, pressure, bar; **g,** Gibbs free energy, kJ/kmol; c_p, specific heat capacity at constant pressure, kJ/kmol-K; **R,** universal gas constant, kJ/kmol-K; **x,** ammonia mole fraction in liquid phase; **y,** ammonia mole fraction in vapor phase.

Superscripts: g, Gas phase; **l,** liquid phase; **o,** ideal gas state.

Subscripts: b, Bubble point; **d,** dew point; **o,** reference state.

REFERENCES

Ahrendts J (1980). Reference states , Energy, 5: 667-668.

Abovsky V (1996). "Thermodynamics of ammonia-water mixture", Fluid Phase Equilibria, 116: 170-176.

Barhoumi M, Snoussi A, Ben EN, Mejbri K, Bellagi A (2004). "Modeling of the thermodynamic properties of the ammonia/water mixture. Int. J. Refrig., 27: 271-283.

Feng X, Yogi GD (1999). "Thermodynamic properties of ammonia-water mixtures for power-cycle applications" Energy, 24: 525-536.

Hasan O, Stanley I, Sandler (1995). "On the combination of equation ofstate and excess free energy models", Fluid Phase Equilibria, 111: 53-70.

Ibrahim OM, Klein SA, (1993). "Thermodynamic Properies of ammonia-water mixture", ASHRAE Trans., 99: 1495-1502

Eric W, Lemmon, Reiner T (1999). "A Helmholtz energy equation of state for calculating the thermodynamic properties of fluid mixtures", Fluid Phase Equlibria, 165: 1-21.

Mejbri KH, Bellagi A (2006). "Modelling of the thermodynamic properties of the water-ammonia mixture by three different approaches, Int. J. Refrig., 29: 211-218.

Mishra RD, Sahoo PK, Gupta A (2006). "Thermoeconomic evaluation and optimization of an aqua-ammonia vapour-absorption refrig. system" Int. J. Refrig., 29: 47-59.

Nowarski A, Friend DG (1998). "Application of the extended corresponding states method to the calculation of the ammonia-water mixture thermodynamic surface", Int. J. Thermophys., 19: 1133-1141.

Patek J, Klomfar J (1995). "Simple functions for fast calculations of selected thermodynamic properties of the ammonia-water system", Refrig., 18: 228-234.

Reid RC, Prausnitz JM, Poling BE (1987). The Properties of Gases and Liquids. Fourth edition. New York, USA: McGraw-Hill. 667. ISBN 0-07-051799-1.

Renon H, Guillevic JL, Richon D, Boston J, Britt H (1985). "A cubic equation of state representation of ammonia-water vapor-liquid equilibrium data" Refrig., 9: 70-73.

Ruiter JP (1990). "Simplified thermodynamic description of mixtures and solutions", 13: 223-236.

Raj S, Diwakar S, Ranjana G, Ashish D (1999). "Potential applications of artificial neural networks to thermodynamics: Vapor-liquid equilibrium predictions, Comput. Chem. Eng., 23: 385-390.

Senthil R, Murugan PMV, Subbarao(2008) "Thermodynamic Analysis of Rankine-Kalina Combined Cycle", Int. J. Thermodyn., 11: 133-141.

Soleimani G, Alamdari (2007). "Simple functions for predicting the thermodynamic properties of ammonia-water mixture", 20(1): 95-104.

Tillner-Roth R, Friend DG (1998). "A Helmholz free energy formulation of the thermodynamic properties of the mixture, American Institute of Physics and American Chemical Society.

"Vapor Absorption Refrigeration Systems Based on Ammonia-Water Pair" (2004). Version 1. ME, IIT.

Weber LA, (1999). "Estimating the virial coefficients of the ammonia-water mixture" Fluid Phase Equlibria, 162: 31-49

Yousef SH, Najjar (1997). "Determination of thermodynamic properties of some engineering fluids using two-consant equations of state, Thermochim. Acta, 303: 137-143.

Ziegler B, Trepp CH (1984). "Equation of state for ammonia-water mixtures" Refrig., 7: 101-106.

Optimal maintenance scheduling of thermal power units in a restructured nigerian power system

Obodeh, O.[1] and Ugwuoke, P. E.[2]

[1]Mechanical Engineering Department, Ambrose Alli University, Ekpoma, Edo State, Nigeria.
[2]Mechanical Engineering Department, Petroleum Training Institute, Effurun, Delta State, Nigeria.

The optimal preventive maintenance schedules of generating units for the purpose of maximizing economic benefits and improving reliable operation of 44 functional thermal generating units of Nigerian power system, subject to satisfying system load demand, allowable maintenance window and crew constraints over 52 weeks maintenance and operational period is presented. It uses HPSO algorithm to find the optimum schedule. The purpose of the algorithm is to orderly encourage moving maintenance outages from periods of low reliability to periods of high reliability, so that a reasonable reliability level is attained throughout the the year. The maintenance outages for the generating units were scheduled to minimize the sum of the squares of reserves and satisfy 2,943.8 MW system peak load with 6.5% spinning reserve of 2,403.8 MW, available manpower for maintenance per week of 22 and maximum generation of 3,028.8 MW. The reliability criterion of the power system was achieved by maximizing the minimum net reserves along with satisfaction of maintenance window, crew and load constraints. The population size of 30 particles and 2500 iterations were chosen. These were chosen as a trade-off between computational time and complexity. It was shown that the HPSO algorithm is not as time efficient as the standard PSO but it provides more consistent and reliable results. In the periods of low maintenance activities, with the PSO algorithm, the maximum generation is 2,753.8 MW while the HPSO produce 2,943.8 MW. It is glaring from the comparison that the HPSO algorithm shows better performance and produce optimal maintenance scheduling framework for the Nigerian power system that will achieve better utilization of available energy with improved reliability and reduction in energy cost.

Key words: Generator maintenance, deregulated market, optimization.

INTRODUCTION

Adequate power supply plays very important role in the socio-economic and technological development of every nation. The electricity demand in Nigeria far outstrips its supply which is epileptic (Sambo, 2007; Efenedo and Akalagboro, 2012). By the year 2020, the Government's policy objective is that Nigeria should posses a generating capacity of at least 40,000 MW (FGN, 2005). Given Nigeria's population growth rate, 40,000 MW would still be less than 25% of South Africa's generating capacity per capita; but it would be sufficient to allow for a significant growth in manufacturing industry and ultimately Gross Domestic Product (GDP) per capita (Sambo, 2007). However, the investments required to finance an increase in total power station capacity from 12,000 to 40,000 MW is huge. On a conservative estimate, this growth in capacity would require 36 billion US dollar (Presidential Action Committee on Power, 2010) which the Government can ill-afford. Hence, the need to incentivize the private sector to partner with government in this endeavour. The unbundling of the Power Holding Company of Nigeria (PHCN) has been an important step. Restructured and liberalized power sectors promote increased competition through unbundling of generation, transmission and privatization of distribution or retailing function (Billinton and Abdulwahab, 2003; Conejo et al., 2005; Jin-ho et al., 2005). The purpose was to break the monopoly of the traditional electric power industry and encourage a competitive power industry. This can bring about reduced generation cost and retail price.

In this context, both generation company (GENCO) and distribution company (DISCO) or retailer may have open access to the transmission grid for negotiated power transfer. The coordination between the GENCO, transmission company (TRANSCO) and DISCO for technical operation of these sub-entities and the commercial arbitration among them will be carried out by Nigeria Electricity Regulatory Commission (NERC). In such environment, GENCO submit their maintenance plans and constraints to the NERC including maintenance time windows, available maintenance resources and generation price offers. The TRANSCO also submit their respective maintenance plans and constraints to NERC. The NERC will be responsible for the optimal coordination of maintenance for generation units and transmission lines to ensure security of power systems and maximize reliability (Presidential Action Committee on Power, 2010). The primary goal of generator maintenance scheduling (GMS) is the effective allocation of generating units for maintenance while ensuring high system reliability, reducing production cost, prolonging generator lifetime subject to some unit and system constraints. GMS problem is a hard combinational optimization problem and is classified as a deterministic cost-minization problem (Cagnina et al., 2011).

Several optimization methods have been applied to solve the problem, which could be grouped into three categories namely, heuristic methods, mathematical programming methods and artificial intelligent methods. Heuristic methods provides the most primitive solution based on trial and error principles. Mathematical programming methods includes mixed integer programming (MIP), mixed integer linear programming (MILP), decomposition, branch and bound and dynamic programming (Kitayayama and Yasuda, 2006). The main problem with the exact mathematical methods is that the number of combinations of states that must be searched increases exponentially with the size of the problem and becomes computationally prohibitive (Talukder, 2011). Furthermore, these techniques are generally unsuitable for the nonlinear objective functions and constraints in their standard form and several assumptions are required to make the problem solvable using reasonable computational resources (Del Valle et al., 2008). Artificial intelligent methods include neural networks, artificial immune systems, genetic algorithm, fuzzy optimization, ant colony optimization and particle swarm optimization (PSO) algorithm. The PSO is a novel population-based stochastic search algorithm and an alternative solution to the complex nonlinear optimization problem (Yasuda et al., 2010).

The PSO algorithm basically learned from animal's activity or behaviour to solve optimization problems. In PSO, the population is called a swarm and each member of the population is called a particle. Each particle has three main characteristics: an adaptable velocity with which it moves in the search space, a memory where it stores the best position it ever visited in the search space (that is, the position with lowest function value), and the social sharing of information, that is, the knowledge of the best position ever visited by all particles in its neighbourhood. Starting with a randomly intialized population and moving in randomly chosen directions, each particle goes through the search space and remembers the best previous positions of itself and its neighbours. Particles of a swarm communicate good positions to each other as well as dynamically adjust their own position and velocity derived from the best position of all particles. The next step begins when all particles have been moved. Finally, all particles tend to fly towards better and better positions over the searching process until the swarm move close to an optimum of the fitness function $f: R^n \rightarrow R$.

Standard PSO performs well in the early iterations, but it has problems approaching a near-optimal solution (Yasuda et al., 2010). If a particle's current position accords with the global best and its inertia weight and previous velocity are different from zero, the particle will only fall into a specific position. If their previous velocities are very close to zero, then all the particles will stop moving around near-optimal solution, which may lead to premature convergence of the algorithm. In this situation, all the particles have converged to the best position discovered so far which cannot be the optimal solution. This is known as stagnation (Del Valle et al., 2008). Using hybrid particle swarm optimization (HPSO) by adding the mutation operator used in genetic algorithm (GA), the aforementioned problems can be solved (Yare et al., 2008; Yare and Venayagamoorthy, 2010). The aim of mutation is to introduce new genetic material into existing individual; that is, to add diversity to the genetic characteristics of the population.

The objective of this study is to schedule generator maintenance that will reduce the operational cost of generator which includes, the maintenance cost while satisfying all necessary constraints involved using PSO.

MATERIALS AND METHODS

The test problem is to schedule the maintenance of 44 functional generating units (Table 1) over planning period of 52 weeks. The

Table 1. Nigerian thermal grid system.

S/N	Plant	Plant type	Location (state)	Installed capacity (MW)	Available capacity (MW)	Installed units	Units available
1.	Egbin	Thermal	Lagos	1320	880	6	4
2.	Egbin AEs	Thermal	Lagos	270	270	9	9
3.	Sapele	Thermal	Delta	1020	102	10	1
4.	Okapi	Thermal	Cross River	480	320	3	2
5.	Afam	Thermal	Rivers	702	105.3	20	3
6.	Delta	Thermal	Delta	840	560	18	12
7.	Omoku	Thermal	Rivers	150	100	6	4
8.	Ajaokuta	Thermal	Kogi	110	110	2	2
9.	Geregu	Thermal	Kogi	414	414	3	3
10.	Omotosho	Thermal	Ondo	335	83.75	8	2
11.	Olorunsogo/ Papalanto	Thermal	Ogun	335	83.75	8	2
Total				5,976	3,028.8	93	44

Source: Sambo (2007).

Nigerian thermal grid system comprising of 44 generating units spread across 11 generation stations is as depicted in Table 1. Table 2 gives the generating capacities, maintenance allowed periods, maintenance duration, available manpower, and the crew needed for each generator. The GMS problem has a number of units and system constraints to be satisfied. The constraints include: the maintenance window, crew, demand and reserve constraints. The objective function to be minimized is given by:

$$Min\left[\sum_{t=1}^{T}\left(\sum_{i=1}^{I} g_{i,t} - \sum_{i\in I_t}\sum_{r\in s_{i,t}} x_{i,t} g_{i,t} - D_t\right)^2\right] \quad (1)$$

Subject to: maintenance window constraint:

$$\sum_{t\in T_i} x_{i,t} = 1, \text{ for all } i = 1,2,\dots,I \quad (2)$$

Crew constraint:

$$\sum_{i\in I_t}\sum_{k\in s_{i,t}} x_{i,r} M_{i,r} \le A_t, \text{ for all } t = 1,2,\dots,T \quad (3)$$

Load constraint:

$$\sum_{i=1}^{I} g_{i,t} - \sum_{i\in I_t}\sum_{k\in s_{i,t}} x_{i,r} g_{i,r} \ge D_t, \text{ for all } t = 1,2,\dots,T \quad (4)$$

Where t = index of period, $t = 1,2,\dots,T$; T = total number of planned horizons; I = index of the generators number, $i = 1,2,\dots,I$; I = total number of generators; $g_{i,t}$ = generating capacity for each generator (MW); I_t = the set of indices of generators in maintenance at time t; r = index of start periods of maintenance for each generator, $r = t,\dots,s$; $S_{i,t}$ = set of start period r such that if the maintenance generator i starts at period r that generator will be in maintenance at period t, $s_{i,r} = [r \in T_i : t - N_i \le r \le t]$; T_i = set of periods when maintenance of generator i may start, $T_i = [t \in T : e_i \le t \le l_i - N_i + 1]$; e_i = earliest period for generator i to start maintenance; l_i = latest period for generator i to start maintenance; $x_{i,r}$ = variable for the start of maintenance for each generator i at time r; if generator i is on maintenance $x_{i,r} = 1$, otherwise $x_{i,r} = 0$; D_t = demand per period; $M_{i,r}$ = number of crew used for maintenance of generator i at time r; A_t = available number of crew at every time t; N_i = duration of maintenance on each generator i.

The augmented objective function $F(x)$ formulated for this study is a weighted sum of the objective function and the penalty functions for violations of the constraints; hence:

$$F(x) = \omega_o SSR + \omega_m TMV + \omega_l TLV \quad (5)$$

Where SSR is the sum of squares of reserves as in Equation 1, TMV is the total manpower violation as in Equation 3 and TLV is the total load violation as in Equation 4.

The weighting coefficients ω_o, ω_m and ω_l were chosen so that the violation of the relatively hard load constraint (Equation 4) gives a greater penelty value than for soft crew constraint (Equation 3). As in Khan et al. (2010), ω_o, ω_m and ω_l were chosen as 10^{-5},

Table 2. Maintenance data of 44 functional units.

S/N	Power station	Capacity (MW)	Earliest period	Latest period	Outage (weeks)	Available manpower	Required manpower
				Egbin			
1	ST1	190	7	23	5	22	6+5+5+4+2
2	ST2	190	29	45	5	22	6+5+5+4+2
3	ST3	190	36	52	5	22	6+5+5+4+2
4	ST4	190	24	50	5	22	6+5+5+4+2
5	ST5	190	39	52	5	22	6+5+5+4+2
6	ST6	190	1	20	5	22	6+5+5+4+2
7	GT1	30	42	52	2	7	4+3
8	GT2	30	8	21	2	7	4+3
9	GT3	30	1	20	2	7	4+3
10	GT4	30	1	20	2	7	4+3
11	GT5	30	13	36	2	7	4+3
12	GT6	30	16	39	2	7	4+3
13	GT7	30	16	41	2	7	4+3
				Sapele			
14	ST1	120	20	45	4	12	4+3+3+2
15	ST2	120	1	14	4	12	4+3+3+2
16	ST3	120	1	20	4	12	4+3+3+2
17	ST4	120	1	15	4	12	4+3+3+2
	Okapi						
18.	GT1	160	1	9	5	20	6+5+4+3+2
19.	GT2	160	1	16	5	20	6+5+4+3+2
	Afam						
20	GT18	138	30	45	4	20	7+6+4+3
21	GT19	138	14	36	4	20	7+6+4+3
22	GT20	138	7	27	4	20	7+6+4+3
				Delta			
23	GT3	19.6	11	26	2	7	4+3
24	GT4	19.6	1	19	2	7	4+3
25	GT6	19.6	35	51	2	7	4+3
26	GT7	19.6	35	51	2	7	4+3
27	GT8	19.6	1	23	2	7	4+3
28	GT15	85	33	52	4	14	4+4+3+3
29	GT16	85	40	52	4	14	4+4+3+3
30	GT17	85	30	52	4	14	4+4+3+3
31	GT18	85	29	52	4	14	4+4+3+3
				Omuku			
32	GT1	25	40	52	2	7	4+3
33	GT3	25	26	52	2	7	4+3
34	GT5	25	20	40	2	7	4+3
35	GT6	25	11	31	2	7	4+3
				Ajaokuta			
36	GT1	50	2	22	4	15	5+4+3+3
37	GT2	50	31	51	4	15	5+4+3+3

Table 2. Contd.

				Geregu			
38	GT1	138	2	20	4	20	7+6+4+3
39	GT2	138	3	15	4	20	7+6+4+3
40	GT3	138	2	17	4	20	7+6+4+3
				Omotosho			
41	GT1	40	5	25	3	8	3+3+2
42	GT2	40	40	52	3	8	3+3+2
				Olorunsogo			
43	GT1	40	37	52	3	8	3+3+2
44	GT2	40	42	52	3	8	3+3+2

4 and 2 respectively.

Particle swarm optimization (PSO)

The PSO algorithm is an iterative optimization process and repeated iterations will continue until a stopping condition is satisfied. Within one iteration, a particle determines the personal best position, the local or global best position, adjusts the velocity, and a number of function evaluations are performed. If N is the total number of particles in the swarm, then N function evaluations are performed at each iteration. The algorithm is terminated when there is no significant improvement over a number of iterations. That is, if the average change of the particles' positions are very small or the average velocity of the particles is approximately zero over a number of iterations. It can also be terminated when the maximum number of iterations has been attained. The updating rule of the positions and velocities are given as:

$$x_{jd}(k+1) = x_{jd}(k) + v_{jd}(k+1) \tag{6}$$

Where:

$$v_{jd}(k+1) = \omega v_{jd}(k) + c_1 rand(\ )(P_{jd}(k) - x_{jd}(k)) + c_2 rand(\ )(P_{gd}(k) - x_{jd}(k)) \tag{7}$$

and $v_{jd}(k)$ is the velocity vector of particle j in dimension d at iteration k; $x_{jd}(k)$ is the position vector of particle j in dimension d at iteration k; $P_{jd}(k)$ is the personal best position of particle j in dimension d found from initialization through iteration k. ω is the inertia weight used to weigh the last velocity; c_1 is a variable used to weigh the particle's knowledge and c_2 is a variable used to weigh the swarm's knowledge. $rand(\ )$ are uniformly distributed random numbers between zero and one. The best position of each particle is updated at each iteration by setting $P_{jd}(k+1) = x_{jd}(k+1)$, if $f(x_{jd}) < f(P_{jd})$, otherwise it remains unchanged.

An update of the index g is also required at each iteration. The recommended relationship between c_1 and c_2 (Talukder, 2011) is:

$$c_1 + c_2 \leq 4 \tag{8}$$

The parameters c_1 and c_2 are usually set fixed and equal, so that the particle is equally influenced by its best position P_{jd}, as well as the best position of its neighbourhood P_{gd}. ω often decrease linearly from 0.9 to 0.4 during each iteration (Yasuda et al., 2010). Its values are set according to:

$$\omega(k) = \omega_{max} - \frac{k(\omega_{max} - \omega_{min})}{k_{max}} \tag{9}$$

Where ω_{max} and ω_{min} are the initial and final values of the inertia weight respectively.

Herein, the mutation operator is introducted into the PSO algorithm. The main goal is to increase the diversity of the population by preventing the particles from moving too close to each other thus converging prematurely to local optima (Song et al., 2012). This in turn improves the PSO's search performance. The mutation operator is defined as follows:

$$\alpha_m(k) = \alpha_m^{max} - \left(\frac{\alpha_m^{max} - \alpha_m^{min}}{k_{max}}\right) \times k \tag{10}$$

Table 3. Parameters used in this study.

Parameter	c_1	c_2	ω_{max}	ω_{min}	α_m^{max}	α_m^{min}	N	k_{max}
Value	2	2	0.9	0.4	0.01	0.001	30	2500

Where k_{max} is the maximum iteration number and k is th current iteration number (Table 3).

GMS procedure using standard PSO are as follows:

Step 1: the initial parameters, the number of particles j, where $j = 1,2,...,N$ the maximum number of iterations k_{max} and the iteration counter $k = 1$ were set. The position x_j and volcity v_j at random for every particle were set;
Step 2: set iteration counter $k = 1$
Step 3: evaluate the augmented objective function of the GMS problem (Equation 5) for every particle;
Step 4: determined P_j and P_g. Here P_j is the value of x particle i that gives the minimum evaluated GMS result from initial iteration to the present iteration. P_g is the value of x that gives the minimum evaluated result in the whole swarm from initial to present iteration;
Step 5: increase the iteration counter $k = k + 1$;
Step 6: compared the iteration counter with the preset maximum number k_{max}. When $k < k_{max}$, the algorithm returns to step 3. Otherwise P_g is output as the optimal solution and the search terminated.

The procedure for the hybrid PSO (HPSO) method is as follows:

Step 1: initialize the volocity, position, local best position and global best position;
Step 2: set iteration counter $k = 1$;
Step 3: evaluate the objective function;
Step 4: update $\omega(k)$ and j^{th} particle velocity using Equations 9 and 7 respectively;
Step 5: update the local best position and global best position;
Step 6: calculate the mutation operator using Equation 10;
Step 7: if rand ( ) < $\alpha_m(k)$, then 7(a): if $k < k_{max}$, then new global best position = $x_{j\min} + rand(\ )\times(x_{j\max} - x_{j\min})$, where $x_{j\min}$ represents the lower bound and $x_{j\max}$ represents the upper bound of the parameter if objective value at global best position > objective value at new global best position, then $P_g = P_{gnew}$ and the search terminated; otherwise $P_j = P_{jnew}$
Step 8: the iteration number k was increased to $k = k + 1$ and algorithm returns to step 3.

RESULTS AND DISCUSSION

The PSO algorithm was simulated with different number of population sizes and iterations over the 52 weeks period. The algorithm was implemented using Matlab on a Intel (R) core (TM)[2] Duo 2.10 GHZ personal computer. Matlab does not have any PSO programming function and as such the mixed integer PSO programme was written using the M-file. An M-file was created and edited in the Editor/Debugger Window of the Matlab programme (Zimmerman and Gan, 2005). The effect of population size on time of computational time is shown in Figure 1. The population size of 30 particles has a lower computational time than that of 40 particles population but generates the same results. It is noteworthy, that the number of iteration was used as a stopping criterion for the optimization and as such the smaller the iteration value used to obtain the optimal results the better. In view of the forthgoing, the population size of 30 particles and 2500 iterations were chosen for this study. The chosen population and iteration ensure that the search space was fully utilized without putting strain on the computation time and complexity. The variation of objective function with the number of iteration is depicted in Figure 2. The population sizes of 30 and 40 particles gave relatively lower values under each iteration. But population size of 30 particles has egde over that of 40 particles because it has a lower computation time.

Choosing the population size determines the diversity and search space for each particle. More particles in the swarm provide a good uniform initialization scheme but at the expense of the computational complexity (Yasuda et al., 2010), as a result the search degrades to a parallel random search. The convergence iteration is a little late in the HPSO algorithm because it takes more time for particles to mutate around feasible areas (Song et al., 2012). In other words, a particle has a higher possibility than PSO algorithm to find better optimized solution. The HPSO algorithm has a better performance related to maintenance scheduling problem having higher accessibility to search the optimal solution. As maintenance scheduling is an annual planning problem in a long horizon over one year, emphasis has to be not on convergence time but a set of optimal solution. Hence, the HPSO is not time efficient as the standard PSO but it provides more consistent and reliable results. Figure 3 shows the reserve margin. The reserve margin is non negative because the load constraint (Equation 4) is satisfied. The addition of the generation limits (Equation 4) ensures that the load constraint is never violated and thus reduces the SSR. NERC may employ penalty factor to patronize unit not to have maintenance in peak load. By this strategy, NERC will have more effect on unit

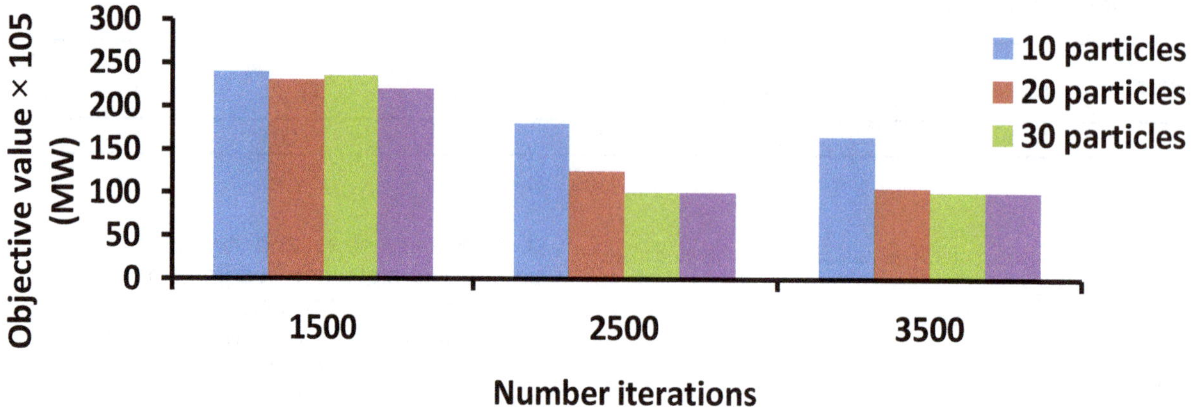

Figure 1. Variation of objective function with number of iteration.

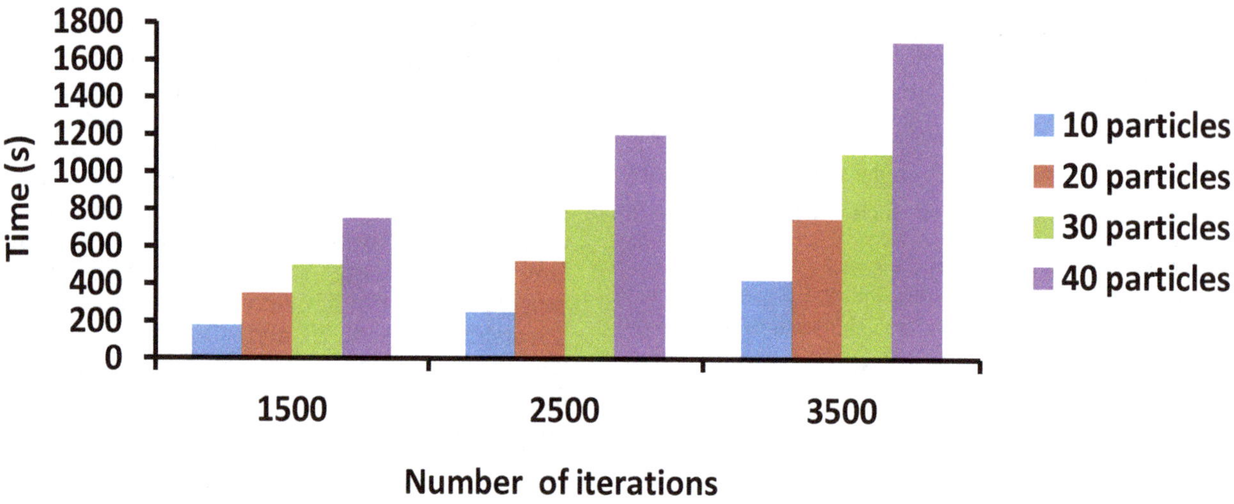

Figure 2. Comparison of population size with computational time.

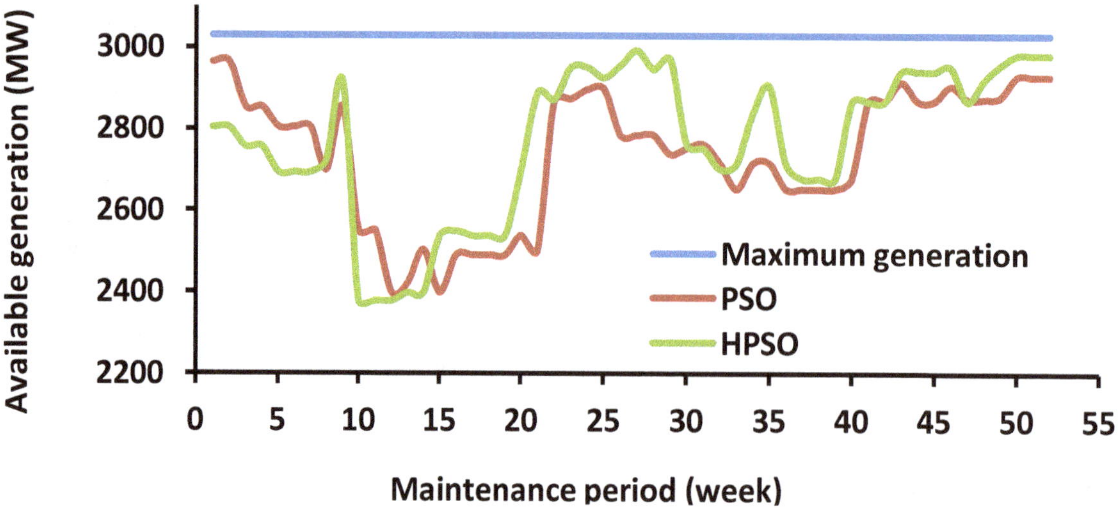

Figure 3. Available generation versus maintenance period.

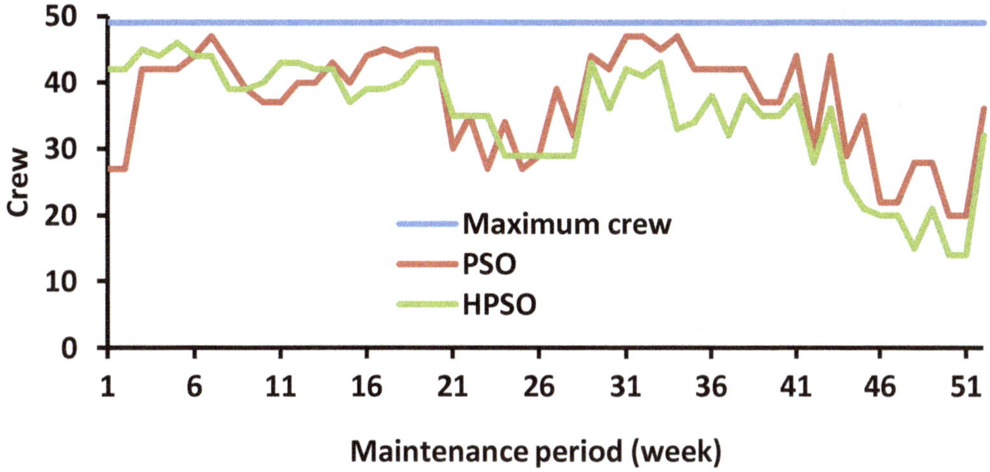

Figure 4. Crew requirement against maintenance period.

Table 4. Generator maintenance schedule.

Week No.	Generating unit scheduled for maintenance		Week No.	Generating unit scheduled for maintenance	
	PSO	HPSO		PSO	HPSO
1	10, 18	6, 15, 18	27	14, 21, 33	4, 33
2	10, 18	6, 15, 18	28	14, 21	4, 35
3	6, 18, 36	6, 15, 18, 24	29	4, 14, 21	2, 31, 35
4	6, 18, 36	6, 15, 18, 24	30	4, 21	2, 31
5	6, 16, 18	6, 16, 18, 36	31	4, 20, 30	2, 31, 37
6	6, 16, 17	9, 16, 17, 36	32	4, 20, 30	2, 31, 37
7	6, 16, 17	9, 16, 17, 36	33	4, 20, 30	2, 20, 37
8.	8, 16, 17, 40	16, 17, 36	34	2, 20, 30	13, 20, 37
9	8, 17, 40	10, 17, 22	35	2, 28, 37	13, 20, 25
10	11, 19, 40	10, 22, 39	36	2, 28, 37	3, 20, 25
11	11, 19, 40	1, 22, 39	37	2, 28, 37	3, 26, 43
12	15, 19, 39	1,19, 22, 39	38	2, 28, 37	3, 26, 43
13	15, 19, 39	1, 19, 39	39	5, 26, 43	3, 28, 43
14	9, 19, 23, 39	1,19, 40	40	5, 26, 43	3, 28, 42
15	9, 23, 39	1, 19, 40	41	5, 29, 43	5, 28, 42
16	1, 24, 38	8, 19, 40	42	5, 44	5, 28, 42
17	1, 24, 38	8, 21, 40	43	5, 29, 44	5, 14, 30
18	1, 27, 38	11, 21, 38	44	7, 29, 44	5, 14, 30
19	1, 27, 38	11, 21, 38	45	7, 29	5, 14, 30
20	1, 34, 41	12, 21, 38	46	25, 32, 42	7, 14, 44
21	12, 34, 41	12, 38, 41	47	25, 32, 42	7, 30, 44
22	12, 22, 41	27, 34, 41	48	3, 31, 42	32, 44
23	13, 22	27, 34, 41	49	3, 31	29, 32
24	13, 22, 35	4, 23	50	3, 31	29
25	22, 35	4, 23	51	3, 31	29
26	14, 33	4, 33	52	3, 31	29

maintenance schedules. Figure 4 illustrates the effectiveness of the generator limits constraint. The reliability criterion of the power system was achieved by maximizing the minimum net reserves along with satisfaction of maintenance window, crew and load constraints.

The complete maintenance schedules obtained by PSO and HPSO are presented in Table 4. It presents the corresponding crew availability needed to carry out the scheduled shutdown maintenance of the generating units. The proposed algorithm attains the maximum generation in weeks 49, 50, 51 and 52. However, the PSO algorithm has also low maintenance activities in those weeks resulting in high available generation. With the PSO algorithm, the maximum generation is 2,753.8 MW in those weeks whilst HPSO algorithm produce 2,911.8 MW in week 49 and 2,943.8 MW in others. It is evident from the comparison that the HPSO algorithm shows great potential for effective energy management, short and long term generation scheduling, system planning and operation.

Conclusion

Annual maintenance scheduling of generating units of GENCOs in deregulated power system was formulated and validated. It uses HPSO algorithm to find the optimum schedule. GENCO's objective is to sell electricity as much as possible and the goal of NERC is to maximize the reserve of the system at every time interval, provided the energy purchase cost is smaller than a pre-determined amount when the units of GENCOs are out for maintenance. In fact, NERC may employ penalty factor to patronize unit not to have maintenance in peak loads. By this strategy, NERC will have more effect on maintenance schedules. The purpose of the algorithm is to orderly encourage moving maintenance outages from periods of low reliability to periods of high reliability, so that a reasonable reliability level is attianed throughout the year. The algorithm was applied to GMS problem with 44 generating units. The maintenance outages for the generating units were scheduled to minimize the sum of the squares of reserves and satisfy 2,943.8 MW system peak load with 6.5% spinning reserve of 2,403.8 MW, available manpower for maintenance per week of 22 and maximum generation of 3,028.8 MW. The reliability criterion of the power system was achieved by maximizing the minimum net reserves along with satisfaction of maintenance window, crew and load constraints. The population size of 30 particles and 2500 iterations were chosen. The chosen population size and iteration ensure that the search space was utilized to the fullest without putting strain on computation time and complexity. It was shown that the proposed algorithm satisfied the load and crew requirements. It was evident that the HPSO algorithm has a better performance related to maintenance scheduling problem having higher accessibility to search the optimal solution. Although, the HPSO is not as time efficient as the standard PSO, it provides more consistent and reliable results.

In the periods of low maintenance activities, with the PSO algorithm, the maximum generation is 2,753.8 MW while the HPSO produce 2,943.8 MW. It is glaring from the comparison that the HPSO algorithm shows better

performance and produce optimal maintenance scheduling framework for the Nigerian power system that will achieve better utilization of available energy with improved reliability and reduction in energy cost. The proposed algorithm can be modified to accommodate the maintenance unit requirements of emerging independent power producers and future generation additions.

REFERENCES

Billinton R, Abdulwahab A (2003). Short-term Generating Unit Maintenance in a Deregulated Power System using a Probabilistic Approach. IET Proceeding- Gen. Trans. Distrib. 150(4):463-468.

Cagnina LC, Esquivel SC, Coello CA (2011). Solving Constrained Optimization Problems with a Hybrid Particle Swarm Optimization Algorithm. Eng. Optim. 43(8):843-866.

Conejo AJ, García-Bertrand R, Díaz-Salazar M (2005). Generation Maintenance Scheduling in Restructured Power Systems. IEEE Trans. Power Syst. 20(2):984-992.

Del Valle Y, Venayagamoorthy GK, Mohagheghi S, Hernandez J, Harley RG (2008). Particle Swarm Optimization: Basic Concepts, Variants and Applications in Power System. IEEE Trans. Evol. Computat. 12(2):171-195.

Efenedo GI, Akalagboro IO (2012). The Challenges of Power Generation, Utilization and Supply: The Nigeria Vision for the 21st Century. J. Emerging Trends Eng. Appl. Sci. 3(5):886-891.

FGN (2005). Electric Power Sector Reform Act, 2005. Federal Republic of Nigeria Official Gazzete Vol. 92, No. 77 Government Notice No. 150.

Jin-ho K, Jong-Bae, P, Jong-Keun P, Yeung-Han C (2005). Generating Unit Maintenance Scheduling under Competitive Market Environments. Elect. Power Energy Syst. 27:189-194.

Khan L, Mumtaz S, Javed K (2010). Generator Maintenance Schedule for WAPDA System using Meta- Heuristic Paradigms. Austral. J. Basic Appl. Sci. 4(7):1656-1667.

Kitayayama S, Yasuda K (2006). A Method for Mixed Integer Programming Problems by Particle Swarm Optimization. Elect. Eng. Japan. 157(2):813-820.

Presidential Action Committee on Power (2010). Roadmap for Power Sector Reform: A customer-Driven Sector-Wide Plan to Achieve Stable Poweform: A customer-Driven Sector-Wide Plan to Achieve Stable Power. Accessed from www.nigeriaelectricityprivatisation.com/wp-content/plugins/download-monitor/download.php?id=43 on December 12, 2012.

Sambo AS (2007). Achieving the Millennium Development Goals (MDGs): The Implication for Energy Infrastructure in Nigeria. Proceedings of COREN 16th Engineering Assembly, August 28-29, Abuja, Nigeria, pp. 121-141.

Song Y, Zhijian W, Wang H, Zhangxin C, Zhong H (2012). A Hybrid Particle Swarm Optimization Algorithm based on Space Transformation Search and A Modified Velocity Model. Int. J. Num. Anal. Model. 9(2):371-377.

Talukder S (2011). Mathematical Modelling and Applications of Particle Swarm Optimization. M. Sc. Dissertation, School of Engineering, Blekinge Institute of Technology, Sweden.

Yare Y, Venayagamoorhty GK (2010). Optimal Generator Maintenance Scheduling using Multiple Swarms- MDPSO Framework, Eng. Appl. Artif. Intell. 23(6):895-910.

Yare Y, Venayagamoorthy GK, Aliyu UO (2008). Optimal Generator Maintenance Scheduling using a Modified Discrete PSO. IET J. Gen. Trans. Distrib. 2(6):834-846.

Yasuda K, Yazawa K, Motoki M (2010). Particle Swarm with Parameter Self-adjusting Mechanism. IEEJ Trans. Elect. Electr. Eng. 5(2):256-257.

Zimmerman R, Gan D (2005). MATPOWER: A Matlab Power System Simulation Package.

Effect of nozzle type, angle and pressure on spray volumetric distribution of broadcasting and banding application

Nasir .S. Hassen, Nor Azwadi .C. Sidik and Jamaludin .M. Sheriff

Department of Thermo fluid, Faculty of Mechanical Engineering, University Technology Malaysia, UTM Johor Bahru, Malaysia.

The ultimate goal of agricultural spraying application system is to put the correct amount of pesticides, in the correct place, at the correct time to reduce the pest to a level below the economic threshold in order to improve agricultural production. A spray patternator was fabricated for the selection of a suitable nozzle to have uniform distribution of the spray liquid. Experiments were conducted on a spray patternator through two types of spray nozzles (even flat fan nozzle TPE for banding application and standard flat fan nozzle TP for broadcasting application). Spray distribution was determined and compared by using single nozzle, at a height of 0.5 m under laboratory conditions. In addition, this paper examined the effect of spray fan angles 65 and 80° and liquid pressures 200 and 300 kPa on the spray distribution. The best distribution of the spray application was obtained by using banding nozzles, whereas the broadcasting nozzle gave an uneven spray distribution with a high peak just below the nozzle centre and taper off towards the edges of the spray pattern. For the two nozzle types tested, results revealed that increasing nozzle angle and pressure reduce the value of the coefficient of variation CV%.

Key words: Static spray distribution, nozzle, patternator, coefficient of variation (CV).

INTRODUCTION

Agricultural chemical can be applied according to the American Society of Agricultural and Biological Engineers (ASABE) Standards (2006) by broadcast application that uses spray over an entire field and band application that uses spray in parallel bands leaving areas between the bands free of chemical. Chemical control in row crops is typically carried out as a broadcast application by using standard flat fan nozzles and most farmers use this kind of application because it is the easiest and preferred method. In fact, standard flat-fan nozzles are not recommended for banded application as result to the following reasons: (1) These nozzles should be overlapped (a array of nozzles) to achieve spray uniform distribution across the entire width of the boom but using the overlapped spray in the row crop fields will cause losing of the spray among the rows or strips. (2) If these nozzles use without overlapping or as single nozzles above the parallel bands, the spray distribution will be uneven. However, management of the precision agriculture encourages reducing the use of pesticides in fields. The question is whether it is possible to use the existing even flat-fan nozzles for spray distribution in very narrow bands instead of standard flat fan nozzles to achieve more efficient spray distribution in the fields.

The technology of band spraying application has not been developed because broadcast applications are the preferred method. Very few publications on band applications and their equipment can be found and those available are with limited information as compared with broadcast application, (Williams, 1981). In modern agriculture, improving quality food production and reducing farming cost can be obtained by using suitable and correct applications. Band spraying is more economical as compared to broadcast spraying (Nigar et al., 2011) because it is targeting a specific area of the field such as rows or strips, amount of chemical used on a specific portion of the field area and reducing water use per area. Spraying application nozzle is designed to achieve good spray distribution and uniformity (Bahadir and Saim, 2011). According to the Spraying System Co (2011), even flat-fan nozzles are used to apply a uniform distribution across the entire spray pattern width. These nozzles should be used for banding applications over the strip or the row and should not be used for broadcast applications, (Spraying System Co, 2011; Vern and Elton, 2004). Wang et al. (1995) reported that uniformity of spray volumetric distribution is the most important indicator of the nozzle performance. Measuring the volumetric distributions of liquid from individual nozzle or group of nozzles have been investigated and identified by researchers using patternator under laboratory conditions (Krishnan et al., 1988; Ozkan et al., 1997; Lebeau et al., 2000; Womac et al., 2001; Sidahmed et al., 2004; Bayat and Yarpuz-Bozdogan, 2005; Javier et al 2008; Jean et al 2012). An even distribution of spray liquid is obtained by selecting suitable nozzles and calibrating them correctly (Lardoux et al., 2007). The objective of the present study was to examine the spray volumetric distribution and coefficient of variation (cv) for standard flat-fan nozzles and even flat- fan nozzles when used as a single nozzles for narrow-band spraying application using different nozzle angles and different liquid pressures.

MATERIALS AND METHODS

The nozzles selected for the study are standard flat-fan nozzles for broadcasting application and even flat- fan nozzles for banding application, these nozzles are classified according to International Organization for Standardization (ISO) of size 03 (0.3 gpm). This current research was carried out in laboratory of the university technology Malaysia (UTM). Before spray distribution measurements, the flow rate of each nozzle was tested by collecting the amount of water directly from the nozzle on a graded container at a pressure 300 kPa for one minute and measuring nozzle output with precision electric balance. Measurements were carried out at 26°C and 75% RH. Tests were repeated three times and the maximal deviation of all nozzles with nominal flow rate was ± 2.5%.

Static spray distribution test

Obtaining an even liquid-spray distribution is considered important in connection with row crop field application of pesticides, whether considering broadcast application or band spraying. Spray volumetric distribution of nozzle was determined in the laboratory by using a spray pattern analyzing system or patternator. The system was fabricate in UTM's workshop and contains a 300 cm long × 100 cm wide spray table with fifty (6 cm wide × 3 cm deep) V-shaped gutters. During the tests, the spray table was inclined 6° from the horizontal plane. Spray liquid was tap water. Water discharged from the nozzle was supplied from a 140 L pressured bottle, the pressured bottle was pressurized by a compressor and the pressure was adjusted by a pressure regulator. Static single nozzle was mounted on heights 50 cm above the spray table. In front of the table, a set of cylinders (250 mL) was used to collect the liquid from each channel. The weighting method was used to determine the transversal volumetric distributions collected during one minute by using a precision balance. Results of spray volumetric distribution were presented as (ml/min) at two nozzle pressures 200 and 300 kPa. This test was repeated three times.

Statistical analysis

The data from spray tests were collected to analyse the variance using a mathematical model for calculating the mean, standard deviation. The calculations made use of the statistical package of applications Microsoft Excel. The coefficient of variation was estimated by using the standard equation. $CV = (SD/X) \times 100$ Where: CV presents the coefficient of variation%, SD standard deviation and X the mean data. The analysis of variance (ANOVA) was performed on coefficient of variation to determine the effects of nozzle types, angles and pressures, and their interactions at the confidence interval was set at α = 0.05 using SPSS software version 20.

RESULTS

One of the most important requirements on agricultural boom sprayers is to produce a uniform distribution of the applied chemical on the target area. A series of laboratory measurements were conducted to examine the spray uniformity distribution. In total, 8 combinations of nozzle type, angle and pressure were tested corresponding with 24 measurements for standard flat-fan nozzles and even flat- fan nozzles.

Effect of nozzle type, angle and pressure on coefficient of variation (C.V %)

According to results of coefficient of variation in the Table 1 and analysis of variance Table 2 indicated that nozzle type, angle and pressure affect significantly on the spray uniformity distribution. The decreasing nozzle angle tends to increase the coefficient of variation of spray. As well as, increasing of nozzle pressure tend to give a good uniformity of dose. We can observe from Table 1 that the best value of coefficient of variation 42.73% was achieved by using nozzle TPE with angle 80° and at pressure 300 kPa. In general, even flat fan nozzle provided uniform distribution better than standard flat fan

Table 1. Effect of two nozzle types, angles and pressures on coefficient variation (CV).

Type of nozzle	Nozzle angle (°)	Nozzle pressure (kpa)	Coefficient of variation, CV (%)
TPE	80	200	44.10
		300	42.73
	65	200	58.60
		300	57.96
TP	80	200	68.16
		300	65.80
	65	200	74.44
		300	69.70

Table 2. Variance analysis of the effect of two nozzle types, angles and pressures on coefficient variation (CV).

Source	Type III sum of squares	df	Mean square	F	Sig.	Partial Eta squared
Corrected model	2876.926[a]	7	410.989	1481710.385	.000	1.000
Intercept	86962.874	1	86962.874	313520950.952	.000	1.000
nozzle	2091.469	1	2091.469	7540223.201	.000	1.000
angle	596.854	1	596.854	2151794.986	.000	1.000
pressure	31.061	1	31.061	111980.443	.000	1.000
nozzle * angle	143.253	1	143.253	516458.348	.000	1.000
nozzle * pressure	9.605	1	9.605	34628.735	.000	1.000
angle * pressure	1.017	1	1.017	3667.340	.000	.996
nozzle* angle* pressure	3.667	1	3.667	13219.643	.000	.999
Error	.004	16	.000			
Total	89839.804	24				
Corrected total	2876.930	23				

a. R Squared = 1.000 (Adjusted R Squared = 1.000).

nozzle throughout the flat spray pattern.

Effect nozzle type on spray distribution

According to Figure 1, it is clear that the type of nozzle has an important influence on the spray distribution. Banding nozzle TPE achieved the best spray volumetric distribution than broadcasting nozzle TP because the peak under the nozzle center became less acute as the result of increasing the height and size of the neighboring peaks around it and bind them together to reach the same height.

Effect nozzle type and pressure on spray volume distribution

The use of a high pressure instead of the low pressure improves the spray distribution. The spray nozzles TPE8003 and TP8003 at pressure 300 kPa gave better distribution than spray nozzles TPE8003 and TP8003 at pressure 200 kPa respectively as shown in Figure 2.

Effect of nozzle type and angle on spray volumetric distribution

Results in Figure 3, show that using nozzles having spray fan angles of more than 65° reduce spray peak under nozzle center. The spray nozzles TPE8003 and TP8003 gave better distribution than spray nozzles TPE6503 and TP6503 respectively and at the same pressure. Increasing nozzle angle from 65 to 80° reduced coefficient of variation for TPE from 58.60 to 44.10% more than the nozzle TP from74.44 to 68.16% as result to the reducing of the heights of the spray peaks under nozzle center.

Effect of nozzle type, angle and pressure on spray volume distribution

The effect of interaction of nozzle type, angle and pressure were also investigated. The interaction among these three factors affect the results of the spray

Figure 1. Spray volumetric distribution of two nozzle types TPE and TP at pressure 300 kPa.

Figure 2. Spray volumetric distribution of two nozzle types TPE and TP at two nozzle pressures 200 and 300 kPa.

distribution and was noted from the static spray distribution test Figure 4 that even-spray flat-fan nozzles TPE were better than that of standard flat-fan nozzles TP in the spray distribution at the same nozzle angles and pressures.

DISCUSSION

A spray patternator was fabricated for the selection of a suitable nozzle type, its angle and pressure to provide uniform distribution of spray liquid above the plant. A spray analysis system or patternator measurement would probably be sufficient to accurately evaluate the static spray volumetric distribution. The two nozzles selected for this study were standard flat-fan nozzle normally recommended for broadcast spraying of pesticides in row-grown crops and even-spray flat-fan nozzle which was specially developed for band spraying. From laboratory experiments results, it was noted that the

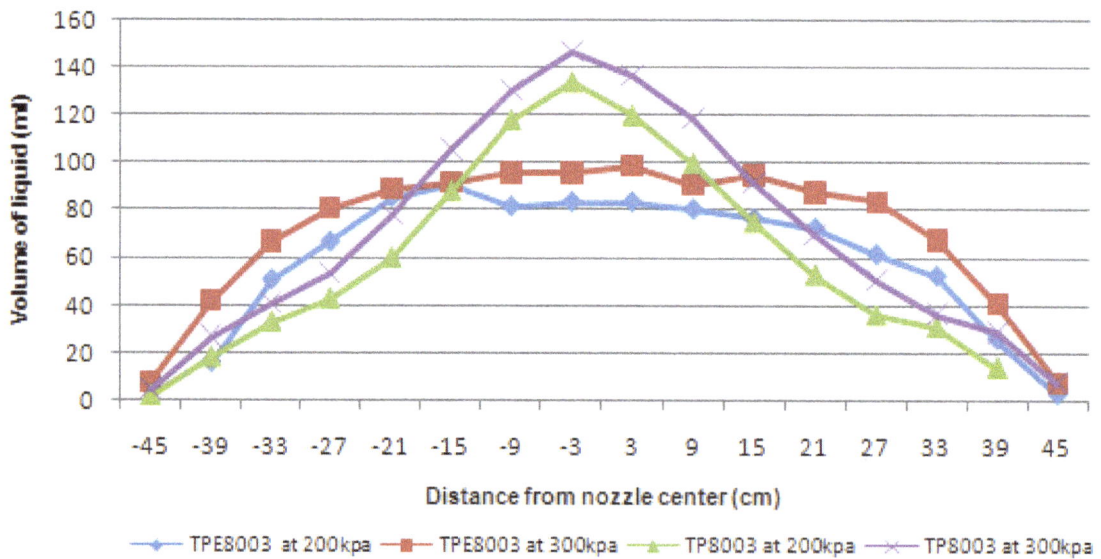

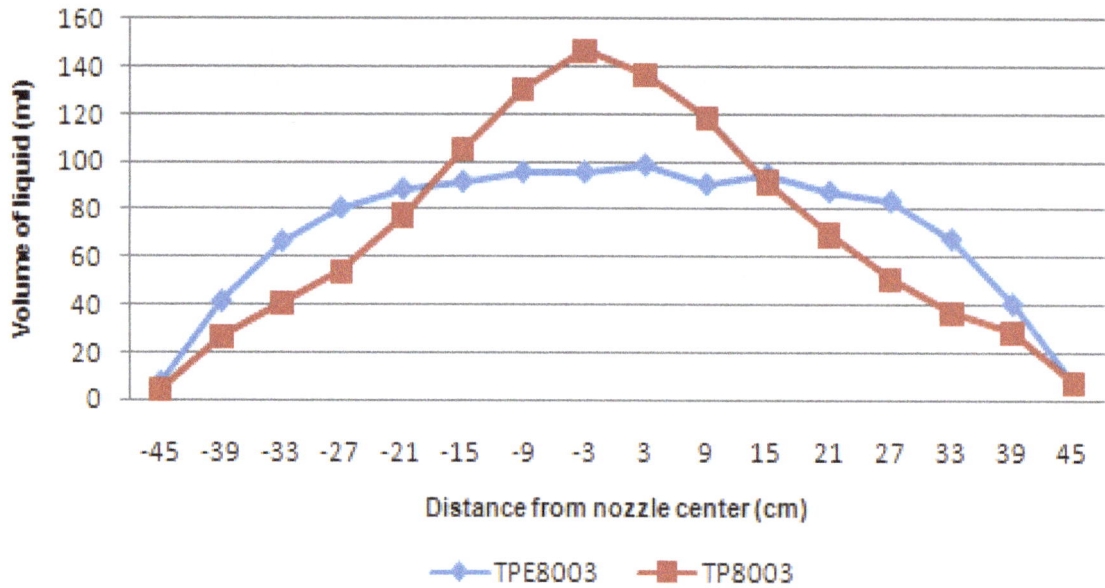

Figure 3. Spray volumetric distribution of two nozzle types TPE and TP with two nozzle angles 65 and 80° at pressure 200 kPa.

Figure 4. Spray volumetric distribution of two nozzle types TPE and TP with angle 65 and 80° and at two pressures 200 and 300 kPa.

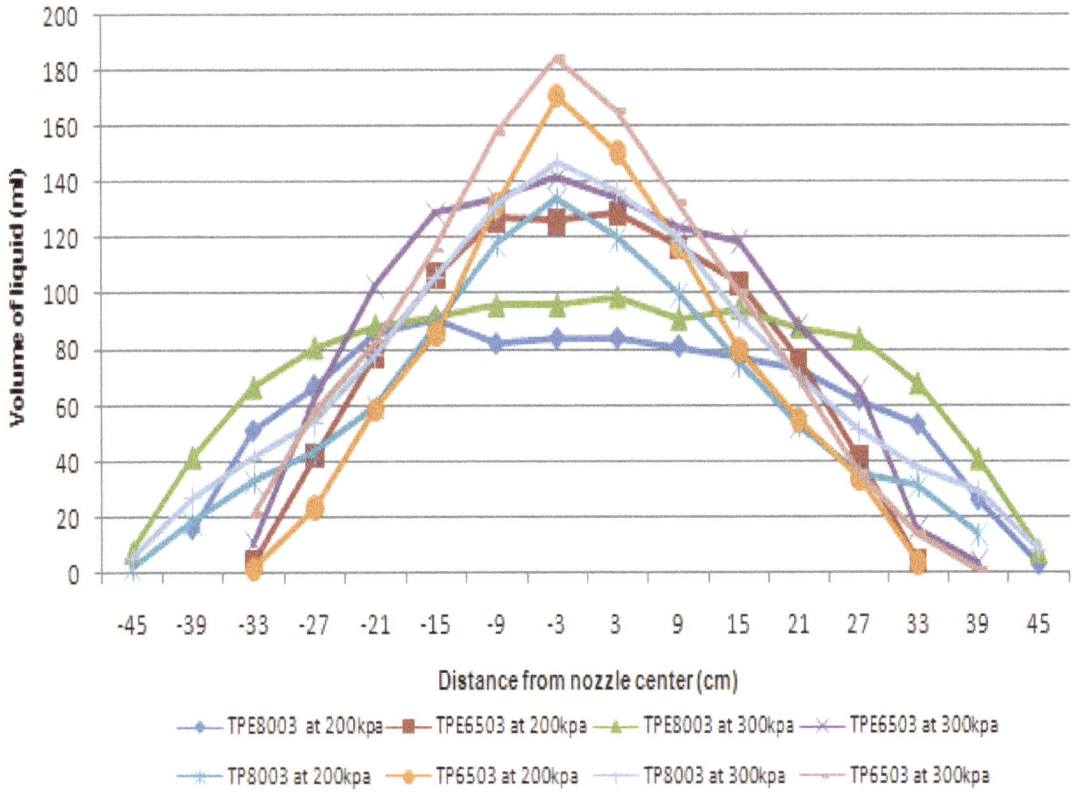

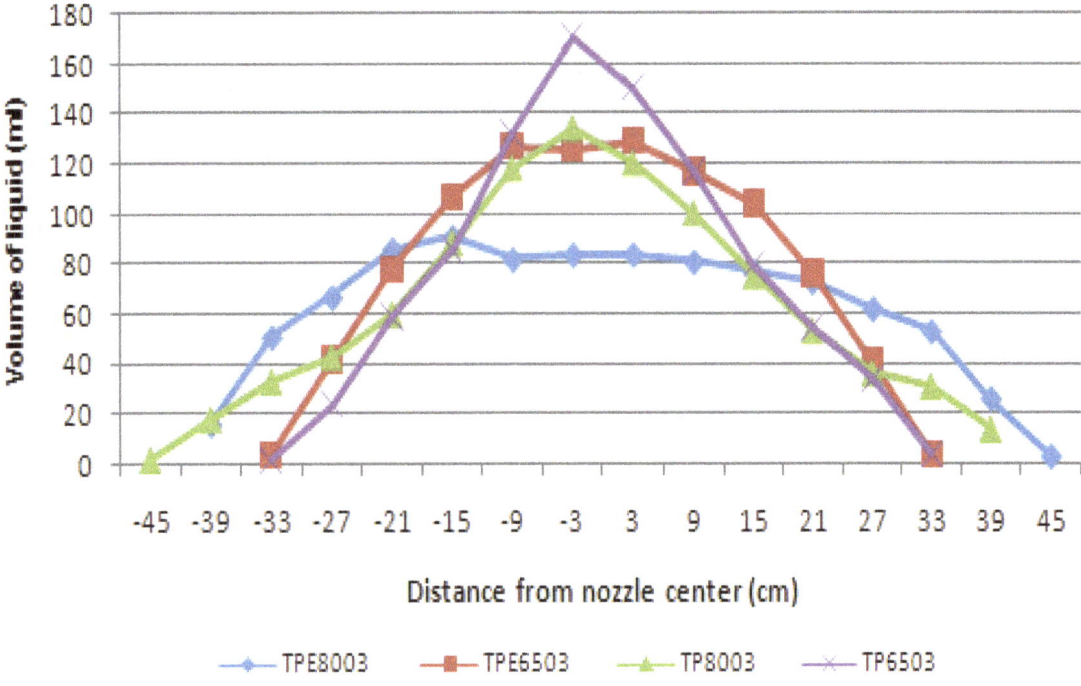

banding nozzle gave the best spray uniformity with the minimum coefficient of variation at all nozzles angles and pressures. The results of the study showed that the combination of the nozzle type TPE, nozzle angle 80° and nozzle pressure 300 kPa gave the best spray volumetric distribution and minimum coefficient of variations 42.73%. Banding nozzle provide spray volumetric distribution better than broadcasting nozzle because they reduce height of the peak under the nozzle center by transferring part of the size of the liquid from the nozzle center and then distribute it at the nozzles sides. Increasing nozzle angle and pressure improve spray uniformity of all broadcasting and banding nozzles. This study supports the use of even flat nozzles in row crop spraying application as a means for improving spray distribution.

ACKNOWLEDGMENT

The authors are thankful to all members of the Aeronautics Laboratory of Mechanical Engineering Faculty, University of Technology, Malaysia for providing facilities for this research.

REFERENCES

ASABE Standards (2006). Terminology and definitions for agricultural chemical application. 53rd ed. The American Society of Agricultural and Biological Engineers. 2950 Niles Road, St. Joseph, MI 49085-9659, USA, pp. 198-201.

Bahadir S, Saim B (2011). Spray distribution uniformity of different types of nozzles and its spray deposition in potato plant. Afr. J. Agric. Res. 6(2):352-362.

Bayat A, Yarpuz-Bozdogan NY (2005). An air-assisted spinning disc nozzle and its performance on spray deposition and reduction of drift potential. Crop Prot. 24:651-960.

Javier AV, Gilberto C, Casimiro DG, Luiz RPT (2008). Effectiveness of the standard evaluation method for hydraulic nozzles employed in stored grain protection trials. Revista. Colombiana de Entomología. 34(2):182-187.

Jean PD, Antoine P, Pierre F (2012). Simulating cov from nozzles spray distribution: A necessity to Investigate spray distribution quality with drift reducing surfactants. International Conference on Agricultural Engineering. CIGR-Ageng, 8-12 Juillet. Valence. Spain.

Krishnan P, Williams TH, Kemble LJ (1988). Technical note: Spray pattern displacement measurement technique for agricultural nozzles using spray table. T. ASAE 31(2):386 -389.

Lardoux Y, Sinfort C, Enfält P, Sevila, F (2007). Test method for boom suspension influence on spray distribution, Part I: Experimental study of pesticide application under a moving boom. Biosyst. Eng. 96(1):29-39.

Lebeau F, Hamza E, Destain M (2000). Automation of a patternator to measure liquid distribution of nozzles. Cahiers Agricultures 9(6):505-509.

Nigar Y, Ekrem A, Ali M, Huseyin Y, Nebile D, Tunahan E, Ebru K (2011). Effect of different pesticide application methods on spray deposits, residues and biological efficacy on strawberries. Afr. J. Agric. Res. 6(4):660-670.

Ozkan HE, Miralles A, Sinfort C, Zhu H, Fox RD (1997). Shields to reduce spray drift. J. Agric. Eng. Res. 67:311-322.

Sidahmed MM, Awadalla HH, Haidar MA (2004). Symmetrical multi-foil shields for reducing spray drift. Biosyst. Eng. 88(3):305-312.

Spraying System CO. (2011). TeeJet Technologies. Catalog 51-m., Wheaton, Illinois, USA, pp. 5-35.

Vern H, Elton S (2004). Spray equipment and calibration. NDSU Extension Service, Agricultural and Biosystems Engineering, North Dakota State University, pp. 1-44.

Wang L, Zhang N, Slocombe JW, Thierstein GE, Kuhlman DK (1995). Experimental analysis of spray distribution pattern uniformity for agricultural nozzles. Appl. Eng. Agric. 11(1):51-55.

Williams K (1981). A guide to band spraying. Beet Rev. 49:33–34.

Womac A, Etheridge R, Seibert A, Hogan D, Ray S (2001). Sprayer speed and venture-nozzle effects on broadcast application uniformity. T. ASAE 44(6):1437-1444.

Review of internal cooling augmentation using baffles

R. M. Majumdar* and V. M. Kriplani

Mechanical Engineering Department, G. H. Raisoni College of Engineering, Nagpur, 440016, India.

Heat transfer augmentation techniques refer to different methods used to increase rate of heat transfer without affecting much overall performance of the system. These techniques are used in heat exchangers. Like ribs, jet impingement and other passive heat transfer enhancement methods, insertion of baffle in a cooling system has been used for various types of industrial applications such as internal cooling systems of gas turbine blades, electronic cooling devices, shell and tube type heat exchangers, thermal regenerators and labyrinth seals for turbo-machines. These techniques broadly are of three types namely: passive, active and compound techniques. This paper reviews the internal cooling augmentation using different type of baffles.

Key words: Heat transfer augmentation technique, baffles, active method, passive method, compound method.

INTRODUCTION

Generally, heat transfer augmentation techniques are classified in three broad categories:

Active method

This method involves some external power input for the enhancement of heat transfer; some examples of active methods include induced pulsation by cams and reciprocating plungers, the use of a magnetic field to disturb the seeded light particles in a flowing stream, etc.

Passive method

These methods generally use surface or geometrical modifications to the flow channel by incorporating inserts or additional devices. For example, use of inserts, use of rough surfaces etc.

Compound method

This is the combination of the aforementioned two methods.

PASSIVE HEAT TRANSFER AUGMENTATION METHODS

Passive heat transfer augmentation methods as stated earlier does not need any external power input. In the convective heat transfer one of the ways to enhance heat transfer rate is to increase the effective surface area and residence time of the heat transfer fluids. The passive methods are based on the same principle. Use of this technique causes the swirl in the bulk of the fluids and disturbs the actual boundary layer so as to increase effective surface area, residence time and consequently heat transfer coefficient in existing system. The following methods are used generally used:

1) Inserts
2) Extended surface
3) Surface modifications
4) Use of additives

INSERTS

Inserts refer to the additional arrangements made as an obstacle to fluid flow so as to augment heat transfer as

*Corresponding author. E-mail: reena101@rediffmail.com

explained earlier. Different types of inserts are:

1) Twisted tape and wire coils.
2) Ribs, baffles and plates.

This paper contributes for review of baffles.

BAFFLES

There are several techniques available to enhance the heat transfer coefficient of gases in internal cooling. One of the common internal cooling enhances techniques is the placement of internal flow swirls, tape twisters or baffles. The swirl insert and tape twister techniques create a significant amount of bulk flow disturbance, and the pressure drop penalties are much higher compared to the gain in heat transfer coefficient. Baffles also create bulk flow disturbance, but unlike tapes or swirls, baffles are discrete objects. Therefore, the flow disturbance created by baffles may be localized, but more intense. Usually the baffle plate is attached to the thermally active surface to augment heat transfer by providing additional fin-link surface area for heat transfer and better mixing.

Use of baffles

The main roles of a baffle in a shell and tube heat exchanger are:

i) Hold tubes in position (preventing sagging), both in production and operation.
ii) Prevent the effects of vibration which is increased with both fluid velocity and the length of the exchanger.
iii) Direct shell-side fluid flow along tube field. This increases fluid velocity and the effective heat-transfer co-efficient of the exchanger.

In a static mixer, baffles are used to promote mixing. In a chemical reactor, baffles are often attached to the interior walls to promote mixing and thus increase heat transfer and possibly chemical reaction rates.

Types of baffles

Implementation of baffles is decided on the basis of size, cost and their ability to lend support to the tube bundles and direct flow. Often this is linked to available pressure drop and the size and number of passes within the exchanger. Special allowances/changes are also made for finned tubes. The different types of baffles include:

i) Segmental baffles (of which single segment is the most common).
ii) Rod or bar baffles (giving a uniform shell-side flow).
iii) Helical baffles (similar to segmental with less pressure drop for same size exchanger).
iv) Longitudinal flow baffles (used in a two-pass shell).
v) Impingement baffles (used for protecting bundle when entrance velocity is high).
vi) Orifice baffles.

Installation of baffles

Baffles deal with the concern of support and fluid direction in heat exchangers. In this way it is vital that they are spaced correctly at installation. The minimum baffle spacing is the greater of 50 mm or one fifth of the inner shell diameter. The maximum baffle spacing is dependent on material and size of tubes. The Tubular Exchanger Manufacturers Association sets out guidelines. There are also segments with a "no tubes in window" design that affects the acceptable spacing within the design. An important design consideration is that no recirculation zones or dead spots form - both of which are counterproductive to effective heat transfer. Some of the common "baffle configurations" are shown in Figures 1 to 4. Examples of baffle installation are shown in Figures 5 to 6.

REVIEW OF WORK CARRIED OUT

Duttaa et al. (2004) investigated the local heat transfer characteristics and the associated frictional head loss in a rectangular channel with inclined solid and perforated baffles. The main objective of the study was to augment both local and global heat transfer behavior of a gaseous fluid (air) by placement of two inclined baffles. Since the flow disturbances and wakes generated by the upstream inclined baffle can potentially affect the performance of the downstream baffle, an average heat transfer performance is considered to cover the entire heated length. The local Nusselt number ratio with two inclined baffles significantly depends on the arrangement (orientation, perforation, and position of the baffles) used. Two inclined baffles augment the local heat transfer coefficient for a longer region of interest. The overall heat transfer coefficient is much higher with two inclined baffles than that with a single baffle placed in the same channel. The average Nusselt number can be as high as 5.0 times the average Nusselt number of a smooth channel. Localized high heat flux zones can be effectively cooled with properly designed perforated baffles in those regions. The local Nusselt number ratio is not a strong function of flow Reynolds number. However, in a particular arrangement the friction factor ratio increases with increase in the flow Reynolds number. For two inclined baffle cases, the frictional head loss is much higher than that of a single baffle arrangement. Moreover, in two baffle cases the friction factor ratio is larger if the second baffle is attached to the bottom plate instead of the top heated surface. Pulvirenti et al. (2010) experimentally studied on saturated flow boiling heat

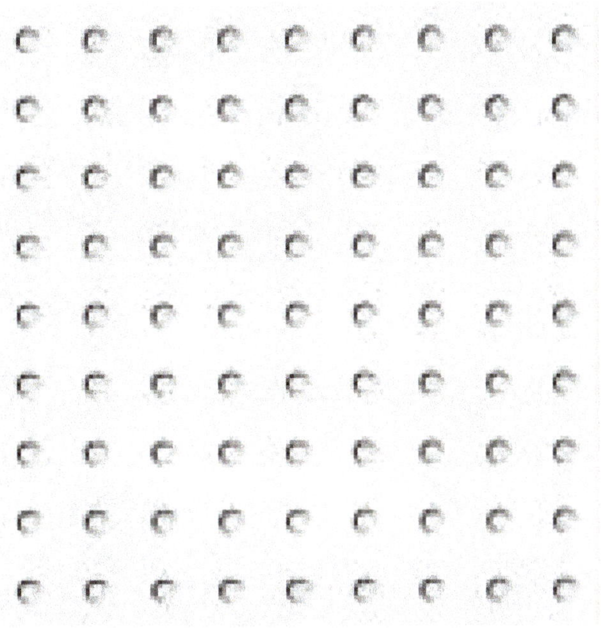

Figure 1. Perforated baffles.

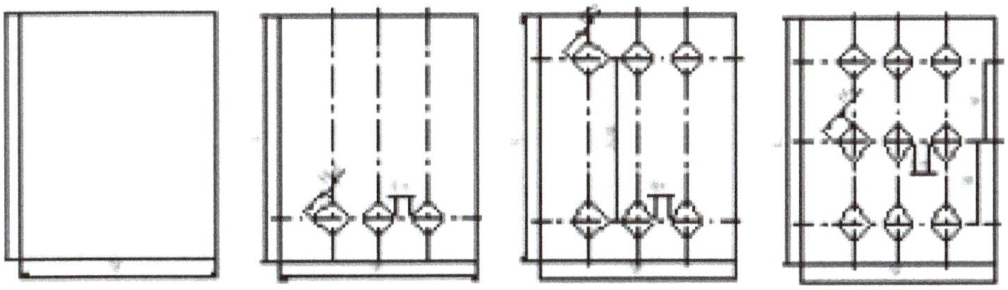

Figure 2. Baffle plate configuration.

Figure 3. Porous baffle.

Figure 4. Solid baffles.

Figure 5. Tape with attached baffle.

Figure 6. Baffles in rectangular duct.

transfer of HFE-7100 in vertical rectangular channels with offset strip fins was presented. The experiments had been carried out at atmospheric pressure, over a wide range of vapour quality and heat fluxes up to 1.8 × 10 W/m^2. The local boiling heat transfer coefficient has been obtained from experiments and analyzed by means of Chen superposition method. In the present study, experimental investigations have been performed to analyze the flow boiling heat transfer of HFE-7100 in vertical rectangular channels with offset strip fins. Two different regimes have been detected.

The first is convective boiling regime, where heat transfer coefficient depends on quality and mass flux but is independent of heat flux. Under this regime, the correlation found by Feldman et al. (2000), together with as hoc single-phase Colburn factor, give a good agreement with our experimental results. The second is nucleate boiling regime, with high heat flux, where the heat transfer coefficient depends on heat flux but is independent of quality and mass flux. Under this regime, correlations found by Kim and Sohn (2006) gives a good agreement with these experimental results. Warrier et al. (2002) performed experiments with the channels oriented horizontally and uniform heat fluxes applied at the top and bottom surfaces. The parameters that were varied during the experiments include the mass flow rate, inlet liquid sub cooling, and heat flux. Additionally, in these experiments, the single and two-phase pressure drop across the channels was also measured. A correlation has been developed for two-phase flow pressure drop under sub cooled and saturated nucleate boiling conditions. Furthermore, two new correlations are proposed one for sub cooled flow boiling heat transfer and the other for saturated flow boiling heat transfer.

Single-phase forced convection

The variation of Nusselt No. for single-phase forced convection as a function of the axial distance. The numerical results in the textbook by Kays and Crawford (1993); similar results were obtained for all the other test cases. It can be seen that the measured values agree quite well with the numerical results, especially near the fully developed region. A comparison of the measured single-phase pressure drop with that predicated for laminar flow in rectangular channels. The fanning friction factor used in predicting the pressure drop was f = 17.25/Reynold No. It was seen that there is good agreement between the experimentally measured and analytically predicted values, especially at lower mass fluxes.

Two-phase pressure drop

The two-phase flow conditions in the channel were assumed to exist downstream of the location where onset of nucleate boiling was observed to occur. The location for onset of nucleate boiling was established as the location where a large increase in the heat transfer coefficient beyond the single-phase value was observed. Aroon et al. (2006) numerically predicted the hydrodynamic and thermally developed turbulent flow for a stationary duct with square ribs aligned normal to the main flow direction. The capability of the detached eddy simulation (DES) version of the 1988 k-w model has been validated in predicting the turbulent flow field and the heat transfer in a complete two pass channel. Results of mean flow quantities, secondary flows, friction and heat transfer were compared with experiments and large-eddy simulations (LES). In the current study, detached eddy simulation (DES) is carried out in a complete two-pass channel with 12 ribs in the first and second passes, connected by a 180° bend. The flow in the first pass of the duct is compared with LES computations by Sewall and Tafti (2004a). Experiments by Rau et al. (1988), the flow in the 180° bend is compared with LES results in the bend by Sewall and Tafti (2005) and experimental results by Nan et al. (1988) and Sewall et al. (?) all physical phenomena characterized by mean and turbulent rms quantities were reproduced with excellent quantitative accuracy. Byongjoo et al. (2006) experimentally studied on saturated flow boiling heat transfer of R113 was performed in a vertical rectangular channel with offset strip fins. Two-phase pressure gradients and boiling heat transfer coefficients in an electrically heated test section were measured for the quality range of 0 to 0.6, mass flux range of 17 to 43 kg/m^2.

Two-phase frictional multiplier was determined as a function of marginally parameter. A superposition method for the flow boiling heat transfer coefficient that included the contribution of saturated nucleate boiling was verified also for flow boiling in a channel with offset strip fins. Two-phase frictional multiplier and the local boiling heat transfer coefficients were measured and correlated. Measured values of two-phase frictional multiplier in a channel with offset strip fins were higher by as much as 80% than those in round tubes. Kang-Hoon et al. (2003) carried out experiment to measure module average heat transfer coefficients in uniformly heated rectangular channel with wall mounted porous baffles. Baffles were mounted alternatively on top and bottom of the walls. Heat transfer coefficients and pressure loss for periodically fully developed flow and heat transfer were obtained for different types of porous medium (10, 20 and 40 pores per inch) with two window cut ratios and two baffles thickness to channel hydraulic diameter ratios. The experimental procedure was validated by comparing the data for the straight channel with no baffles with those in the literature [Publications in Engineering, vol. 2, University of California, Berkeley, 1930, 443, Int. Chem. Eng. 16(1976): 359. The use of porous baffles resulted in heat transfer enhancement as high as 300% compared to heat transfer in straight channel with no baffles.

Experimental procedure was validated by making heat transfer measurements for flow through a straight channel without any baffles. The variation of average Nusselt number with Reynolds number for fully developed flow in straight channels. Experimental data is compared against the correlations available in the literature. The average Nusselt number for fully developed flow in a straight channel was compared with results obtained from correlation of Gnielinski (1976) and Dittus-Boelter (1930). The maximum difference for average Nusselt number obtained in the present work and those correlations in the literature of Dittus-Boelter (1930) and Gnielinski (1976) is 5.9%.

Won et al. (2003) experimentally investigated spatially-resolved, local flow structure and surface Nusselt numbers presented for a stationary channel with an aspect ratio of 4 and angled rib tabulators inclined at 45° with perpendicular orientations on two opposite surfaces. Instantaneous flow visualizations and time-averaged flow structural data show a variety of flow phenomena, including the development of increased numbers of multiple, smaller vortex pairs as the Reynolds number increases, strong span wise secondary flow components, which move in opposite directions in the top and bottom halves of the channel, and result in the formation of other secondary flows and vertical motions. The flow structure results include time-averaged distributions of local stream wise vorticity, different components of local velocity, local total pressure, local static pressure, and secondary flow vectors, as well as instantaneous flow visualization images. Flow visualizations show that increased numbers of multiple, smaller vortex pairs develop in the channel as the Reynolds number increases, which implies that energetic flow structures with a wide range of length scales are present in the fully turbulent channel flow and indicates significant augmentations of local surface heat transfer rates. Tsay et al. (2005) numerically investigated the heat transfer enhancement on a vertical baffle in backward facing step flow channel. The effect of the baffle height, thickness and the distance between the baffle and the backward facing step on the flow was studied. They found that an insertion of a baffle into the flow could increase the average Nusselt number by 190%. They also observed that the flow conditions and heat transfer characteristics are strong function of the baffle position.

REFERENCES

Byongjoo K, Byonghu S (2006). "An experimental study of flow boiling in a rectangular channel with offset strip fins." Int. J. Heat Fluid Flow 27:514-521.

Dittus FW, Boelter LMK (1930). Publication in Engineering, University of Calofornia, Berkeley, CA, USA 2:433.

Duttaa P, Hossainb A (2004). Internal cooling augmentation in rectangular channel using two inclined baffles. Int. J. Heat Fluid Flow 26:223-232.

Feldman A, Marvilletb C, Lebouché M (2000). "Nucleate and convective boiling in plate fin heat exchangers". Int. J. Heat Mass Transf. 43:3433-3442.

Gnielinski V (1976). "New equations for heat and mass-transfer in turbulent pipe and channel flow". Int. Chem. Eng. 16:359-368.

Ko KH, Anand NK (2003). "Use of porous baffles to enhance heat transfer in a rectangular channel". Int. J. Heat Mass Transf. 46:4191-4199.

Kays WM, Crawford ME (1993). A textbook of "Convective Heat and Mass Transfer". McGraw-Hill, New York.

Kim B, Sohn B (2006). "An experimental study of flow boiling in rectangular channel with offset strip fins." Int. J. Heat Fluid Flow 27:514-521.

Pulvirenti B, Matalone A, Barucca U (2010). "Boiling heat transfer in narrow channels with offset strip fins: Application to electronic chipsets cooling". Appl. Therm. Eng. 30(14-15):1-8.

Rau G, Çakan M, Moeller D, Arts T (1998). "The effect of periodic ribs on the local aerodynamic and heat transfer performance of a straight cooling channel. "ASIME J. Turbomach. 120:368-375.

Sewall EA, Tafti DK (2004). Large eddy simulation of the developing region of a rotating ribbed internal turbine blade cooling channel. Paper GT2004–53833, ASME Turbo Expo, Vienna, Austria.

Sewall EA, Tafti DK (2005). Large Eddy Simulation of Flow and Heat Transfer in the 180° Bend Region of a Stationary Ribbed Gas Turbine Internal Cooling Duct. ASME Turbo Expo, GT2005-68518 pp. 481-493.

Tsay YL, Chang TS, Cheng JC (2005). "Heat transfer enhancement of backward-facing step flow in a channel by using baffle installed on the channel wall". Acta Mech. 174:63-76.

Warrier GR, Pan T, Dhir VK (2002). "Heat transfer and pressure drop in narrow rectangular channels". Exp. Therm. Fluid Sci. 26:53-64.

Won SY, Mahmood GI, Ligrani PM (2003). "Flow structure and local Nusselt number variations in a channel with angled crossed-rib turbulators". Int. J. Heat Mass Transf. 46:3153-3166.

15

Design analysis of solar bi-focal collectors

Abdulrahim A. T.[1]*, Diso I. S.[2] and Abdulraheem A. S.[2]

[1]Department of Mechanical Engineering, University of Maiduguri, Nigeria.
[2]Department of Mechanical Engineering, Bayero University Kano, Nigeria.

This study carried out the design analysis of solar bi-focal collectors with the basic units comprising the paraboloid concentrators, receivers and support/connectors. The design of the receivers is such that it works on thermo-siphon principle while the heat energy requirement for each receiver is 650 kJ. Solar energy required to provide the needed power input in the collector's receiver is amounted to 0.967 kJ/s. The results of the analysis revealed that each collector has diameter of the receiver of 0.3 m, aperture diameter of 1.4 m and internal surface area of 1.53 m^2. The materials and dimensions of parts were selected based on the design specifications. The design analysis will serves as basis for the development of a solar tracking bi-focal collector system for possible steam generation.

Key words: Solar, collector, concentrator, receiver, heat energy.

INTRODUCTION

Solar radiation has been identified as the largest renewable resource on earth (Muller-Steinhagen, 2003). The maximum intensity of solar radiation at the earth's surface is about 1.2 kW/m^2 but it is encountered only near the equator on clear days at noon. Under these ideal conditions, the total energy received is from 6 to 8 kWh/m^2 per day (Androsky, 1973; Spillman et al., 1979; Halacy, 1980). Solar energy can be captured in different ways; also many different methods for collecting the solar energy from incident radiation are available. Solar thermal power devices use the sun rays to heat fluid, from which heat transfer system may be used to produce steam. The steam in turn is converted into mechanical energy in a turbine and into electricity from a conventional generator coupled to the turbine (Eastop and Mc Conkey, 1993).

Paraboloidal concentrators have the ability to raise various absorbers and working fluids to high temperatures. Broman and Broman (1996) noted that the concentration factors needed for temperatures of several 100°C were obtained easily with parabolic troughs if the mirrors were made with high precision. They also observed that if one chooses a dish-type mirror that concentrates the light onto a focal spot instead of a focal line, there is a rather large freedom to appropriate the parabolic surface and still obtain the desired concentration ratio. The maximum concentration factor and temperature attainable in practice depends on the aperture size, reflectivity and accuracy of the surface contour, and the degree to which the concentrator approximates a true paraboloidal geometry (Ong, 1974). The basic objective of this study is to design solar bi-focal collectors that can heat water for continuous production of steam at a temperature above 100°C.

*Corresponding author. E-mail: engrabdulrahimat@yahoo.com.

MATERIALS AND METHODS

Energy calculation in order to determine the heat energy required to generate steam at the required temperature in the receiver was carried out using the formulae (Abbott, 1977; Sayigh, 1977):

$$Q = M_w C_w \Delta\theta \quad (1a)$$

$$\text{Power} = \text{Energy}/\text{Time}, \quad (1b)$$

Where Q is the heat energy (J), M_w is the mass of fluid, C_w is the specific heat of fluid (kJ/kg.K) and $\Delta\theta$ is change in temperature.

Determination of solar energy required to provide the needed power input in the collector's receiver using expression given by Sayigh (1977) is:

$$Ei = \tfrac{1}{4}\pi D^2 \times Ha \quad (2)$$

Where Ei is the solar energy, D is paraboloid aperture diameter (m), Ha is intensity of direct solar radiation (W/m^2).

Calculation of collector design parameters and component parts material selections can be done using the following equations (Backhouse et al., 1962; Sayigh, 1977; Stroud, 1988).

Aperture diameter; $D = \sqrt{\frac{4E_{i1}}{\pi H_a}}$ (3)

Equation of a parabola, $x^2 = 4fy$ (4)

Where x and y are coordinates of the points along the curve of the parabola and f is the focal length.

Internal surface area of the parboloid,

$$A_p = \int_{y_1}^{y_2} 2\pi \left[1 + \left(\frac{dx}{dy}\right)^2\right]^{\frac{1}{2}} dy \quad (5)$$

Concentration ratio, $CR = \left(\frac{D}{d}\right)^2$ (6)

Where d is the receiver diameter (m)

Focal length, $f = \frac{D(1+\cos\emptyset)}{4\cos\emptyset}$ (7)

Where ø is the rim half angle

Depth of the dish, $h = f - \frac{D}{2}\cot\emptyset_{opt}$ (8)

Where $ø_{opt}$ is the optimum rim angle

The Length of the arc of the Parabola,

$$S = \int_{x_1}^{x_2} \left[1 + \left(\frac{dy}{dx}\right)^2\right]^{\frac{1}{2}} \quad (9)$$

Determination of the (x, y) coordinates of the points along the curve of the parabola was done by employing parabola calculator version 2.0 program developed by Mike (Light Concentrated From Below, 2004).

Receivers' support design was carried out using the formulae given by Khurmi and Gupta (2004): The Euler formula for the crippling or buckling load W_{cr}:

$$W_{cr} = \frac{\pi^2 EI}{L^2} \quad (10)$$

Where E is Modulus of elasticity (N/m^2), I is moment of inertial ($kg.m^2$) and L is equivalent length of column (m).

The Rankine's formula is expressed as:

$$W_{cr} = \frac{\sigma c \times A}{1 + a\left[\frac{L}{K}\right]^2} \quad (11)$$

Where σ_c is crushing stress or yield stress in compression, a is Rankine's constant, A is area of cross section (m^2), L is equivalent length of column (m), and K is least radius of gyration

RESULTS AND DISCUSSION

Design consideration

For this study, bi-focal solar collector system is conceived for possible generation of steam. Maiduguri, a semi-arid area, located on latitude 11.85° north and longitude 13.08° east with altitude of 354 m and annual mean daily global solar radiation on a horizontal surface of 6.176 kWh/m^2-day (Layi Fagbenle, 1990; Gisilanbe, 1995; Onyegegbu, 1993; Onyegegbu, 2003) has been chosen as the study area. The inlet water temperature to the first collector is taken as 35°C, an overall mean monthly maximum temperature over a period of 7 years for Maiduguri (NIMET, 2003).

Energy calculation

Heat energy required to generate steam at a temperature above 100°C should be useful energy delivered to the working fluid in the absorbers. In order to estimate the required heat energy, the following computation was carried out based on the assumptions that:

1) The first collector is expected to generate steam at a temperature of 100°C from inlet water temperature of 35°C.
2) The second collector is to raise steam temperature from 100 to 235°C.
3) The absorbers are of flat metal plate with arrangement for circulating water pass through a continuous tube bent in a sinusoidal shape fastened to the flat plate facing the reflectors.

For this study, the capacity of the first collector absorber pipe was taken as 0.25×10^{-3} m^3 made of 15 mm diameter and 1 mm thick galvanised iron because of its ability to withstand high temperature and pressure. The main water way to the receiver is a 15 mm galvanised iron pipe from the centre of the reflector along the focal line. A parallel pipe conveys heated water to the second collector system. The design of the receivers is such that they work on thermo-siphon principle. The flat plate receivers' diameters on which the absorbers are fixed are taken as 0.3 m.

The receiver (Figure 1) is made up of flat plate of mild steel sheet of 1 mm thickness and 0.3 m diameters, galvanised iron absorber pipe of 15 mm diameter and 1.414 m length in coil form, mild steel sheet of 1 mm thickness to serve as cylindrical part enclosure, insulator of 20 mm thickness behind the plate, insulator lagging from mild steel sheet of 1 mm thickness, and glass cover

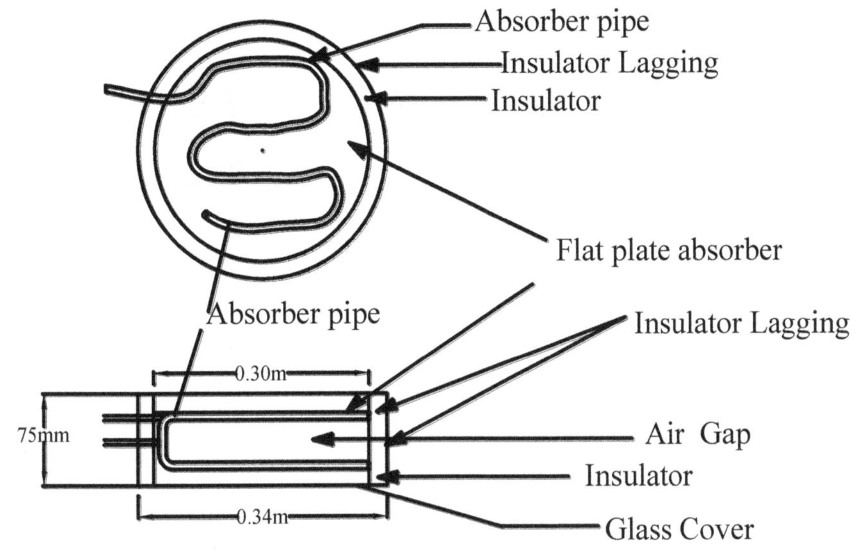

Figure 1. Receiver's components.

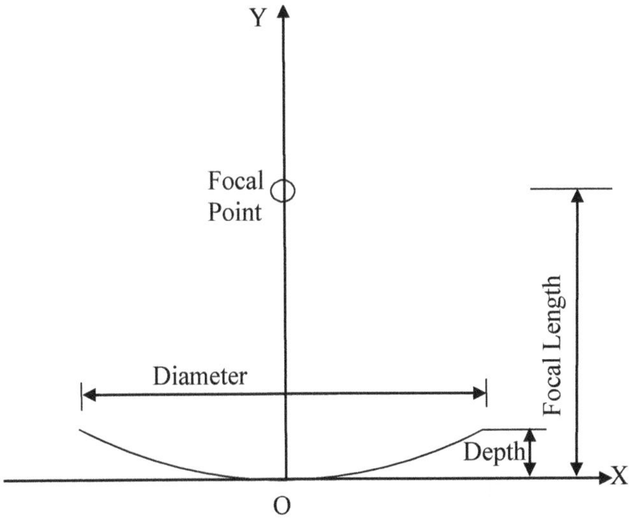

Figure 2. Parabola curve and coordinates
x -700.00, y 145.00; x -612.50, y 111.02; x -525.00, y 81.56; x -437.50, y 56.64; x -350.00, y 36.25; x -262.50, y 20.39; x -175.00, y 9.06; x -87.50, y 2.27; x 0.00, y 0.00; x 87.50, y 2.27; x 175.00, y 9.06; x 262.50, y 20.39; x 350.00, y 36.25; x 437.50, y 56.64; x 525.00, y 81.56; x 612.50, y 111.02; x 700.00, y 145.00.

over the absorber leaving an air gap of about 50 mm. Input values used for some quantities are as follows:

Inlet water temperature to the first collector, T_w = 35°C
Density of water, ρ_w = 1000 kg/m^3
Specific heat capacity of water at 35°C, C_w = 4.179 kJ/kg.K
Specific latent heat of steam, L_s = 2260 kJ/kg
Specific heat capacity of steam at 235°C, C_{s1} = 3.69 kJ/kg.K
Specific heat capacity of galvanised iron, $C_{G.I}$ = 0.473 kJ/kg.K
Density of galvanised iron, $\rho_{G.I.}$ = 7, 801 kg/m^3

Design results

A) Heat energy required to convert 0.25 x 10^{-3} m^3 of water at 35°C to steam at 100°C in the absorber of the first collector was calculated to be 650 kJ
B) Power input required, P_1 if heat energy of 650 kJ is to be made available in the absorber under 15 min (resident period) is 0.722 kJ/s
C) Solar energy required to provide the needed power input in the collector's receiver, E_{i1} is 0.967 kJ/s.
D) Dimension of the collectors.

For this design, the dimensions of the first and second collector are the same. These are
1) Aperture diameter of the paraboloid, D = 1.4 m
2) Focal length of the paraboloid, f = 0.8428 m
3) Diameter of the receiver, d = 0.3 m
4) Concentration ratio of the paraboloid, CR = 22
5) Length of the Latus rectum, LR = 3.3712 m
6) Internal surface area of the paraboloid, Ap = 1.53 m^2
7) Length of arc of the parabola, S=0.73 m

E) The (x, y) coordinates of the points along the curve of the parabola are as follows and the schematic diagram is shown in Figure 2.

Material selection and specifications

1) Flat sheet metal of 0.5 mm thickness was selected for the shell (Figure 3),

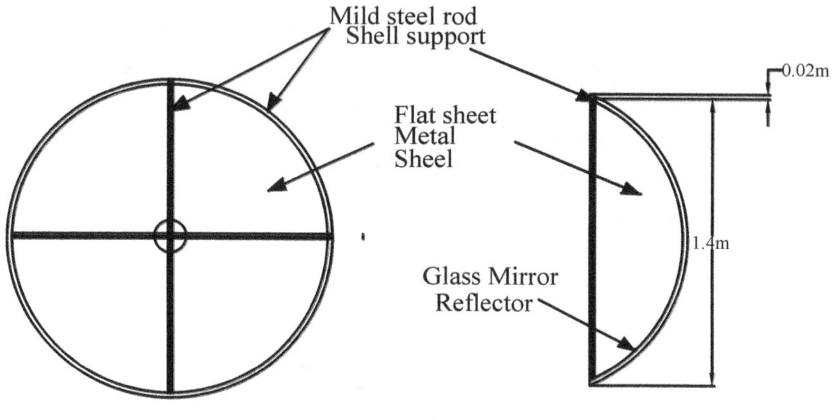

Figure 3. Paraboliod concentrator components.

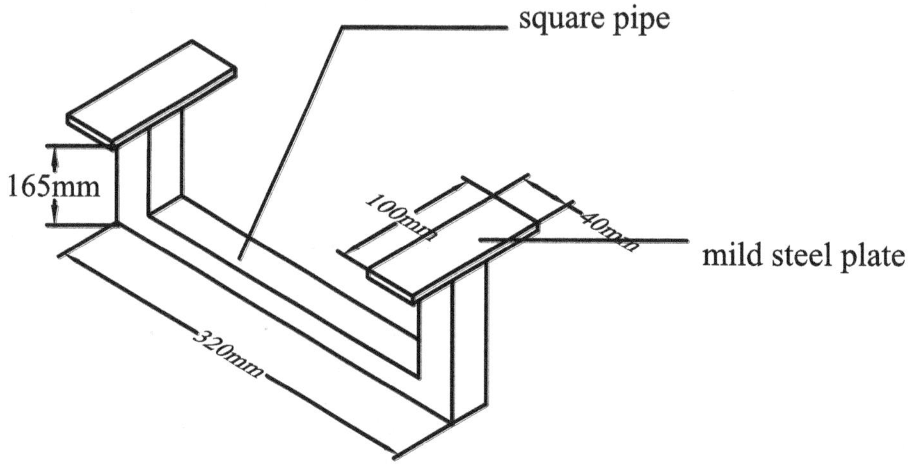

Figure 4. Hinge support.

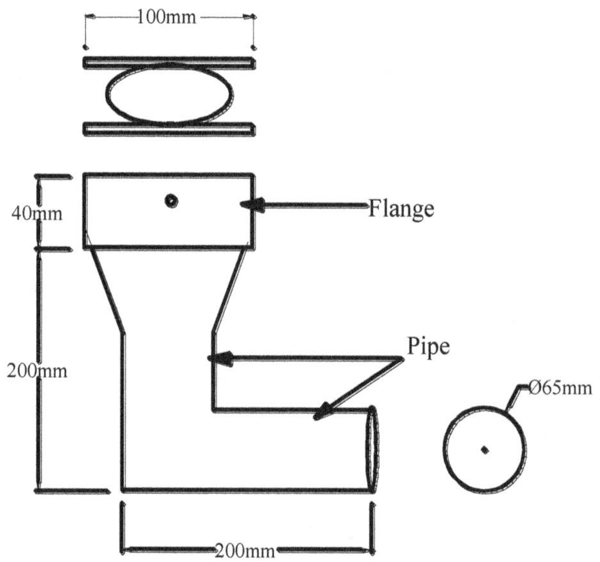

Figure 5. Connector for concentrator to the rotating support.

2) Mild steel rod of 4 mm radius was selected for the shell's support (Figure 3)

3) Glass mirrors reflector (Figure 3) of 3 mm thickness was selected for the reflective surface of the concentrators with a total area of 1.53 m^2.

4) Hinge/support for positioning collector relative to the axis of rotating support (Figure 4). Two numbers mild steel bar of 3 mm thickness, 100 mm length and 40 mm breath to be joined with 40 mm square pipe of 0.65 m length and thickness of 1 mm was selected.

5) Connector for the concentrator to the rotating support (Figure 5) is made up of mild steel pipe of internal diameter 65 mm, 2 mm thickness and length of 0.2 m, and edge flanges (4 numbers) of mild steel flat bar of 3 mm thickness, length of 100 mm and breath of 40 mm.

6) Receiver support

For proper support and stability, three number supports were selected for the receiver (Figure 6). The supports are to be equally spaced and connected between

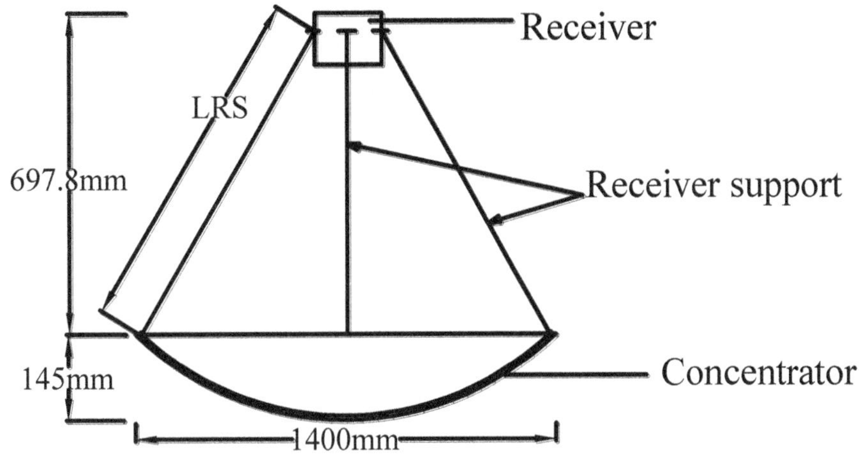

Figure 6. Receiver's support arrangement.

Table 1. Paraboloid concentrator components specifications.

S/No.	Component description	Material specification	Quantity	Mass (kg)
1.	Shell	Mild steel 0.5 mm thickness	1.53 m^2	5.28
2.	Shell support	Mild steel rod of 8 mm diameter	20.866 m	4.10
3.	Hinge/support	Mild steel flat bar of 3 mm thickness and 40 mm breath	0.2 m	0.18
		Square pipe of 40 mm square and thickness 1 mm.	0.65 m	0.23
4.	Connector	Mild steel pipe of 65mm internal diameter and 2 mm thickness	0.2 m	0.63
5.	Reflector	Mild steel flat bar of 3 mm thickness and 40 mm breath	0.3 m	0.9
		Back-silvered glass mirrors of 3 mm thickness	1.53 m^2	11.38
6.	Other: Bolts and nuts screw	Mild steel	Lot	0.5

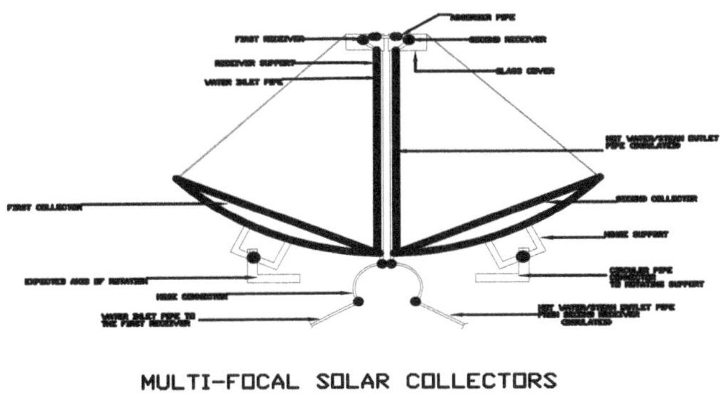

Figure 7. Solar bi-focal collectors assembly.

concentrator aperture diameter edge and the base of the receiver. One of the supports is designed to extend from the concentrator edge to the rotating support.

Taking the larger of the two values arrived at using Euler formula and Rankine formula gives the rod diameter, d = 4.257 mm. For this design, mild steel rod of 8 mm diameter was selected for the support of the receiver since it may be subjected to deflection due to varying collector position and wind actions. Table 1 shows the collectors component parts specification and the assembly of the solar bi-focal collectors is shown in Figure 7.

Conclusion and Recommendations

Solar collectors comprising basic units such as the paraboloid concentrators, receivers and support/connectors have been analysed in this study. The results indicated that two number solar parabolic concentrators each having diameter of the receiver as 0.3 m, aperture diameter of 1.4 m and internal surface area of 1.53 m^2 will be adequate to collect the solar energy required to provide the needed power input of 0.967 kJ/s in the collector's receiver. The analysis will serve as a basis in designing the other component parts such as the tracking and support systems of a solar tracking bi-focal collector system. It is therefore recommended that the procedures used in this work should be applied in the design of solar collectors of parabolic type of other magnitude.

REFERENCES

Abbott AF (1977). Ordinary Level Physics. 3rd edition, Heinemann Educational Books, London. pp. 203-219.

Androsky A (1973). Uses of the Sun in the Service of Man. Aerospace Corporation El Segundo, CA, Report No: ATR –74 – (9470).

Androsky A (1973). Uses of the Sun in the Service of Man. Aerospace Corporation El Segundo, CA, Report No: ATR –74 – (9470).

Backhouse JK, Houldsworth SPT, Cooper BED (1962). Pure Mathematics. A Second Course, Longmans Green and Co. Ltd., U. K. pp. 175-186.

Broman L, Broman A (1996). Parabolic Dish Concentrators Approximated by Simple Surfaces. Solar Energy, 57(4): 317-321.

Eastop TD, McConkey A (1993). Applied Thermodynamics for Engineering Technologists. Addison Wesley Longman Limited, 5th Edition, England. pp. 234-259

Gisilanbe AM (1995). Weather, Health and Development in the Semi-arid Tropics. A Paper Presented at the International Conference on Climate change, Global Warming and Environment Degradation in Africa, Lagos – Nigeria. pp. 1-10.

Halacy DS (1980). Solar Energy and Biosphere. In Solar Energy Technology Handbook, Part A: Engineering Fundamentals, ed. W. C. Dickinson and P.N. Cheremisinoff. Marcel Dekkar, New York. pp. 1-8.

Khurmi RS, Gupta JK (2004). A Text book of Mechine Design. Eurasia Publishing House (Put.) Ltd., Ram Nagar, New Delhi – 110005. pp 537-558.

Layi FR (1995). Fourier Analysis of Climatological Data Series in a Tropic Environment. International J. Energy Res., 19: 117-123.

Muller SH (2003). Concentrating Solar Power: A Vision for Sustainable Electricity..Generation. www.ecm.auckland.ac.nz/conf/Auckland-2.pdf . pp. 1-31.

NIMET (2003). Nigeria Metrological Agency, Maiduguri Zonal Office. Former Zonal Metrological Inspectorate, Federal Ministry of Aviation, Maiduguri, Nigeria. pp. 1-13

Ong KS (1974). A Finite Difference Method to Evaluate the Thermal Performance of a Solar Water Heater. Solar Energy, 16: 137-147.

Onyegegbu SO (1993). General Overview of Solar Energy Availability and Applications Paper presented at the National Centre for Energy Research and Development, University of Nigeria Nsukka.

Onyegegbu SO (2003). Renewable Energy Potentials and Rural Energy Scenario in Nigeria. Renewable Energy for Rural Industrialization and Development in Nigeria. UNIDO. pp. 5-27.

Sayigh AAM (1977). Solar Energy Engineering. Academic Press Inc., New York. pp. 233-262.

Spillman CK, Robbins FV, Hines RH (1979).Solar Energy for Reduction Fossil Fuel Usage. Farrowing Houses Agricultural Experiment Station Paper No: 79-179A, Kansas State University, Manhattan, KS.

Stroud KA (1988). Engineering Mathematics. Macmillan Education Ltd, England.

A new experimental technique to determine heat transfer coefficient and pressure drop in smooth and micro-fin tube

S. N. Sapali[1]* and Pradeep A. Patil[2]*

[1]Department of Mechanical Engineering, Government College of Engineering, Shivaji Nagar, Pune, Maharashtra 411005, India.
[2] Mechanical Engineering Department, Aissms College of Engineering, Pune University, Kennedy Road, Near R. T. O, Pune, Maharashtra, 411001 India.

An experimental test facility is designed and built to calculate condensation heat transfer coefficients and pressure drops for HFC-134a in a 10.21 mm ID smooth and 8.56 mm ID micro-fin tube. The main objective of the experimentation is to investigate the enhancement in condensation heat transfer coefficient and increase in pressure drop using micro-fin tube for different condensing temperatures and further develop an empirical correlation for heat transfer coefficient and pressure drop, which takes into account, variation of condensing temperature and mass flux of refrigerant. The experimental setup has a facility to vary the different operating parameters such as condensing temperature, cooling water temperature, flow rate of refrigerant and cooling water etc. and study their effect on heat transfer coefficients and pressure drops. The hermetically sealed reciprocating compressor is used in the system, thus the effect of lubricating oil on the heat transfer coefficient is taken in to account. This paper reports the detailed description of design and development of the test apparatus, control devices, instrumentation, experimental procedure and data reduction technique. It also covers the comparative study of experimental apparatus with the existing one from the available literature survey. The condensation and pressure drop of HFC-134a in a smooth tube are measured and the values of condensation heat transfer coefficients for different mass flux and condensing temperatures were obtained using modified Wilson plot technique with correlation coefficient above 0.9. The condensation heat transfer coefficient and pressure drop increases with increasing mass flux and decreases with increasing condensing temperature. The results are compared with existing available correlations for validation of test facility. The experimental data points have good association with few available correlations. The condensation and pressure drop of HFC-134a in a micro-fin tube are also measured and the values of condensation heat transfer coefficients obtained. The enhancement and penalty factors of HFC-134a are 1.24 - 2.42 and 1 - 1.77 respectively.

Key words: Experimental technique, micro-fin tube, condensation heat transfer, pressure drop, heat transfer enhancement.

INTRODUCTION

In-tube condensation is quite common in refrigeration and air-conditioning applications. It is the binding choice for air-cooled and evaporative condensers. In-tube condensation is often thought of as a process of film-wise condensation (less effective than drop-wise condensation) (Kern, 2003) of vapor inside a tube, hence air-cooled condensers are less effective. Another draw back of air-cooled condenser is that it operates at a greater condensing temperature than water-cooled condenser; hence the compressor (and the refrigeration system) delivers 15 to 20% lower capacity (Arora, 2004). Therefore one has to use a larger compressor to meet the requirement. At the same time, the compressor consumes

*Corresponding author. E-mail: papatil73@yahoo.co.in, padu_patil@rediffmail.com.

greater power. Hence the air-cooled system has a lower ratio of overall energy efficiency. The augmentation of In-tube evaporation and condensation heat transfer can result in smaller and more efficient evaporators and condensers. Micro-fin tubes (Figure 3) have been successfully implemented in the air-conditioning and refrigeration industries for effectively improving tube-side performance. This success is because of their ability to significantly improve heat transfer coefficient with only moderate increase in pressure drop; hence this augmentation technique shows great potential as an energy saving technique. An experimental program designed to investigate potential augmentation technique has been carried out worldwide as part of a large study of In-tube condensation.

The range of operating parameters used in experimental test facilities developed by different researchers is given in Table 1. It is found that in many test setups, refrigerant pump is used as a circulating device instead of compressor and used for small range of operating conditions. No study found higher condensing temperatures such as 55 - 60°C. Also in very few studies new refrigerants are used. The present test facility overcomes these deficits of the literature survey and achieved the following range:

Mass flux (Gr) = 50 - 800 $kg/s.m^2$
Condensing pressure (Pd) = 7.5 - 16.5 bar (gauge)
Condensing temperature (Th) = 35 - 60°C
Cooling water temperature (Tci) = 2 - 40°C

As for cooling water supply for test, condenser evaporator tank is utilized, no separate chilled water plant is required and heating is achieved with the help of 8 kW capacity heaters which are immersed in the evaporator tank. The test is carried out with HFC-134a refrigerant.

DESCRIPTION OF THE TEST APPARATUS

The test apparatus, as shown schematically in Figure 1 consist of four circuits namely, refrigerant main, auxiliary, cooling water and chilled water circuit. Details of these circuits are given below.

The refrigerant main circuit links compressor to main condenser to expansion valve to evaporator and back to compressor. Compressor used is of hermetically sealed reciprocating type with a cooling capacity of 7.6 kW and suitable for HFC-134a, R-404A, R-407C, R-507A refrigerants. Main condenser is shell and tube type with refrigerant through shell and cooling water through tubes. Thermostatic expansion valve is used as an expansion device. The evaporator is of tank and coil type; with refrigerant flowing through coil and surrounded by water in the tank, heaters are immersed in the tank to provide heat source for evaporator as well as maintain desired water temperature in the tank.

The refrigerant auxiliary circuit links compressor to test condenser to expansion valve to evaporator and back to compressor. All the devices in this circuit are common with main circuit except test condenser. The test condenser is a shell and U bend tube exchanger with the refrigerant flowing inside the inner tube (di = 10.21 mm) and chilled water flowing through the shell of diameter 50.8 mm. Table 2 provides the dimensions of smooth and micro-fin tube. In order to induce turbulence and direct the water flow outside the tubes, baffles are employed. The center to center distance between baffles is called baffle spacing (B). The baffle spacing is not usually greater than shell ID and not less than one-fifth the shell ID. For desired effect it is generally taken as 0.2 Ds or 2 inches whichever is greater. Considering that (B = 2 inches = 50.8 mm) (Kern, 2003), baffles will be of segmental type, also known as 25% cut baffles. The test condenser is designed for maximum loading capacity. The maximum loading condition occurs for 35°C condensing temperature with mass flux of 800 $kg/m^2.s$.

The chilled water is used in test rig which flows in close cycle between evaporator and test condenser. The circuit mainly joins components such as, pump, Rota meter, test condenser evaporator and back to pump. This circuit allows increasing or decreasing the chilled water flow rate with the help of valve according to cooling required in test condenser. The heat absorbed in test condenser is rejected at evaporator. To match the cooling capacity of refrigeration unit extra arrangement of heaters are used. The pump is selected on the basis of maximum flow rate and maximum pressure drop. The pump selected to meet the requirements is 3000 Lph and 28 m head.

The cooling water circuit as shown in Figure 2 is used to cool water circulating from the main condenser; the heat absorbed in the main condenser by cooling water is ejected in the force drought cooling tower and circulated back from the main condenser with the help of pump of capacity 1500 Lph and 2 m head. Plate type valves are used in lines to regulate the flow of refrigerant and water.

Instrumentation

The measurements taken in the system are pressure, temperature and flow at various locations in the apparatus. These measurement points are as follows.

Temperature measurements

1. Before and after the test condenser (refrigerant circuit), in order to measure the degree of superheating and sub cooling during condensation process.
2. Before and after the test condenser (chilled water circuit), to measure chilled water temperatures used for the calculation of heat absorbed by water.
3. To measure the temperature of chilled water in the evaporator thus monitoring the steady state.
4. Before and after evaporator, to measure the refrigerant temperatures, to ensure state of refrigerant.

Pressure measurements

1. At the inlet and outlet of the test condenser, to measure the refrigerant pressures required to calculate the pressure drop across the test condenser, consequently used to calculate the friction factor.
2. At the inlet of compressor, to measure the suction pressure required during analyzing system performance.
3. Mounted on main condenser, to measure condenser pressure, monitor the condensing temperature and to ensure the system balancing when the refrigerant flow rate is changed.

Flow measurements

1. In the auxiliary refrigerant circuit, to measure the refrigerant flow rate in the test condenser, required to calculate Reynolds number and heat rejected by refrigerant.

Table 1. Range of operating parameters used in various test facilities.

S. No	Authors (Year)	Range of experimental parameters covered	Working fluids	Circulating device
1	Yirong Jiang, Srinivas Garimella (2003)	Gr: 200 - 500 kg/m^2.s Pd: 7.5 - 10.5 bar Tci: not given	R-404A, water coolant, steam	Refrigerant pump
2	L.M.Schlager, M. B. Pate, Bergles (1990)	Gr: 75 - 400 kg/m^2.s Pd: 15 - 16 bar Tci: not given	R-22, water-glycol, water	Refrigerant pump
3	J. C. Khanpara, Bergles (1986)	Gr: 197 - 594 kg/m^2.s Pd: fixed pressure 2.41 bar Tci: not given	Refrigerant, water, coolant	Refrigerant pump
5	Wang Fazio (1985)	Gr: 17.14 - 85.55 kg/m^2.s Pd: -6.8 - 11.4 bar Tci: city water at constant temperature	R-12,R-22,cold water, hot water	Open type reciprocating compressor
6	Said and Azer (1982)	Gr: 14.14 - 305.89 kg/m^2.s Pd: 1.32 - 3.05 bar Tci: 11.7 - 35.9°C	R-113, water	Refrigerant pump
7	Stoecker and Kornota (1985)	Gr: fixed flow rate of 0.023 kg/s was maintained. Pd: 4.78 - 6.09 bar Tci: city water at constant temperature	R-114,R-12, cooling water	Refrigerant pump
8	Tichy, Macken and Duval (1985)	Gr: 94.44 - 944.44 kg/m^2.s Pd: 4.8 - 9.3 bar Tci: city water at constant temperature.	R-12, cooling water	Open type reciprocating compressor
9	Keumnam and Sang-Jin Tae (2000)	Gr: 100 - 400 kg/m^2.s Pd: fixed pressure Tci: 11.7 - 35.9°C	R-407C, R-12,	Refrigerant pump
10	Steve J. Eckels and Brian A. Tesene (1999)	Gr: 125 - 600 kg/m^2.s Pd: 8.8 - 11.6 bar Tci: contant temperature water	R-22, R-134a, R-410a	Refrigerant pump
11	Minh Luu And Bergles (1980)	Gr: 86 - 760 kg/m^2.s Pd: 2.41 - 6.55 bar Tci: 10 - 104°C	R-113, water, steam	Refrigerant pump
12	Smit and Meyer (2002)	Gr: 100 - 600 kg/m^2.s Pd: fixed pressure of 24.3 bar Tci: 10 - 85°C	R-22, water-glycol, water	Open type reciprocating compressor
13	Tandon,varma and Gupta (1985)	Gr: 175 - 560 kg/m^2.s Pd: 1.4 - 8 bar Tci: fixed temperature water	R-22, water-glycol, water	Open type compressor
14	Steve J. Eckels Doerr and Pate Brian A. Tesene (1994)	Gr: 86 - 375 kg/m^2.s Pd: fixed 8.3 bar Tci: not given	R-134a	Refrigerant pump

Table 1. Contd.

15	Eckels and Pate (1991)	Gr:130 - 400 kg/m^2.s Pd: 6.2 - 11.5 bar Tci: not given	HFC-134a, CFC-12, water-glycol mixture	Refrigerant pump
16	Agrawal,Kumar and Varma (2004)	Gr: 210 - 372 kg/m^2.s Pd: 14.4 - 21.9 bar Tci: 20 - 30°C	R-22, water	Open type compressor
17	Chato and Dobson (1998)	Gr: 25 - 800 kg/m^2.s Pd: 7.5 - 10.5 bar Tci: constant temperature water	R-134a, R-22, R-32/R-125	-

Gr: mass flux of refrigerant; Pd: condensing pressure; Tci: temperature of cooling water used in condenser.

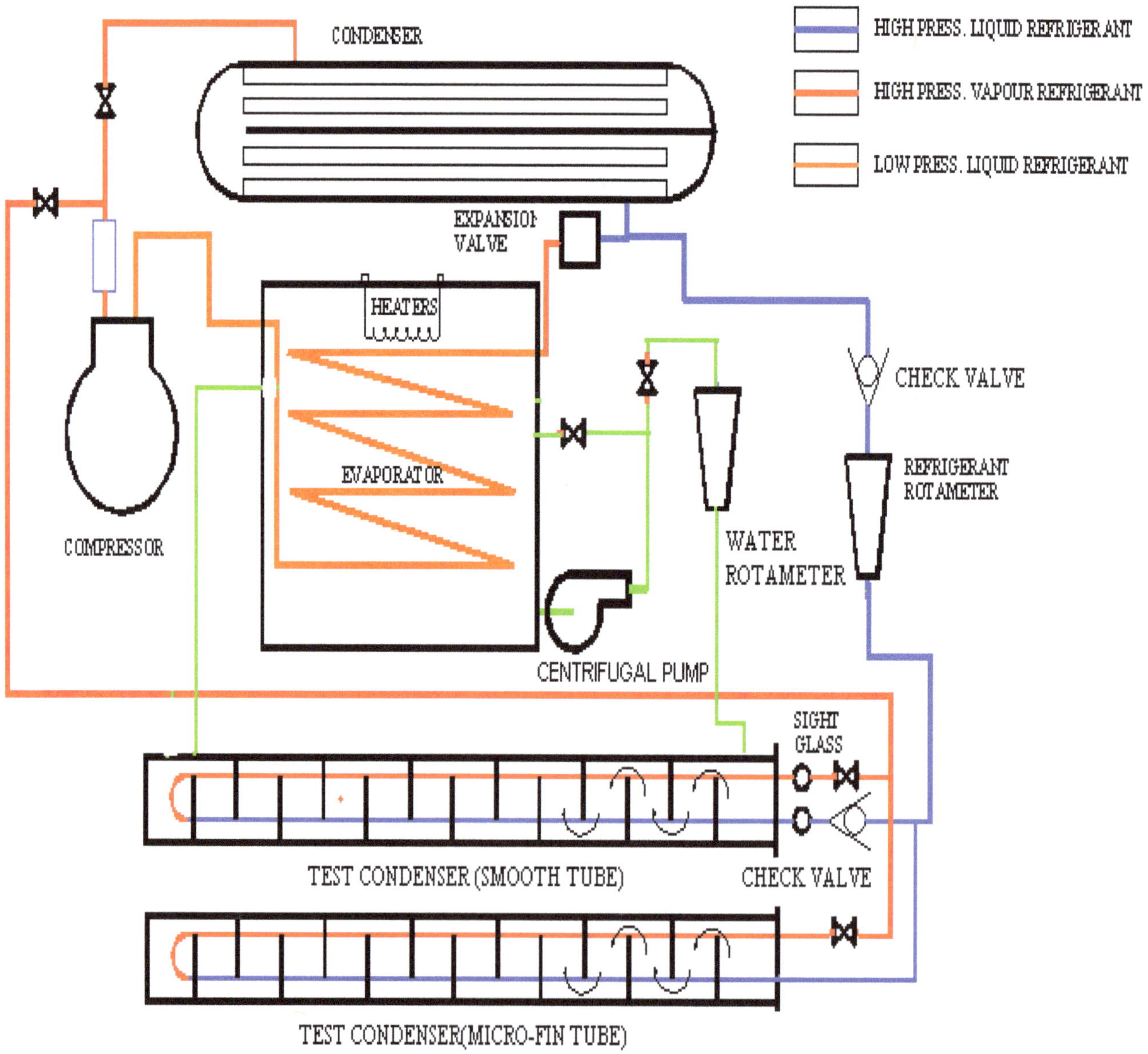

Figure 1. Experimental test facility.

Table 2. Smooth and micro-fin tube dimensions.

Parameter	Smooth tube	Micro-fin tube
Outside diameter, do (mm)	9.42	9.52
Bottom thickness, t (mm)	0.64	0.28
Number of fins, N	--------	60
Spiral angle, β , degree	--------	18
Apex angle, γ, degree	--------	45
Fin height, ef (mm)	--------	0.2
Fin tip diameter, dt (mm)	--------	8.56
Max. inside diameter, di (mm)	8.14	8.96
Length of tube, L (m)	4.5	4.5
Cross sectional area, Ac (mm^2)	52.04	63.053

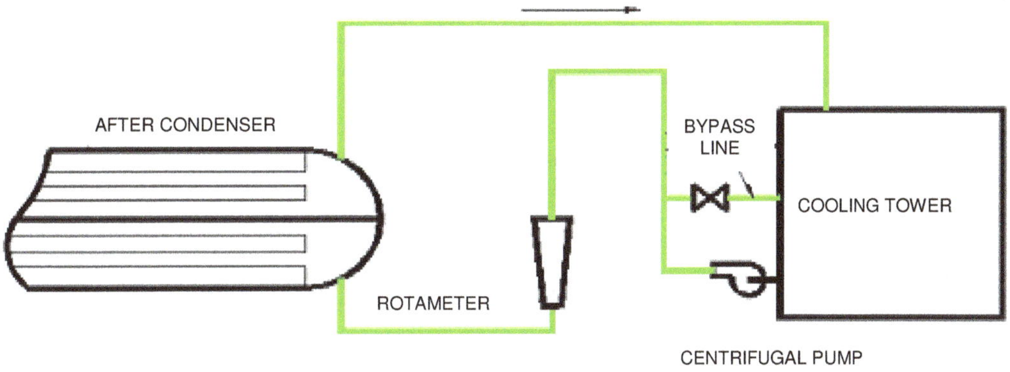

Figure 2. Cooling water circuit for main refrigerant circuit.

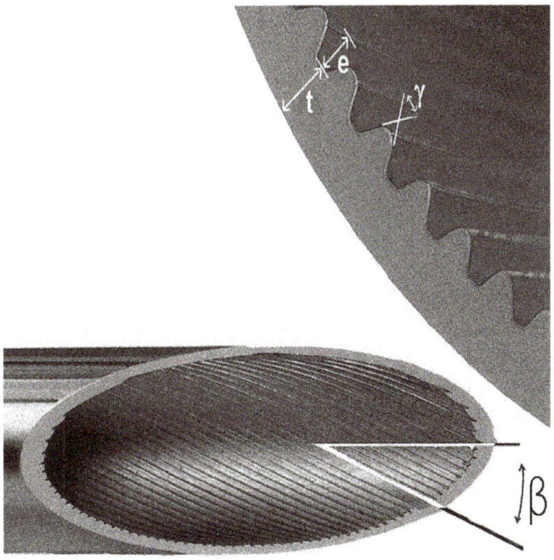

Figure 3. Micro-fin tube.

2. In the chilled water circuit, to measure the water flow rate in the test condenser, required to calculate the heat absorbed by chilled water in the test condenser.
3. In the cooling water circuit to measure the water flow rate, used during analyzing system performance.

PT100 (Resistance Temperature Detector made of platinum with a base of 100 Ω at 0°C) with 1% accuracy is used for temperature measurements. Pressure transmitters with 0.25% accuracy and 13% uncertainties are used to measure pressure difference across the test condenser, while Bourdon pressure gauges are used in other locations. Rota meters with 1% accuracy are used to measure all flow rates. All measuring instruments are calibrated from recognized calibration centers.

EXPERIMENTAL PROCEDURE

The experimentation is carried out for different mass flow rate and different condensing temperature of refrigerant. One particular condensation process (for a particular mass flow rate and condensing temperature) is also achieved for different flow rate and temperature of chilled water.

The following are steps for carrying out experimentation for 100 Lph (refrigerant) flow and 40°C condensing temperature:

1. Start refrigerant main and cooling water circuit, auxiliary circuit remains closed.
2. Reduce the temperature of water in the evaporator to 5°C.
3. Adjust the cooling water flow to achieve 40°C condensing temperature in main circuit.
4. Start the chilled water pump and allow the water to flow through test condenser, set the flow rate of chilled water at 1000 Lph.

5. Gradually open the valve of auxiliary circuit until the mass flow rate of refrigerant reaches 100 Lph.
6. Adjust the flow rate of chilled water (say to 700 Lph) to adjust condensing temperature 40°C and achieve the condensation with 10°C sub cooling.
7. Allow the system to stabilize, and record all readings such as test condenser inlet, outlet temperatures of chilled water and refrigerant etc. after steady state.
8. Increase the temperature of water in the evaporator by 5°C with the help of heater.
9. Repeat steps 6 to 8 for different chilled water inlet temperatures say 10, 15, 20, 25 and 30°C respectively.
10. Repeat steps 1 to 9 for mass flow rate of 20, 40, 60, 80,120, 140 and 160 Lph.

Data reduction

The data analysis procedure determines the average convective heat transfer coefficient of pure refrigerant, which also takes into account oil present in the refrigerant. In addition, the data analysis determines the correlation constants required for average convective heat transfer coefficient of water and refrigerant side using modified Wilson plot technique. The following is a brief description of the data reduction equations.

The equations to find rate of heat rejected by refrigerant and rate of heat absorbed by cooling water are as follows. The variation between the heat rejected by refrigerant and heat absorbed by water is within 5%.

$$Q_r = Q_{sv} + Q_c + Q_{sl} \quad (1)$$

$$Q_{sv} = m_r c_{p_v} (T_{r_i} - T_{h_i}) \quad (2)$$

$$Q_{sl} = m_r c_{p_l} (T_{h_o} - T_{r_o}) \quad (3)$$

$$Q_c = m_r (h_{t_i} - h_{t_o}) \quad (4)$$

$$Q_w = m_w c_{p_w} (T_{w_o} - T_{w_i}) \quad (5)$$

The average LMTD value is obtained by using following equations indicated in (Kern, 2003)

$$T_{w_d} = T_{w_i} + \frac{Q_{sv}}{m_w c_{p_w}} \quad (6)$$

$$T_{w_c} = T_{w_d} + \frac{Q_c}{m_w c_{p_w}} \quad (7)$$

$$LMTD_d = \frac{(T_{ri} - T_{wi}) - (T_{hi} - T_{wd})}{\ln \frac{(T_{ri} - T_{wi})}{(T_{hi} - T_{wd})}} \quad (8)$$

$$LMTDc = \frac{(T_{hi} - T_{wc}) - (T_{ho} - T_{wd})}{\ln \frac{(T_{hi} - T_{wc})}{(T_{ho} - T_{wd})}} \quad (9)$$

$$LMTDs = \frac{(T_{ho} - T_{wo}) - (T_{ro} - T_{wc})}{\ln \frac{(T_{ho} - T_{wo})}{(T_{ho} - T_{wc})}} \quad (10)$$

$$LMTD = \frac{Q_r}{\Sigma \frac{Q}{LMTD}} = \frac{Q_r}{\Sigma\left(\frac{Q_d}{LMTD_d} + \frac{Q_c}{LMTD_c} + \frac{Q_s}{LMTD_s}\right)} \quad (11)$$

The overall HTC is determined by using:

$$Uo = \frac{Q_r}{A_o LMTD} \quad (12)$$

The overall thermal resistance of the condensation process in shell and tube condensers (Rov) can be expressed as the sum of the thermal resistances corresponding to external convection (Ro), internal convection (Ri) and the tube wall (Rt) as shown in Eq. (13)

$$Rov = Ri + Ro + Rt \quad (13)$$

The individual resistances can be obtained by using following expressions:

$$R_{ov} = \frac{1}{U_o A_o} \quad (14)$$

$$R_i = \frac{1}{h_i A_i} \quad (15)$$

$$R_o = \frac{1}{h_o A_o} \quad (16)$$

$$R_t = \frac{\ln(\frac{d_o}{d_i})}{2\pi L k_t} \quad (17)$$

For a specific condition of the condensation process (particular condensing pressure and refrigerant flow rate), with different flow rate of cooling water, the overall thermal resistance is varied mainly due to the variation in outside heat transfer coefficient; meanwhile the remaining thermal resistances stay nearly constant. Therefore the thermal resistances due to internal convection and tube wall can be considered constant as indicated in Eq. (18).

$$C1 = Ri + Rt \quad (18)$$

The average heat transfer coefficient for flow across cylinders can be expressed as:

$$ho = CRewmPrw0.33 \left(\frac{k_w}{d_o}\right) \quad (19)$$

Where,

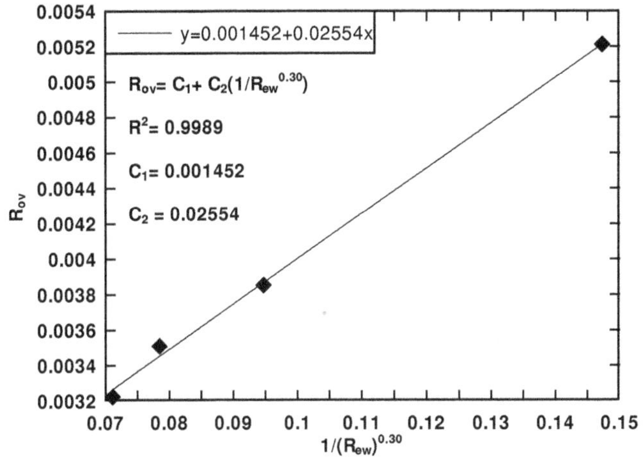

Figure 4. Modified Wilson plot 1.

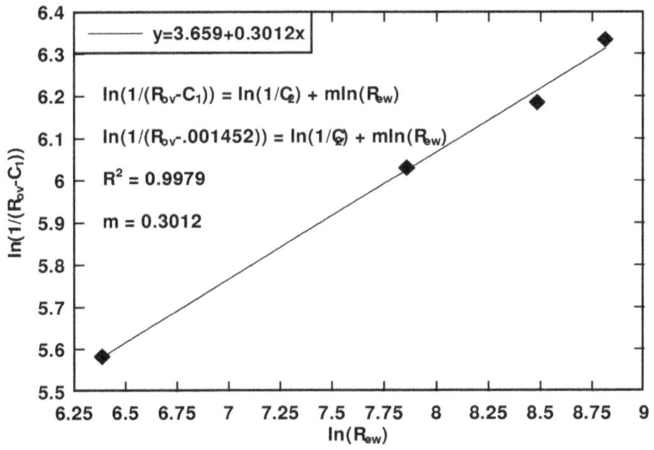

Figure 5. Modified Wilson plot 2.

$$Re_w = \frac{GD_e}{\mu_w} \quad (20)$$

$$Pr_w = \frac{\mu_w C_{pw}}{k_w} \quad (21)$$

Putting Eq. (19) in Eq. (16), we have

$$Ro = C2 \frac{1}{Re_w^m} \quad (22)$$

Where,

$$C2 = \frac{1}{C}\left(\frac{1}{Pr_w^{0.33}}\right)\left(\frac{do}{kw}\right)\left(\frac{1}{A_o}\right) \quad (23)$$

Putting Eq.(18) and (22) in Eq.(13), we have

$$Rov = C1 + C2 \frac{1}{Re_w^m} \quad (24)$$

$$\ln\left(\frac{1}{R_{ov} - C_1}\right) = \ln\left(\frac{1}{C_2}\right) + m \ln(Re) \quad (25)$$

$$\Delta P_{frict} = \Delta P_{total} + \Delta P_{mom} - \Delta P_l - \Delta P_g \quad (26)$$

The values of constants C1 and C2 are obtained according to Eq. (24) using least square technique initially by assuming the value of m and plotting graph as shown in Figure 4. Put the value of C1 in Eq. (25) and determine the value of 'm' again by using the same technique (from plot as shown in Figure 5.) If the value of 'm' obtained is equal to the value initially assumed, then the process is finished and the value of exponent is determined. Otherwise, the iteration process is repeated by assuming new 'm' value. Moreover, the coefficient C and the exponent 'm' of the general dimensionless correlation as indicated in Eq. (19) are also obtained, thus the general correlation is determined assuming only the value of the exponent of the Prantdl number. This technique is known as modified Wilson plot technique (Jose et al., 2005). Obtain the values of ho, Ro and Rt using Eq. (19), (16) and (17) respectively. Putting these values in eq. (13) to determine Ri; consequently determine hi using Eq. (15).

RESULTS AND DISCUSSION

The heat transfer coefficients and pressure drops of HFC-134a are measured in smooth and micro-fin tubes at different condensing temperatures of 35, 40, 45, 50, 55 and 60°C. About 280 data points each are taken during experimentation on smooth and micro-fin tubes. Condensation of refrigerant at specific conditions (mass flow of refrigerant and condensing temperature) is achieved for different flow rates and temperatures of cooling water for obtaining constants of co-relations using modified Wilson plot technique as shown in Figures 4 and 5.

Modified Wilson plot method

The modified Wilson plot method is applied to experimental data according to iteration procedure indicated in experimental procedure. The constants C1 and C2 are obtained as indicated in Figure 4. The Wilson plot is implemented for estimating heat transfer coefficient for every mass flow rate. The experimental data with particular refrigerant flow rate and condensing temperature are considered for each plot.

Figure 5 shows the values of the term ln [1/(Rov-C1)] plotted as a function of ln (Re), taking into account the values of the overall thermal resistance and the constant C1 obtained from least square technique as indicated in Figure 4. If the obtained value of 'm' from regression technique as indicated in Figure 5 is equal to assumed

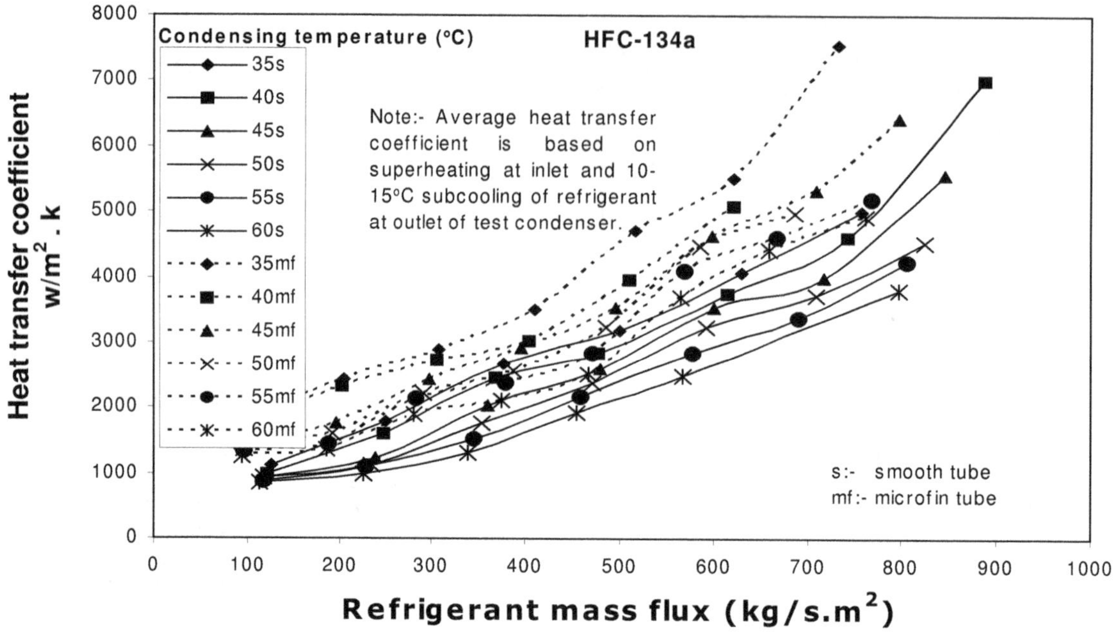

Figure 6. HFC-134a Condensation heat transfer coefficient in a smooth and micro-fin tube.

value of m from Figure 4, the iteration procedure is completed, otherwise repeat the procedure as indicated in Figure 4. This technique is applied for each condensing temperature and for all mass flow rate of refrigerant. Total 42 Wilson plots each are developed with correlation coefficient of above 0.9.

Condensation heat transfer

Condensation heat transfer data for smooth tube and micro-fin tube with HFC-134a are shown in Figure 6. For both tubes, the heat transfer coefficient increases with mass flux but decreases with increasing condensing temperature. The value of heat transfer coefficients is obtained using Eq. (18) and Eq. (15). The heat transfer coefficients obtained for micro-fin tube are greater than that of smooth tube for all condensing temperatures and mass fluxes.

Pressure drop

Frictional pressure drop data obtained using equation (26) during condensation of HFC-134a for smooth tube and micro-fin tube are as shown in Figure 7.

As with heat transfer coefficients, the pressure drop varies considerably with mass flux and condensing temperature.

Enhancement and penalty factors

Another approach for comparing the micro-fin tube heat transfer performance with that of the smooth tube is to form heat transfer enhancement factors, EF, defined as the ratio of micro-fin tube heat transfer coefficient to that of comparable smooth tube at a similar mass flux, heat flux, pressure level, and inlet and oulet quality. Pressure drop performance comparisons between the micro-fin tube and smooth tube can be made by forming ratios of pressures drop in a manner similar to that used to form heat transfer enhancement factors. These ratios are hereafter referred to as pressure drop penalty factors (PF). Figure 8 shows both heat transfer enhancement factors, EF, and pressure drop penalty factors, PF, for the micro-fin tube with HFC-134a. The EFs vary from maximum of 2.42 at low mass flux to a minimum of 1.24 for highest mass flux. The PFs are also shown in Figure 8 and vary from minimum 1 at low mass flux to maximum 1.77 at high mass flux. The penalty factors appear to be nearly constant above 400 kg/s.m^2 mass flux.

Experimental uncertainty

The maximum uncertainties are ±13.2% for the LMTD, ±1.8% for the mass flow rate of water, ±2.81% for the mass flow rate of refrigerant, ±4.72% for the heat dissipation by refrigerant in the test section, ±9.22% for the heat absorbed by the water in the test section, ±13.3 for overall heat transfer coefficient, ±18.2% for refrigerant side heat transfer coefficient and ±13.3% for the pressure drop. A propagation of error analysis (Kline and McClintock, 1953) is used to obtain the uncertainty listed above with a confidence interval of 85 - 90% with a coefficient of correlation above 0.9.

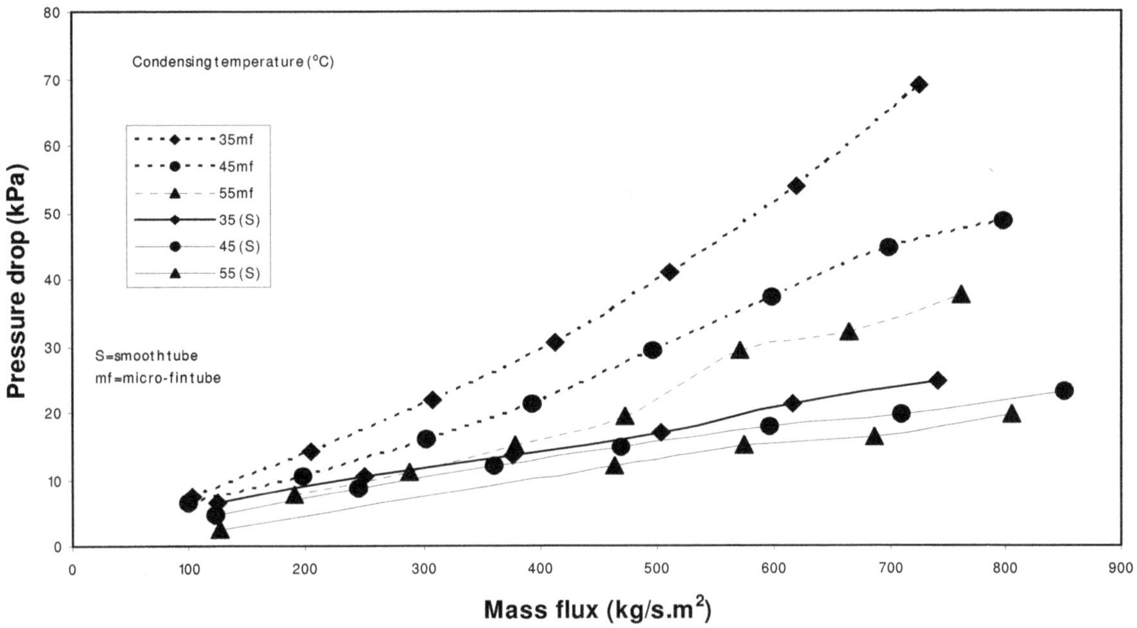

Figure 7. HFC-134a Condensation pressure drop in a smooth and micro-fin tube.

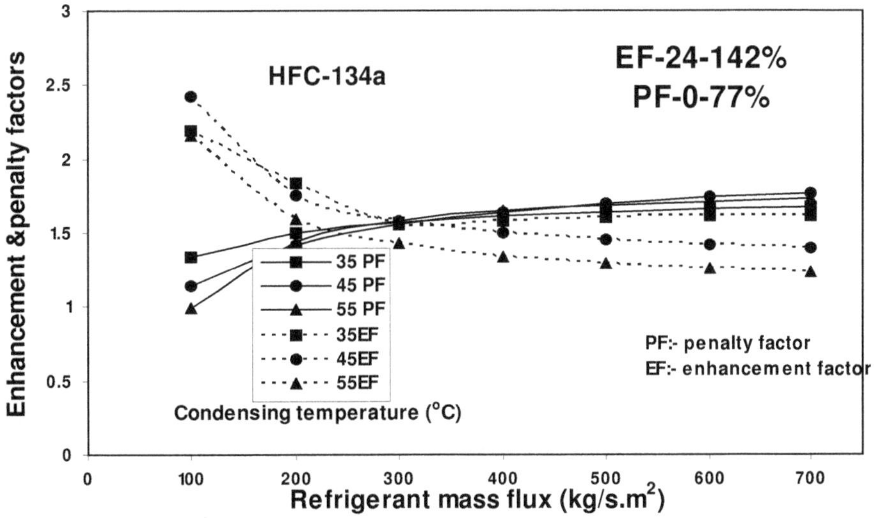

Figure 8. HFC-134a heat transfer enhancement and pressure drop penalty factor.

Correlation comparison

The experimental heat transfer and pressure drop data of smooth and micro-fin tubes are also compared with some available correlations and only the best two correlations for each case is discussed as follows:

Heat transfer

Boyko and Kruzhilin (1967) correlation captures 83.91% HFC-134a data within ±20%. Akers et al. (1959) correlation captures 78.32% HFC-134a data for smooth tube as shown in Figure 9.

For micro-fin tube, Luu and Bergles (1980) correlation captures maximum data points amongst all, capturing 74.64% of HFC-134a data within ±20. Most of the data points corresponding to low mass flux are under predicted, however almost all values corresponding to 60°C condensing temperatures are over predicted by this correlation. Hiroshi Honda, Huasheng Wang and Shigeru Nozu's correlation captures 47.84% of HFC-134a data

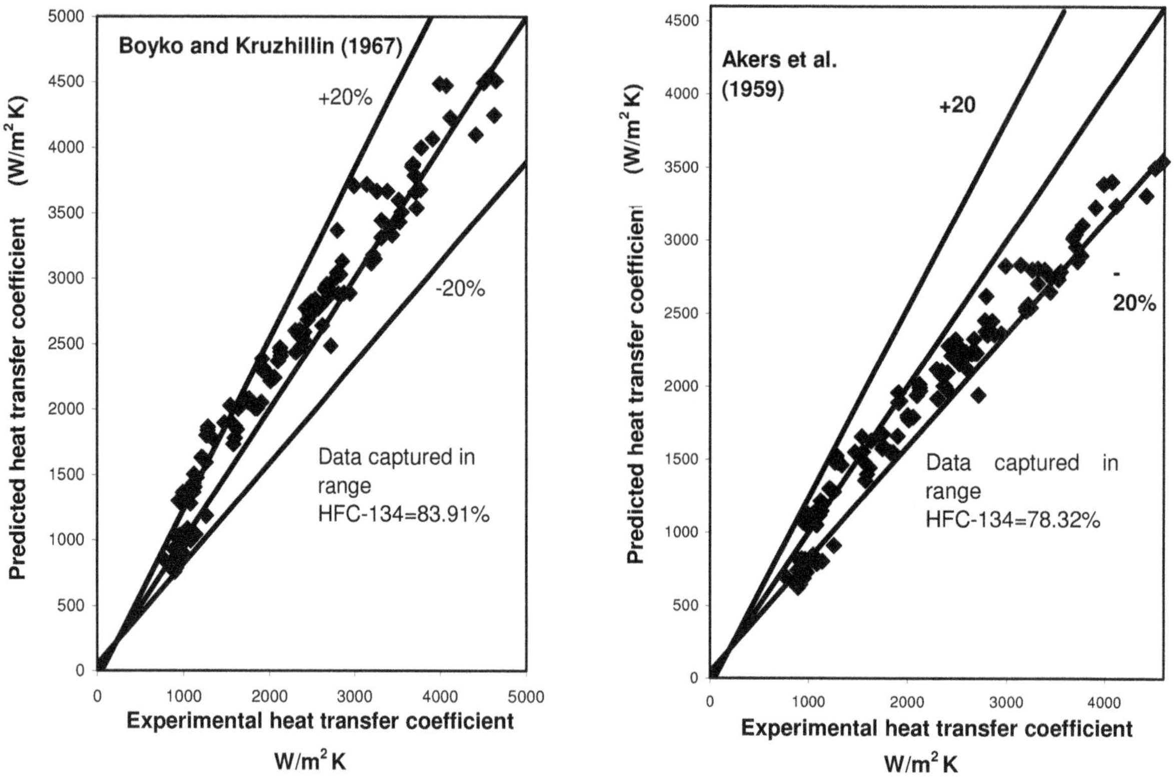

Figure 9. Comparison of smooth tube heat transfer data with existing correlations.

within ±20 (Hiroshi et al., 2002). Most experimental data between 50 and 60°C condensing temperatures are over predicted and low mass flux data between 35 and 55°C is under predicted by this correlation as shown in Figure 10.

Pressure drop

In case of smooth tube, (Friedel, 1979) correlation captures maximum data points amongst all, capturing 75% data of HFC-134a data within ±30%. The experimental data of mass fluxes below 200 kg/s.m^2 are under predicted by this correlation. M¨uller-Steinhagen and Heck (1986) correlation captures 57.57% of HFC-134a data within ±30%. Most of the experimental data from low mass flux area and high condensing temperature are under predicted by this correlation as shown in Figure 11. Choi et al. (2001) correlation captures maximum data points of micro-fin tube amongst all, capturing 69.88% data of HFC-134a within ±30%. The experimental data of mass fluxes below 200 kg/s.m^2 and some of data corresponding to 35 and 40°C condensing temperatures are under predicted by this correlation. Kedzierski and Goncalves (1999) correlation captures 64.2% of HFC-134a data within ±30%. Most of the experimental data from low mass flux area are under predicted, and few data points corresponding to high mass flux are over predicted by this correlation as shown in Figure 12.

Conclusion

The experimental test facility has been designed and developed, which is used to determine the condensation heat transfer coefficient and pressure drop in smooth and micro-fin tubes for various HFC refrigerants namely HFC-134a, R-404A, R-407C, R-507A. As the hermetically sealed compressor used for circulating refrigerant, effect of oil present in the refrigerant during condensation is also taken into account. The experimentation covers wide range of operating parameters such as mass flux and condensing temperatures. The instruments used for measurements are calibrated from recognized calibration centers.

The condensation and pressure drop of HFC-134a in smooth and micro-fin tubes are measured and the values of condensation heat transfer coefficients for different mass flux and condensing temperatures are obtained using modified Wilson plot technique with correlation coefficient above 0.9. The condensation heat transfer coefficient and pressure drop increases with increasing mass flux and decreases with increasing condensing temperature for both smooth and micro-fin tubes. The heat transfer coefficients and pressure drops obtained for micro-fin tube are greater than that of smooth tube for all condensing temperatures and mass fluxes. The EFs obtained varies from 1.24 to 2.42, while PFs varies from 1 to 1.77.

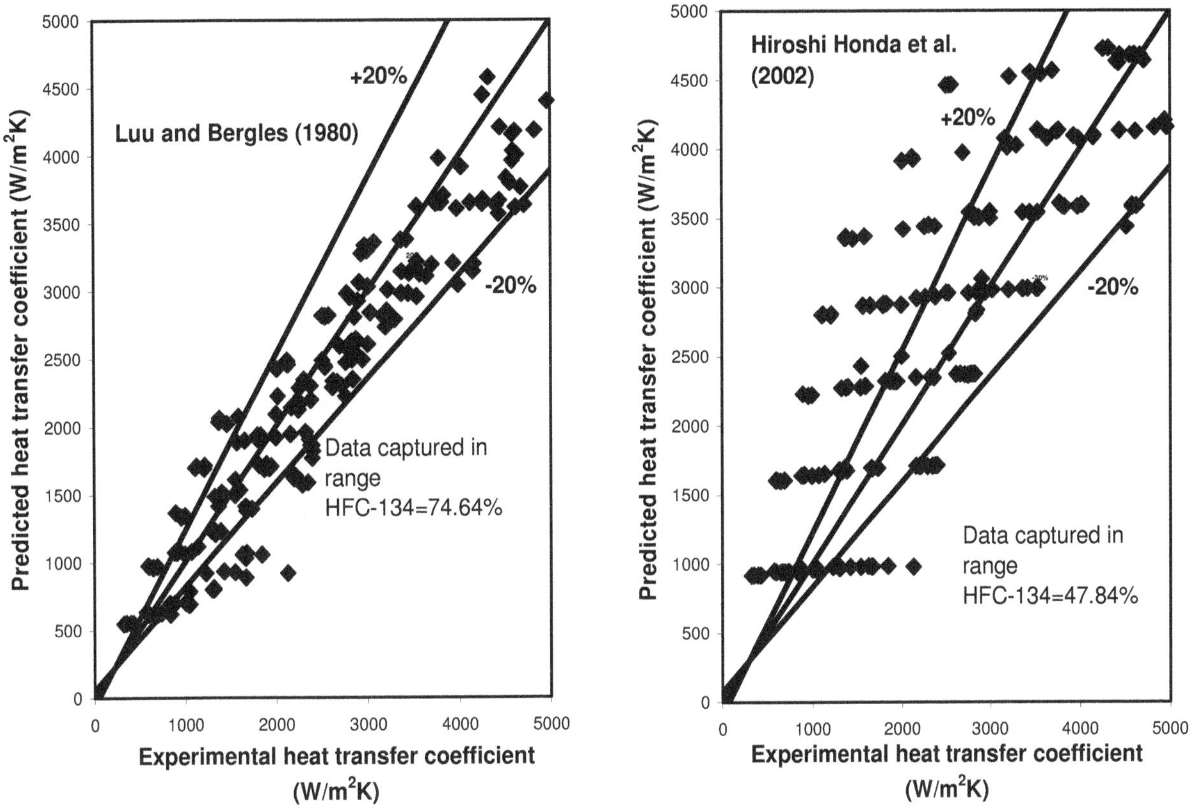

Figure 10. Comparison of micro-fin tube heat transfer data with existing correlations.

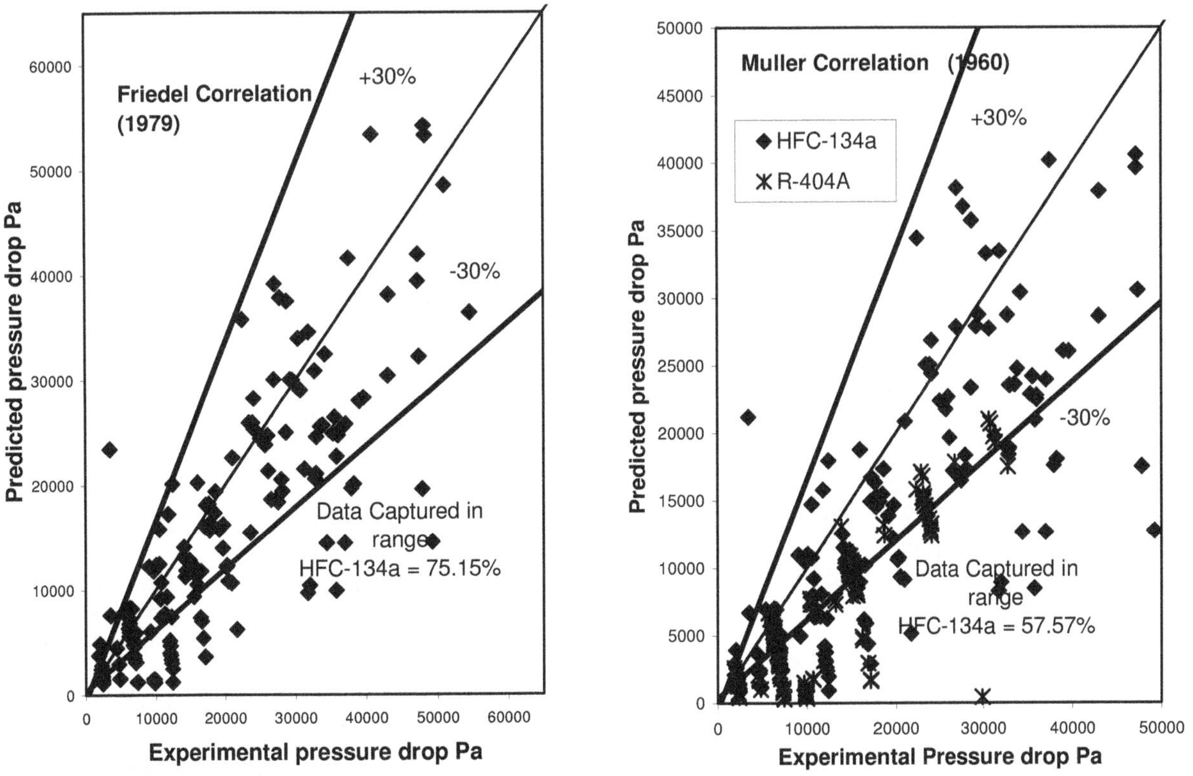

Figure 11. Comparison of smooth tube pressure drop data with existing correlations.

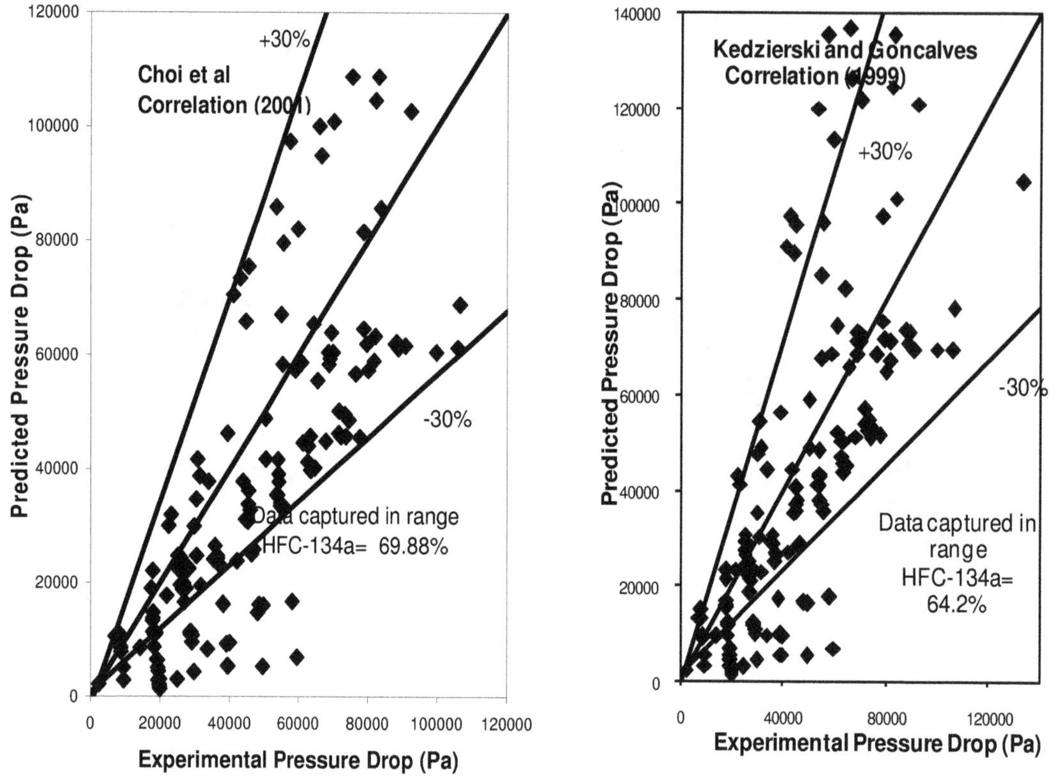

Figure 12. Comparison of micro-fin tube pressure drop data with existing correlations.

The results are compared with existing available correlations for validation of test facility. The experimental data points have good association with few available correlations except some data points from low and high mass flux and data points from higher condensing temperatures, which did not fall within ±20%.

ACKNOWLEDGEMENTS

The authors express their deep appreciation for the financial support provided for this setup by ASHRAE, USA. We would also like to thank our undergraduate students; Mr. Pushkar Natu, Mr. Sagar Shirolkar, Mr. Rahul Piese and Mr. Sachin Pole for their efforts during the fabrication of this test facility, and Mr. Yeole Madhusudan, Mr. Nehete Yatin, Mr. Chaudhari Ashish and Mr. Waykos Yogesh for their efforts during experimentation on smooth tube condensation. We would also like to thank our College authorities Dr. J. D. Bapat and Prof. S. V. Chaitanya for giving support at administration level.

NOMENCLATURE

A_i	inner surface area of tube (m^2) $=\pi d_i L$
A_o	outer surface area of tube (m^2) $=\pi d_o L$
a_f	cross flow area (m2) $=IDxCxB/P_T$
B	baffle space (m)
C	clearance in U-tube (m)
C_{pl}	specific heat of liquid refrigerant (kJ/kg.K)
C_{pv}	specific heat of vapour refrigerant (kJ/kg.K)
C_{pw}	specific heat of water (kJ/kg.K)
D	characteristic diameter of tube (m)
D_e	equivalent diameter of shell (m) $=4x\ (P_T^2 - \pi d_o^2/4)/(\pi do)$
d_i	inner diameter of tube (m)
d_o	outer diameter of tube (m)
G	mass velocity of water ($kg/m^2.s$) $= m_w/a_f$
h_i	film coefficient inner side (refrigerant) (W/m^2K)
h_o	outside heat transfer coefficient (water side) (W/m^2K)
h_{ti}	enthalpy at test condenser inlet (kJ/kg)
h_{to}	enthalpy at test condenser outlet (kJ/kg)
ID	inner diameter of shell (m)
k_t	thermal conductivity of liquid refrigerant (W/m.K)
k_t	thermal conductivity of tube material (W/m.K)
k_w	thermal conductivity of water (W/m^2K)
L	length of U-tube (m)
$LMTD$	average weighted logarithmic mean temperature difference (°C)
$LMTD_c$	logarithmic mean temperature difference (°C) for condensation process
$LMTD_d$	logarithmic mean temperature difference (°C) for

desuperheating process
$LMTD_s$ logarithmic mean temperature difference (°C) for sub cooling process
m_r mass flow rate of refrigerant (kg/s)
m_w mass flow rate of water (kg/s)
Nu Nusselt number
P saturation pressure (bar)
Pr_l Prandtl number for liquid refrigerant
Pr_w Prandtl nuber for water
P_{rc} reduced pressure=(P/Pcr)
P_T pitch of U-tube
Q_c rate of heat rejected by refrigerant during only condensation (kW)
Q_r total rate of heat rejected by refrigerant (kW)
Q_{sl} rate of heat rejected by refrigerant during sub cooling of refrigerant (kW)
Q_{sv} rate of heat rejected by refrigerant during desuperheating of refrigerant (kW)
Q_w rate of heat absorbed by cooling water (kW)
Re_l Reynolds number for liquid refrigerant
Re_g Reynolds number for vapour refrigerant
Re_w Reynolds number for water
R_i thermal resistance due to inner film coefficient (K/W)
R_o thermal resistance due to outer heat transfer coefficient (K/W)
R_{ov} overall thermal resistance (K/W)
R_t thermal resistance due to tube wall (K/W)
T_{hi} = refrigerant saturation temperature at the inlet of condenser (°C)
T_{ho} refrigerant saturation temperature at the outlet of condenser (°C)
T_{ri} refrigerant temperature at the inlet of condenser (°C)
T_{ro} refrigerant temperature at the outlet of condenser (°C)
T_{wc} estimated water temperature at the end of only condensation of refrigerant (°C)
T_{wd} estimated water temperature at the end of desuperheating of refrigerant (°C)
T_{wi} cooling water temperature at the inlet of shell (°C)
T_{wo} cooling water temperature at the outlet of shell (°C)
U_o overall heat transfer coefficient based on outer surface area (W/m^2.K)
X vapour quality of refrigerant
μ_w dynamic viscosity of water (N.s/m^2)
μ_g dynamic viscosity of liquid refrigerant (N.s/m^2)
μ_l dynamic viscosity of vapour refrigerant (N.s/m^2)
ρ_f density of liquid refrigerant (kg/m^3)
ρ_g density of vapour refrigerant (kg/m^3)
ΔP_{total} measured pressure drop during experimentation

$$\Delta P_{mom} \quad G^2\left\{\left[\frac{1}{\rho_l}\right]_{out} - \left[\frac{1}{\rho_g}\right]_{in}\right\}$$

ΔP_l pressure drop occurred during sub cooling process $= \left(\dfrac{4f_l(L_l/d_i)G^2}{2\rho_l}\right)$

ΔP_g pressure drop occurred during desuperheating process $= \left(\dfrac{4f_g(L_g/d_i)G^2}{2\rho_g}\right)$

REFERENCES

Agrawal NK, Ravi Kumar, Varma KH (2004). Heat Transfer Augmentation by Segmented Tape Inserts during Condensation of R-22 inside a Horizontal tube. ASHRAE Transactions, Oct. 143-149.

Arora CP (2004). Refrigeration and Air conditioning, Tata Mc Graw- Hill Publishing Company Ltd., Second edition.

Boyko LD, Kruzhilin GN (1967). "Heat transfer and hydraulic resistance during condensation of steam in a horizontal tube and in a bundle of tubes." Int. J. Heat Mass Transfer, 10: 361–73.

Choi JY, Kedzierski MA, Domanski PA (2001). Generalized pressure drop correlation for evaporation and condensation in smooth and micro fin tubes. In:Proc. of IIF-IIR Commission B1, Paderborn, Germany, B4, p. 9–16.

Dobson MK, Chato JC (1998). Condensation in Smooth Horizontal tubes. ASME Journal of heat transfer, Feb. 193-210.

Eckels SJ, Pate MB (1991). In-tube Evaporation and Condensation of Refrigerant-Lubricant Mixtures of HFC-134a and CFC-12. ASHRAE Transactions, Vol. 97, Part 2.

Eckels SJ, Tesene B (1999). A Comparision of R-22, R-134a, R-410a, and R-407C Condensation Performance in Smooth and Enhanced tubes: Part I, Heat Transfer. ASHRAE Transactions, 428-451

Friedel L (1979). Improved friction drop correlations for horizontal and vertical two-phase pipe flow.In European Two-phase Flow Group Meeting, paper E2, Ispra, Italy.

Hiroshi H, Huasheng W, Shigeru N (2002). A Theoretical study of film condensation in horizontal micro fin tubes, ASME . J. heat Trans., 124: 94-101

Jose FS, Francisco JU, Jaime S, Antonio C (2005). Experimental Apparatus for Measuring Heat Transfer Coefficients by the Wilson Plot Method. Eur. J. Phy., 26, N1-N11.

Kedzierski MA, Goncalves JM (1999). "Horizontal convective condensation of alternative refrigerants within a micro fin tubes." Enhanced Heat Transfer, 6: 161–178.

Kern (2003). Process Heat Transfer. Tata Mc Graw- Hill Publishing Company Ltd., Second edition.

Keumnam Cho, Sang-Jin Tae (2000). Condensation Heat Transfer for R-22 and R-407C Refrigerant-Oil Mixtures in a Micro-Fin tube with a U-Bend. Jan. Intern. J. Heat Trans., 2043-2051.

Khanpara JC, Bergles AE, Pate MB (1986-2B). Augmentation of R-113 in-tube evaporation with micro-fin tubes. ASHRAE Transactions.

Khanpara JC, Pate MB, Bergles AE (1986). Augmentation of R-113 In-tube Condensation with Micro-Fin tubes. Heat Transfer in Air Conditioning and Refrigeration Equipment, HTD-65, New York, American Society of Mechanical Engineers, pp. 21-32.

Kline SJ, McClintock (1953). Describing Uncertainties in Single-Sample Experiments. Mech. Eng, 3.

M¨uller-Steinhagen H, Heck K (1986). A simple friction pressure correlation for two-phase flow in pipes. Chem. Eng. Process, 20:297–308.

Minh Luu, Bergles (1980). Enhancement of Horizontal in-tube Condensation of R-113. ASHRAE Transactions, 293-310.

Minh Luu, Bergles (1980). "Enhancement of Horizontal in-tube Condensation of R-113." ASHRAE Transactions, 293-310.

Said SA, Azer NZ (1982). Heat Transfer and Pressure Drop during Condensation inside Horizontal Finned tubes. ASHRAE Transactions, 114-135.

Schlager LM, Pate MB, Bergles AE (1990). Evaporation and Condensation Heat Transfer and Pressure Drop in Horizontal, 12.7 mm Micro-Fin tubes with Refrigerant 22. Nov. ASME J. Heat Trans.

Smit FJ, Meyer JP, (2002). R-22 and Zeotropic R-22/R-142b Mixture Condensation in Micro-Fin, High-Fin, and Twisted Tape Insert tubes. ASME J. Heat Trans.

Steven J, Eckels TM, Doerr Pate MB (1994-II). In-Tube Heat Transfer and Pressure Drop of R-134a and Ester Lubricant Mixtures in a Smooth tube and a Micro-fin tube: Part II-Condensation. ASHRAE Transactions.

Stoecker WF, Kornota E (1985). Condensing Coefficients when using Refrigerant Mixtures. ASHRAE Transactions, 1350-1367.

Tandon TN, Varma HK, Gupta CP (1985). An Experimental Investigation of Forced Convection Condensation during Annular Flow inside a Horizontal tube. ASHRAE Transactions, 343-355

Tichy JA, Macken NA, Duval WMB (1985). An Experimental Investigation of Heat Transfer in Forced Convection Condensation of Oil-Refrigerant Mixtures. ASHRAE Transactions, 297-309.

Wang JC Y, Kalamchi A, Fazio P (1985). Experimental study on Condensation of Refrigerant-Oil Mixtures: Part I- Design of the Test Apparatus. ASHRAE Transactions, 216-237.

Analysis of potential flow around two-dimensional body by finite element method

Md. Shahjada Tarafder and Nabila Naz

Department of Naval Architecture and Marine Engineering, Bangladesh University of Engineering and Technology, Dhaka-1000, Bangladesh.

The paper presents a numerical method for analyzing the potential flow around two dimensional body such as single circular cylinder, NACA0012 hydrofoil and double circular cylinders by finite element method. The numerical technique is based upon a general formulation for the Laplace's equation using Galerkin technique finite element approach. The solution of the systems of algebraic equations is approached by Gaussian elimination scheme. Laplace's equation is expressed in terms of both steam function and velocity potential formulation. A finite element program is developed in order to analyze the result. The contours of stream and velocity potential function are drawn. The contour of stream function exhibits the characteristics of potential flow and does not intersect each other. The calculated pressure co-efficient shows the pressure decreasing around the forwarded face from the initial total pressure at the stagnation point and reaching a minimum pressure at the top of the cylinder.

Key words: Stream function, velocity potential, number of nodes (NDE), number of elements (NEL).

INTRODUCTION

The flow past two dimensional body such as circular cylinders and hydrofoil has been the subject of numerous experimental and numerical studies because this type of flow exhibits the very fundamental mechanisms. The flow field over both the cylinder and hydrofoil is symmetric at low values of Reynolds number. As the Reynolds number increases, flow begins to separate behind the body causing vortex shedding which is an unsteady phenomenon. To achieve the goal of obtaining the detailed information of the flow field around two dimensional bodies, Finite Element Method (FEM) has been emerged as an attractive, powerful tool in many designing process.

The FEM was originated from the field of structural calculation in the beginning of the fifties and was introduced by Turner et al. (1956). The FEM was introduced into the field of computational fluid dynamics (CFD) by Chung (1977). The first study concerning the steady flow past a circular cylinder was reported by Thom (1933) for Reynolds number of 10 and 20. The works of Kawaguti (1953) and Payne (1958) were restricted to low Reynolds numbers (Re = 40) and relatively low Reynolds numbers (Re = 40~100) respectively. Oden (1969) has presented a theoretical finite element analogue for the Navier-Stokes equations, but without a practical numerical method. Dennis and Chung (1970) introduced finite element method into the field of computational fluid dynamics (CFD) by solving steady flow past a circular cylinder at Reynolds number (Re≤100).

Tong (1971) presented results for steady flow using this method with pressure and velocities as dependent variables. Olson (1974) presented a numerical procedure to investigate steady incompressible flow problems using stream function formulation. Hafez (2004) simulates steady an inviscid flows over a cylinder using both potential and stream functions.

The objective of the present research is to analyze the potential flow around single circular cylinder, NACA 0012 hydrofoil and double circular cylinders by Galerkin technique of finite element method. Due to symmetry of circular cylinders and NACA 0012 hydrofoil, only the upper half portions have been considered as computational domain. Both stream function and velocity potential formulation have been used with definite boundary conditions. Contours of stream and velocity potential function, velocity above crest for both formulations and velocity and pressure distribution along the surface for various discretization are obtained which are compared with the analytical result available from the literature and shown graphically.

MATHEMATICAL FORMULATION

Potential flow around circular cylinder

Let us consider the potential flow of an ideal fluid around the circular cylinder placed with its axis perpendicular to the plane of the flow as shown in Figure 1. Now, the potential flow around circular cylinder can be represented by *Laplace* equation as:

$$\left.\begin{aligned}\nabla^2\Psi = 0\\ \nabla^2\Phi = 0\end{aligned}\right\} \quad (1)$$

The velocity components u and v of the flow field in relation to stream function Ψ or the velocity potential Φ are given by:

$$u = \frac{\partial\Psi}{\partial y},\ v = -\frac{\partial\Psi}{\partial x} \quad \text{or,}$$

$$u = \frac{\partial\Phi}{\partial x},\ v = \frac{\partial\Phi}{\partial x} \quad (2)$$

Boundary conditions for stream function

The half of the fluid domain is taken in the computations as shown in Figure 2 and the boundary conditions that need to be satisfied in order to get the solution of *Laplace* equation: $\nabla^2\Psi = 0$ in Ω are given as follows:

(a) $\Psi = 0$ on the boundary a-e-f-g
(b) $\Psi = yU$ on the boundary a-b
(c) $\Psi = yU$ on the boundary b-h
(d) $\Psi = yU$ on the boundary g-h

Boundary conditions for velocity potential

In case of velocity formulation, the boundary conditions that need to be satisfied in order to get the solution of *Laplace* equation: $\nabla^2\Phi = 0$ in Ω as shown in Figure 3:

(a) On the boundary a-b, $\dfrac{\partial\Phi}{\partial x} = U$

(b) On the boundary a-e-f-g, $\dfrac{\partial\Phi}{\partial y} = 0$

(c) On the boundary b-h, $\dfrac{\partial\Phi}{\partial y} = 0$

(d) On the boundary g-h, $\dfrac{\partial\Phi}{\partial x} = U$

Potential flow around a hydrofoil

Let us consider the flow of an ideal fluid around a hydrofoil placed with its axis perpendicular to the plane of the flow as shown Figure 4.

Boundary conditions for stream function

Now we need to solve the *Laplace* equation: $\nabla^2\Psi = 0$ in Ω as shown in Figure 5 with the following boundary conditions:

(a) $\Psi = 0$ on the boundary a-b-c-d
(b) $\Psi = yU$ on the boundary a-f and e-d
(c) $\Psi = yU$ on the boundary f-e

Boundary conditions for velocity potential

In case of velocity potential formulation, we need to solve the *Laplace* equation: $\nabla^2\Phi = 0$ in Ω as shown in Figure 6 with the following boundary conditions:

(a) On the boundary a-f and d-e, $\dfrac{\partial\Phi}{\partial n} = 1$

(b) On the boundary a-b-c-d and f-e, $\dfrac{\partial\Phi}{\partial n} = 0$

Potential flow around two circular cylinders

Let us consider potential flow around double circular cylinders as shown in Figure 7. The stream function for the flow can be expressed as

$$\Psi(x,y) = \Psi_1(x,y) + a\Psi_2(x,y) + b\Psi_3(x,y) \quad (3)$$

Where a and b are the two constants. Now we need to solve Laplace's equation

$$\nabla^2\Psi_1 = 0;\ \nabla^2\Psi_2 = 0;\ \nabla^2\Psi_3 = 0$$

with the following boundary conditions:

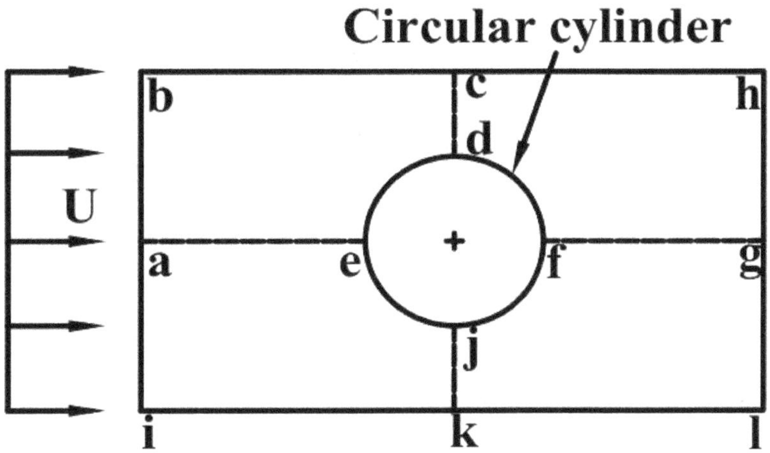

Figure 1. Flow around single circular cylinder.

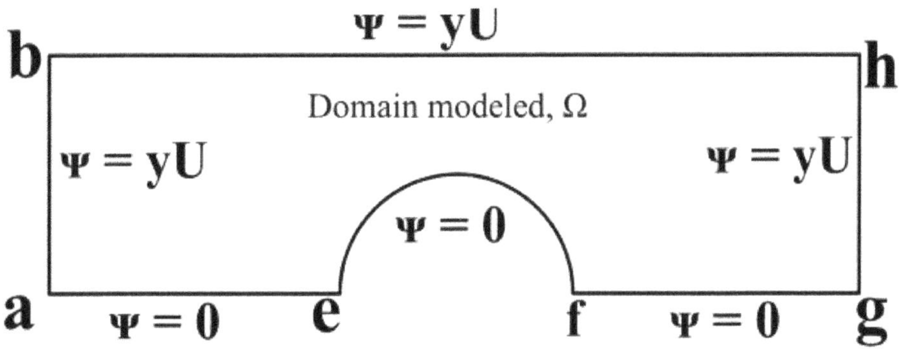

Figure 2. Boundary conditions for the stream function formulation.

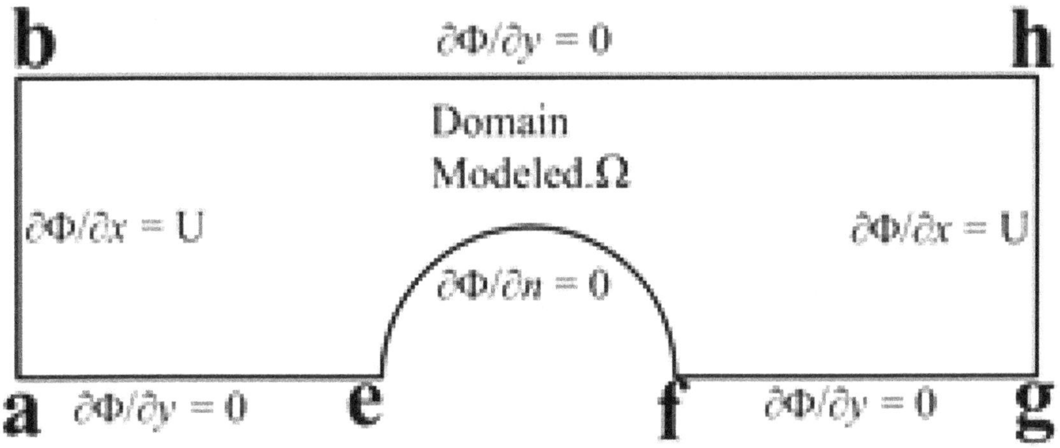

Figure 3. Boundary conditions for the velocity potential function formulation.

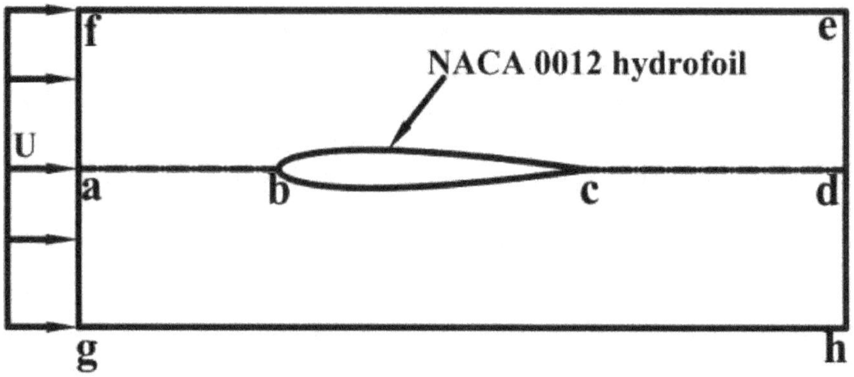

Figure 4. Flow around a NACA 0012 hydrofoil.

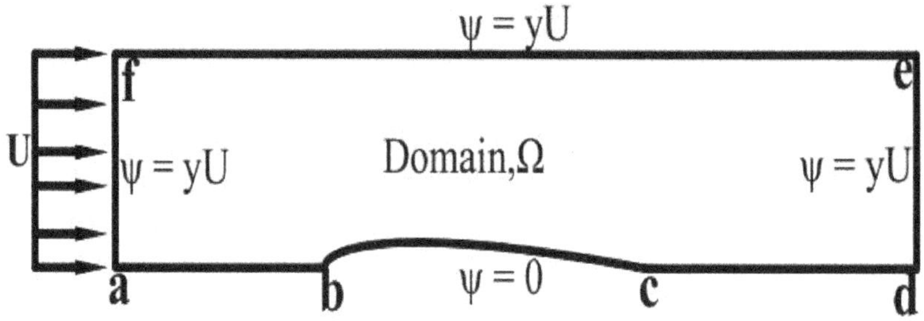

Figure 5. Boundary conditions for the stream function formulation for hydrofoil.

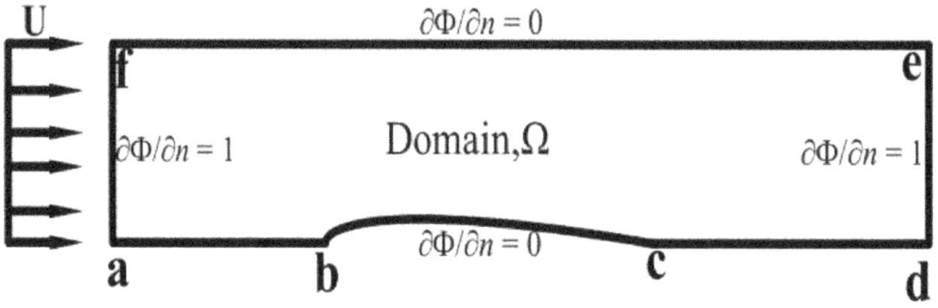

Figure 6. Boundary conditions for the velocity potential function formulation.

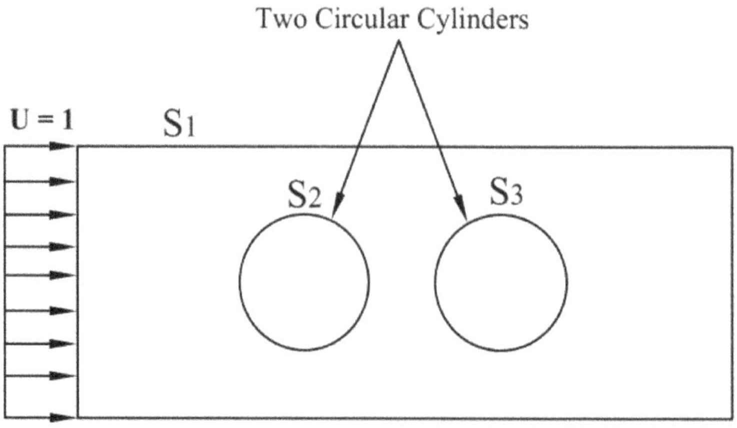

Figure 7. Flow around double circular cylinders with boundary conditions.

(a) $\Psi_1 = U y$ on S_1
(b) $\Psi_1 = 0$ on S_2 and S_3
(c) $\Psi_2 = 0$ on S_1 and S_3
(d) $\Psi_2 = 1$ on S_2
(e) $\Psi_3 = 0$ on S_1 and S_2
(d) $\Psi_3 = 1$ on S_3

NUMERICAL SOLUTION OF POTENTIAL FLOW

Numerical solution by stream function method

The stream function over the domain of interest is discretized into finite elements having M nodes:

$$\Psi(x,y) = \sum_{i=1}^{M} N_i(x,y)\Psi_i = [N]\{\Psi\} \tag{4}$$

Using the Galerkin method, the element residual equations are:

$$\int_{A^e} N_i(x,y)(\frac{\partial^2\Psi}{\partial x^2} + \frac{\partial^2\Psi}{\partial y^2})dxdy = 0, i = 1, M \tag{5}$$

$$or, \int_{A^e} [N]^T(\frac{\partial^2\Psi}{\partial x^2} + \frac{\partial^2\Psi}{\partial y^2})dxdy = 0 \tag{6}$$

Application of the Green-Gauss theorem gives

$$\int_{S^e}[N]^T\frac{\partial\Psi}{\partial x}n_x dS - \int_{A^e}\frac{\partial[N]^T}{\partial x}\frac{\partial\Psi}{\partial x}dxdy + \int_{S^e}[N]^T\frac{\partial\Psi}{\partial y}n_y dS - \int_{A^e}\frac{\partial[N]^T}{\partial y}\frac{\partial\Psi}{\partial y}dxdy \tag{7}$$

Where S represents the element boundary and (n_x, n_y) are the components of the outward unit vector normal to the boundary. Using Equation (4) in Equation (7) and substituting the velocity components into the boundary integrals, results in:

$$\int_{A^e}\frac{\partial[N]^T}{\partial x}\frac{\partial[N]}{\partial x} + \frac{\partial[N]^T}{\partial y}\frac{\partial[N]}{\partial y}dxdy\{\Psi\} = \int_{S^e}[N]^T(un_y - vn_x)ds \tag{8}$$

and this equation is of the form

$$[k^e]\{\Psi\} = \{f^{(e)}\} \tag{9}$$

The $M \times M$ element stiffness matrix is

$$[k^{(e)}] = \int_{A^e}(\frac{\partial[N]^T}{\partial x}\frac{\partial[N]}{\partial x} + \frac{\partial[N]^T}{\partial y}\frac{\partial[N]}{\partial y})dxdy \tag{10}$$

the nodal forces are represented by the $M \times 1$ column matrix

$$\{f^{(e)}\} = \int_{S^e}[N]^T(un_y - vn_x)ds \tag{11}$$

Numerical solution by velocity potential method

The finite element formulation of potential flow of an ideal fluid in terms of velocity potential is quite similar to that of the stream function approach, since the governing equation is *Laplace's* equation in both cases. By direct analogy with Equations (4) to (11) it is obtained as follows:

$$\Phi(x,y) = \sum_{i=1}^{M} N_i(x,y)\Phi_i = [N]\{\Phi\} \tag{12}$$

Using the Galerkin method, the element residual equations are:

$$\int_{A^e} N_i(x,y)(\frac{\partial^2\Phi}{\partial x^2} + \frac{\partial^2\Phi}{\partial y^2})dxdy = 0, i = 1, M \tag{13}$$

$$or, \int_{A^e} [N]^T(\frac{\partial^2\Phi}{\partial x^2} + \frac{\partial^2\Phi}{\partial y^2})dxdy = 0 \tag{14}$$

Application of the Green-Gauss theorem gives

$$\int_{S^e}[N]^T\frac{\partial\Phi}{\partial x}n_x dS - \int_{A^e}\frac{\partial[N]^T}{\partial x}\frac{\partial\Phi}{\partial x}dxdy + \int_{S^e}[N]^T\frac{\partial\Phi}{\partial y}n_y dS - \int_{A^e}\frac{\partial[N]^T}{\partial y}\frac{\partial\Phi}{\partial y}dxdy \tag{15}$$

Utilizing Equation (12) in the area integral of Equation (15) and substituting the velocity components into the boundary integrals, results in:

$$\int_{A^e}\frac{\partial[N]^T}{\partial x}\frac{\partial[N]}{\partial x} + \frac{\partial[N]^T}{\partial y}\frac{\partial[N]}{\partial y}dxdy\{\Phi\} = -\int_{S^e}[N]^T(un_x + vn_y)ds \tag{16}$$

and this equation is of the form

$$[k^e]\{\Phi\} = \{f^{(e)}\} \tag{17}$$

RESULTS AND DISCUSSION

Based on the previous mathematical formulation as outlined as numerical solution by stream function method and numerical solution by velocity potential method a finite element program has been developed in FORTRAN 90 for calculating the potential flow around two dimensional bodies. For all finite element mesh configurations, nodes along the vertical line above the crest of the cylinder are numbered consecutively from top to bottom in order to be compatible with velocity calculations used in the program. The elements are taken in the form of triangle or quadrilateral for the convenience of discretization, thus the edge of the body may not be appeared as a circle or hydrofoil.

Single circular cylinder

Let us consider the flow around the circular cylinder of unit radius confined between two parallel plates having length of 7 m and height 4 m. A fluid of uniform velocity 1.0 m/s is assumed to be flowing from the left to the right of cylinder as shown in Figure 8. The choice of computational domain in the direction of flow is arbitrary and the free stream velocity is considered to prevail at distances sufficiently far from the cylinder. The upper half of the computational domain surrounded by the path (a-b-c-d-e-f) is taken into account for numerical calculation due to symmetry of flow and is discretized by (24×5) triangular elements as shown in Figure 9 for stream function formulation. The contour of stream function has been obtained from stream function formulation and exhibits the characteristics of potential flow as shown in Figure 10. The stream lines have not intersected each other and mean the flow past the cylinder smoothly without any separation at the trailing edge. The upper half of the computational domain for velocity potential formulation is also discretized by (20×7) quadrilateral elements as shown in Figure 11. The contours of velocity potential have also been obtained from velocity potential formulation and exhibit the characteristic of potential flow i.e. no vortices exist at the trailing edge as shown in Figure 12.

The velocities along the vertical line above the crest of the cylinder are calculated and then compared with the analytical result in Figure 13. The average deviation for the velocity profiles between the two cases is less than one percent. Figure 14 is plotted by calculating the velocities above the crest at two points($x = 3.50$, $y = 1.00$) and ($x = 3.50$, $y = 2.00$) against various number of nodes for stream function formulation which shows that computed velocities converges to the analytical solution as number of nodes increases. Figure 15 is obtained by plotting the velocity square along the surface of cylinder against the angular coordinates of nodal points. There are two types of curves of which first type shows a sinusoidal curve for the whole cylinder obtained theoretically and second type consists of four curves for four different finite element mesh configurations.

In Figure 16 the calculated pressure coefficient (C_p) is compared with the theoretical pressure distribution over the surface of the cylinder and the agreement is found to be quite satisfactory. The calculated results show the pressure decreasing around the forwarded face from the initial total pressure at the stagnation point and reaching a minimum pressure at the top of the cylinder.

NACA 0012 hydrofoil

Let us consider the flow around NACA 0012 hydrofoil confined between two parallel plates having length of 10 m and height 4m as shown in Figure 17. A fluid of uniform velocity 1 m/s is flowing from the left to right of the foil. The half of computational domain for stream function formulation is discretized by (16×3) elements as shown in Figure 18 and the contours of the stream lines are given in Figure 19.

Similarly, the half of computational domain for velocity potential formulation is discretized by (16×3) elements as shown in Figure 20 and the contours of the stream lines are given in Figure 21. Figure 22 depicts a comparison of pressure distribution pressure over the surface of the foil with the results obtained from constant strength source method[11] and shows very close agreement both at the leading and trailing edge of the foil.

Double circular cylinders

The flow around two circular cylinders of unit radius is confined between two parallel plates having length of 10 m and height 4m. The distance between the cylinders is unit length and the height above the cylinder is also unit length. A fluid of uniform velocity 1 m/s is flowing from left to right of cylinders as shown in Figure 23. The half of the computational domain is discretized by (16×3) elements for stream function formulation and (20×7) elements for velocity potential as shown in Figures 24 and 25, respectively. The contours obtained from these two formulations are drawn in Figures 26 and 27 respectively.

Conclusions

The paper presents a numerical method of calculating the potential flow around two dimensional bodies by finite element method. The following conclusions can be drawn from the present numerical analysis:

(i) The present method can be an efficient tool for evaluating the potential flow characteristics of two dimensional body.
(ii) The contour of stream function exhibits the characteristics of potential flow and does not intersect each other.

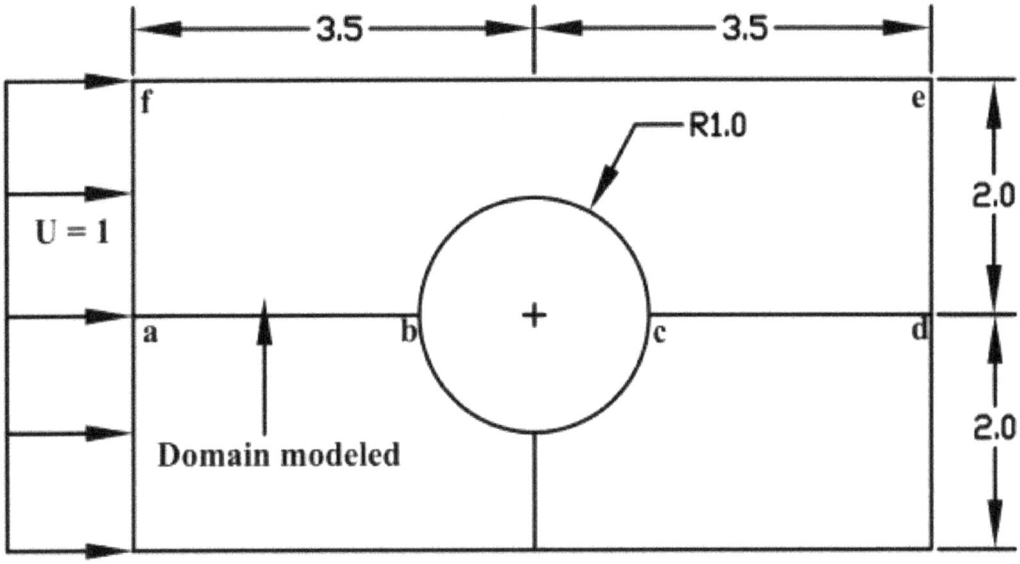

Figure 8. Computational domain for stream function and velocity potential formulation.

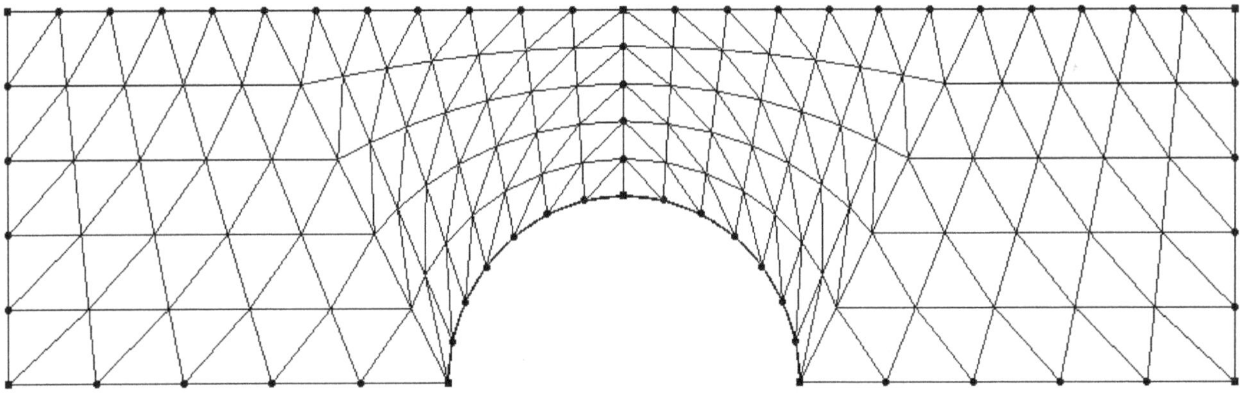

Figure 9. Discretization of domain by 240 triangular elements for stream function formulation.

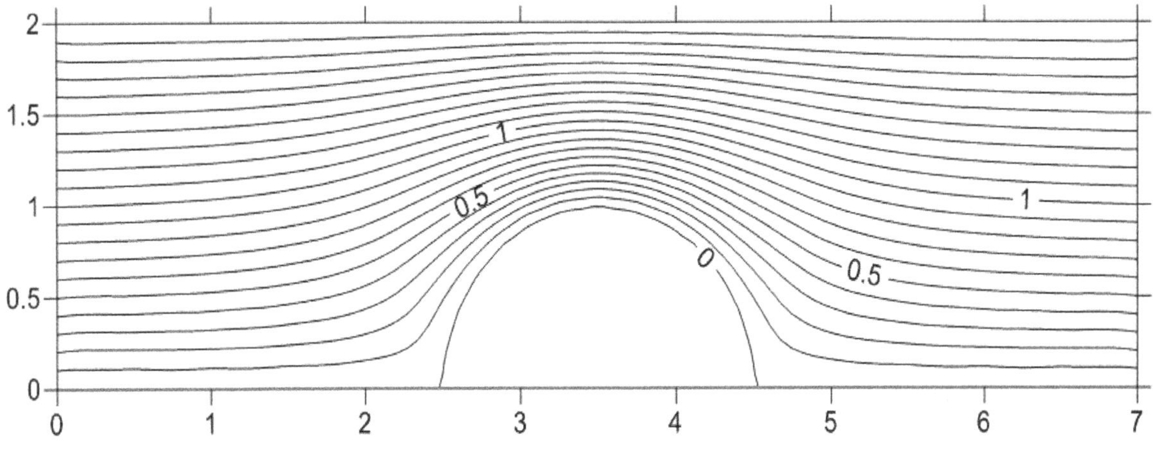

Figure 10. Stream function contours around the half circular cylinder.

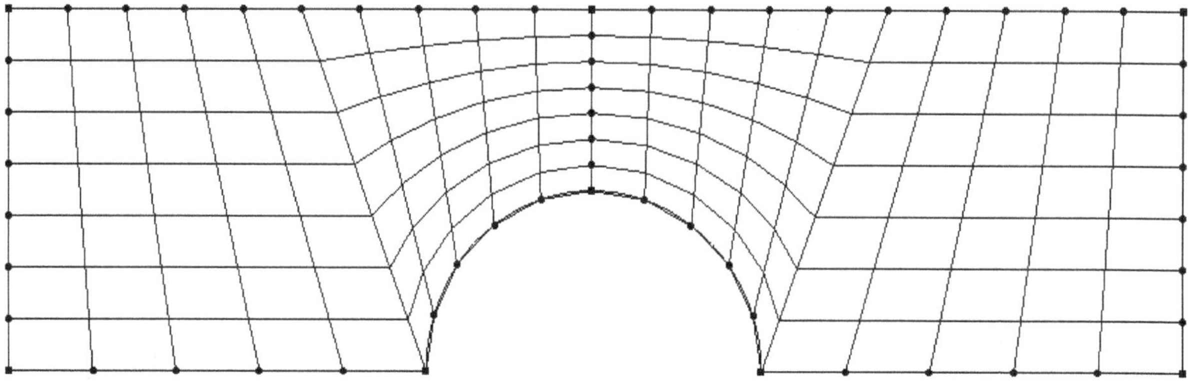

Figure 11. Discretization of computational domain by 140 quadrilateral elements for velocity potential.

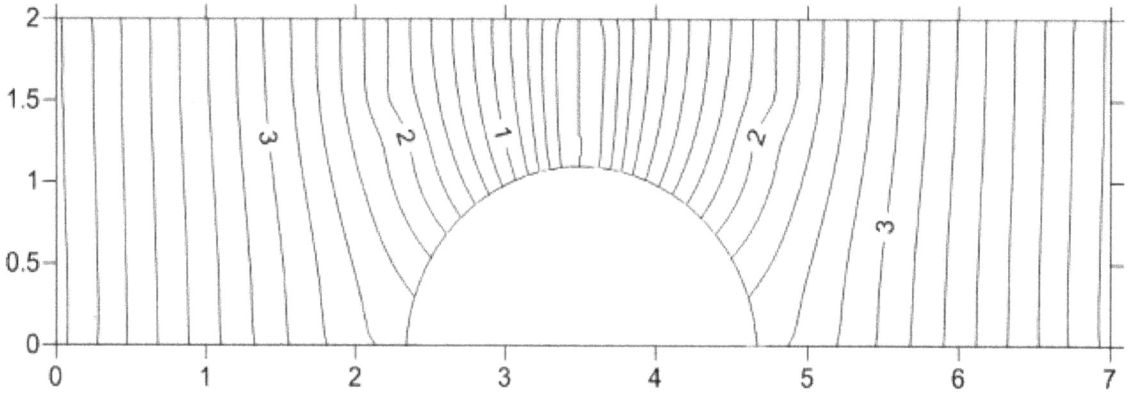

Figure 12. Velocity potential contours around the half circular cylinder.

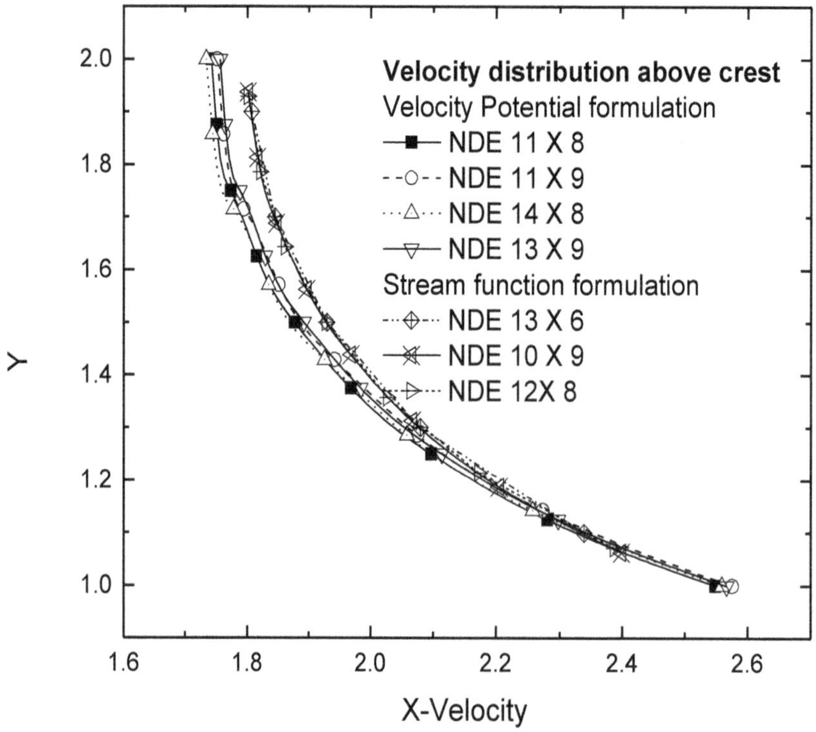

Figure 13. Velocity distributions above the crest of cylinder.

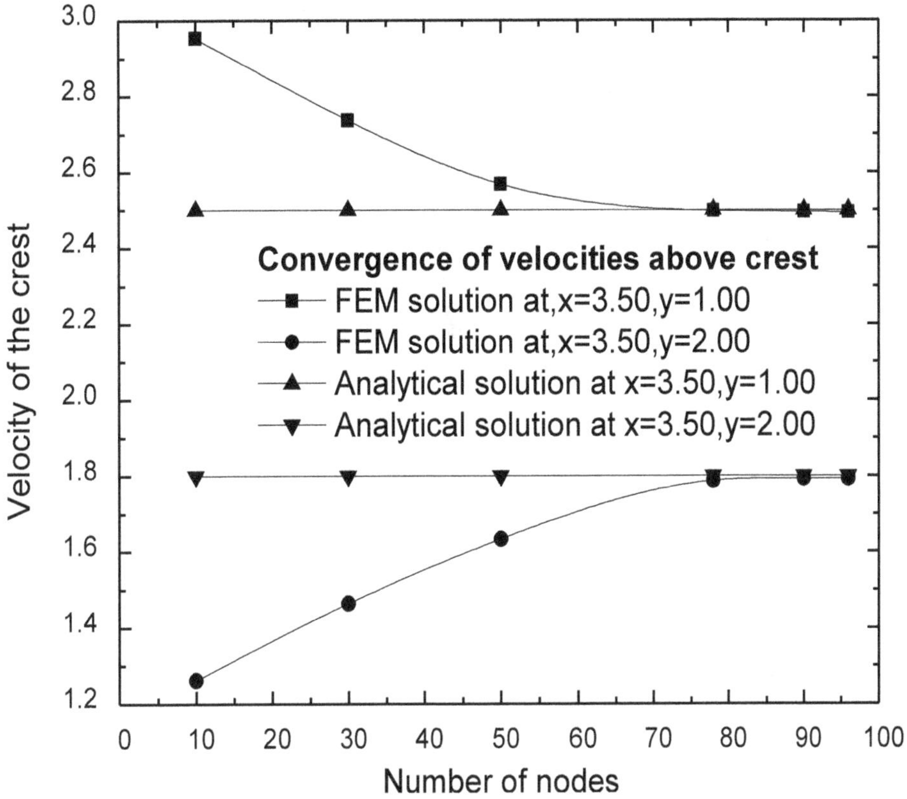

Figure 14. Error analysis for Ψ formulation showing convergence of velocities above the crest.

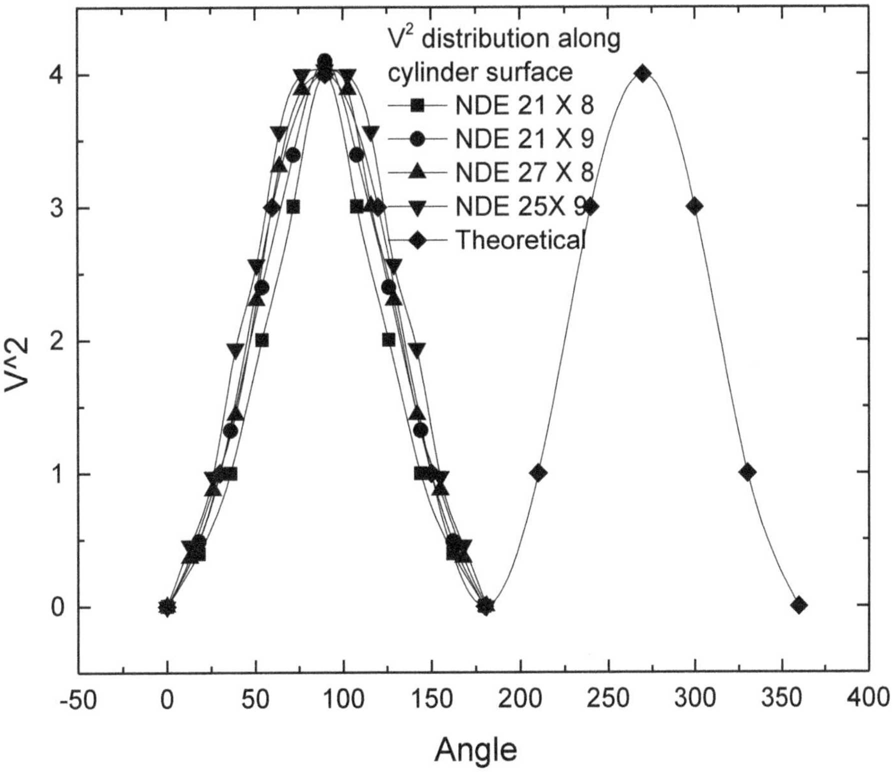

Figure 15. Velocity profile along cylinder surface.

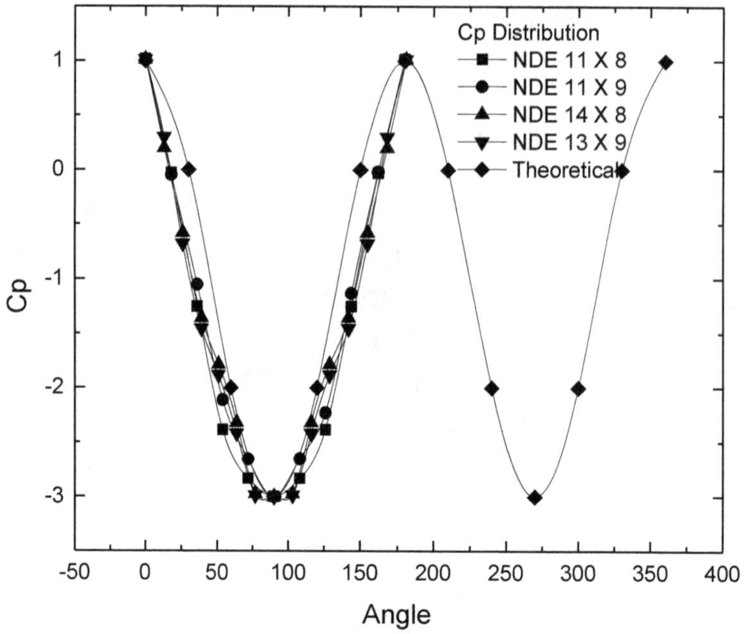

Figure 16. Distribution of pressure coefficient (C_p).

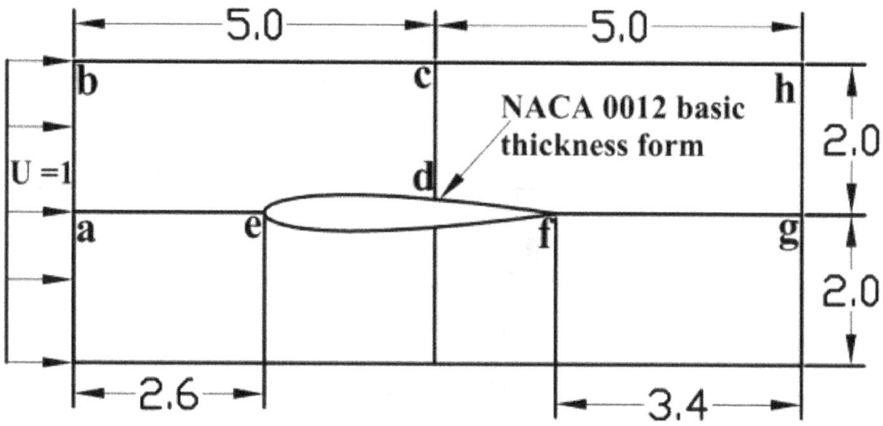

Figure 17. Computational domain for flow around NACA 0012 hydrofoil.

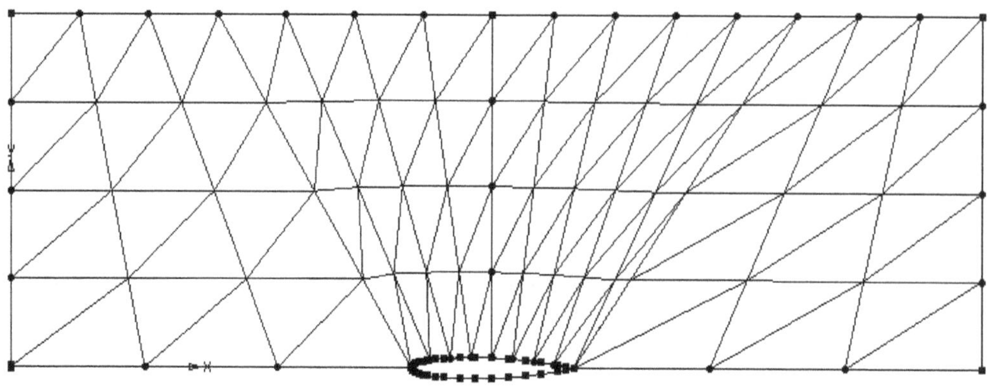

Figure 18. Discretization of computational domain by 120 triangular elements for the hydrofoil.

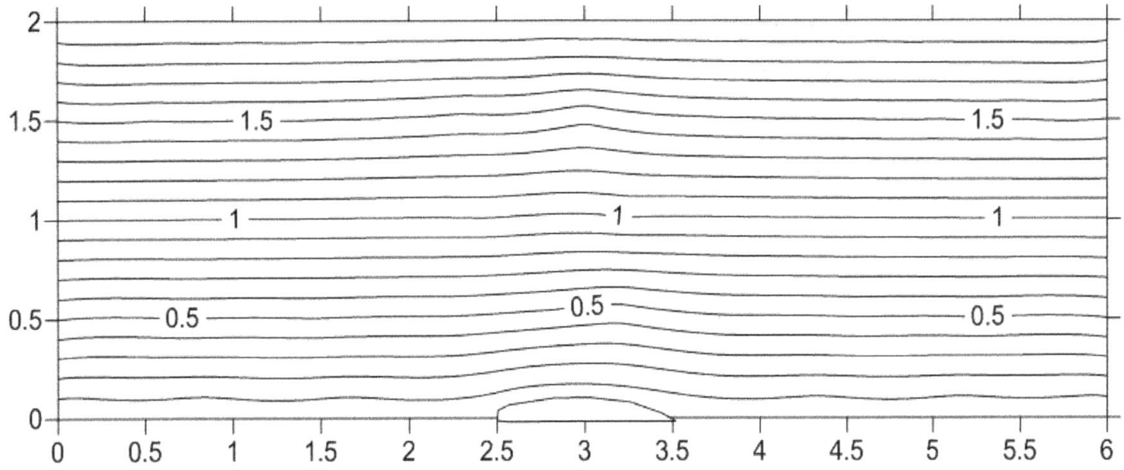

Figure 19. Stream function contours for flow around NACA 0012 hydrofoil.

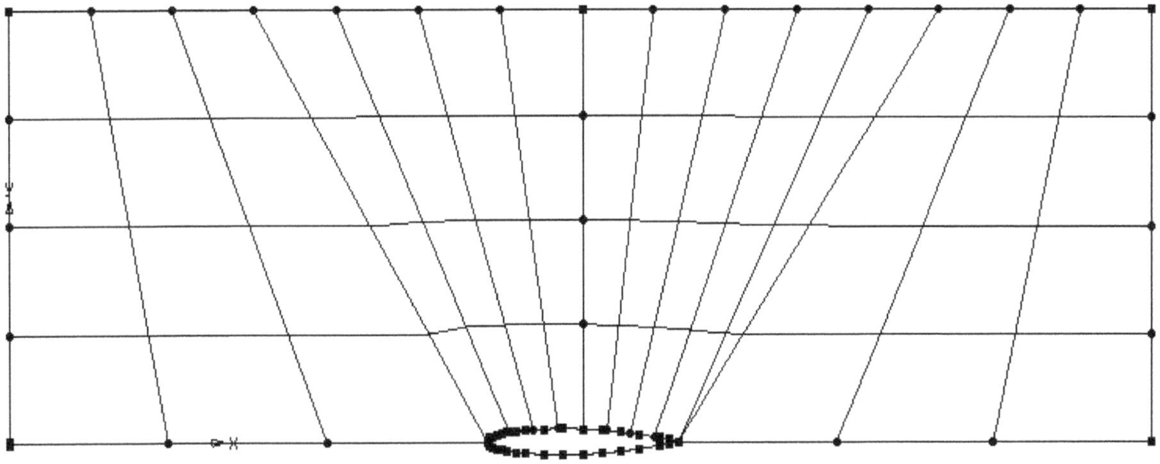

Figure 20. Discretization of computational domain by 60 quadrilateral elements for the hydrofoil.

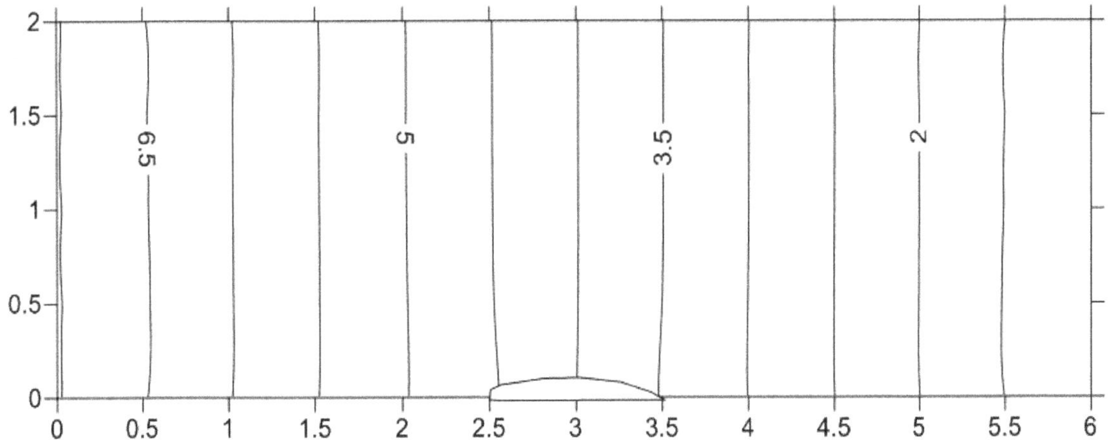

Figure 21. Contours of velocity potential.

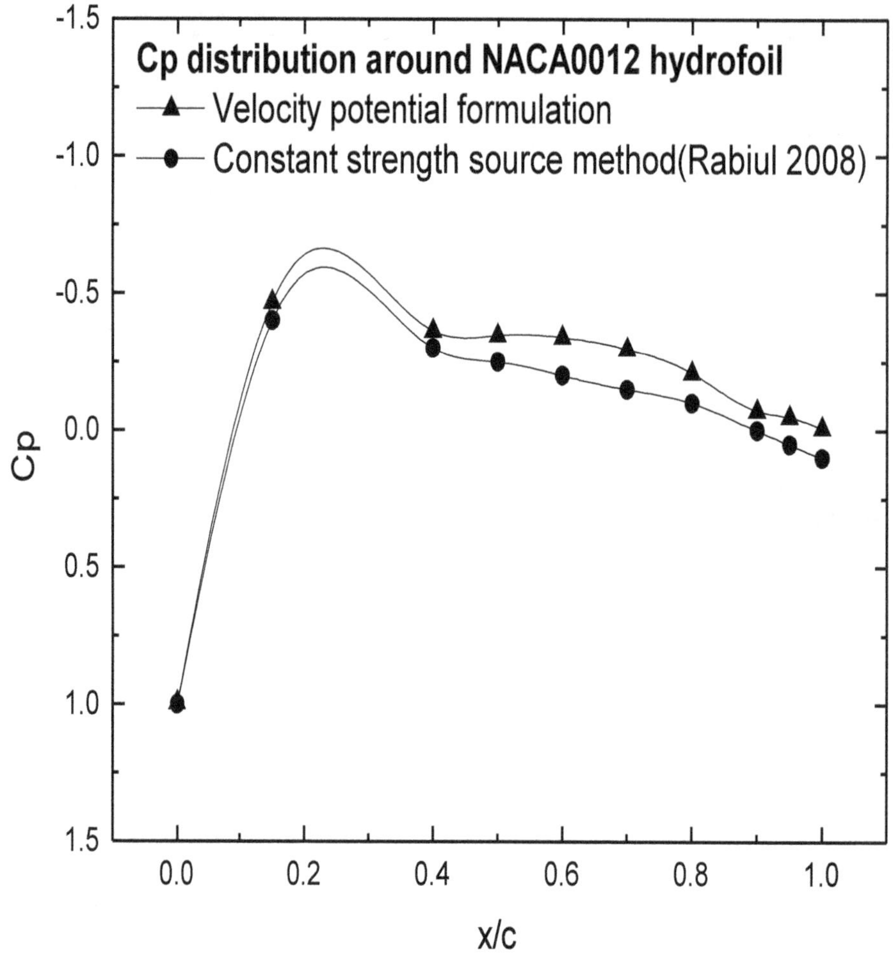

Figure 22. Chord wise pressure variations.

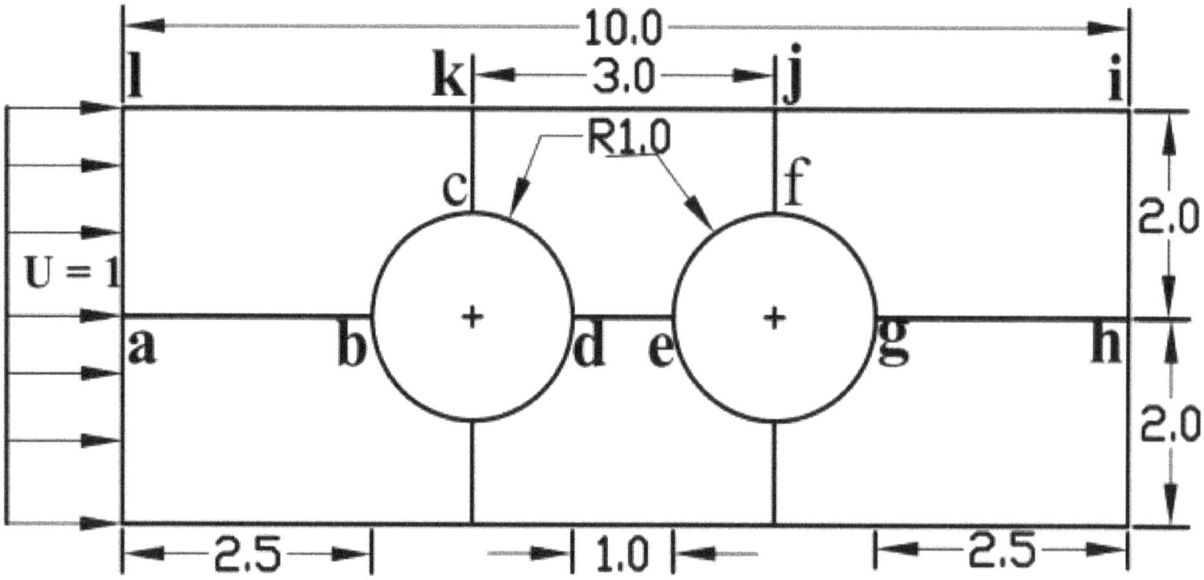

Figure 23. Computational domain for flow around two circular cylinders.

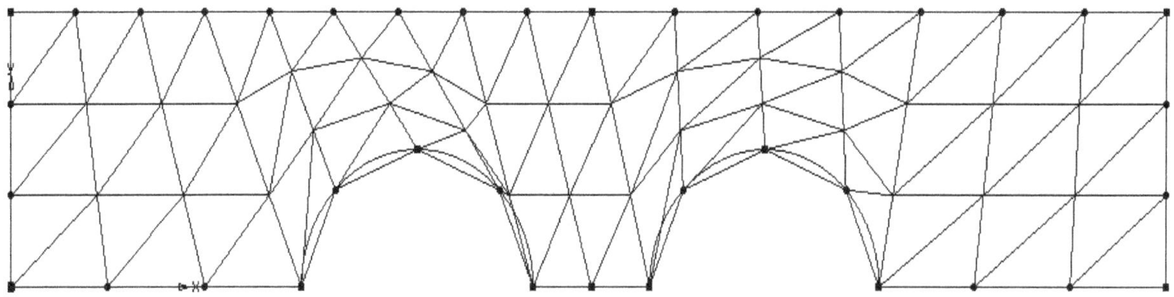

Figure 24. Discretization of domain by 48 triangular elements for flow around two circular cylinders.

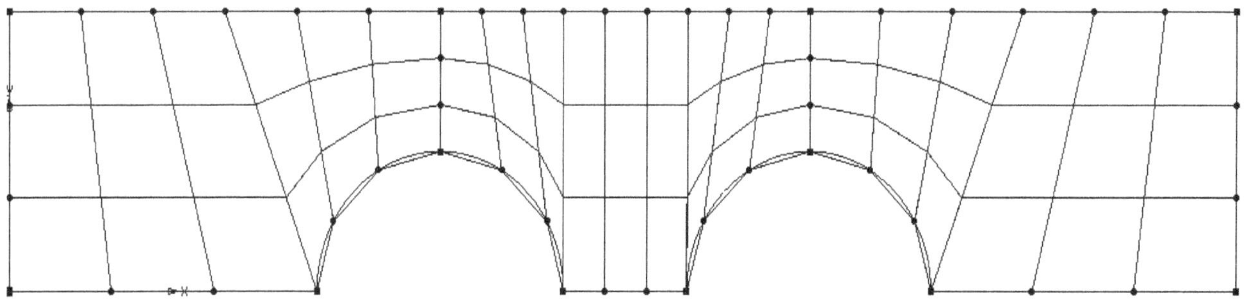

Figure 25. Mesh arrangement for velocity potential formulation.

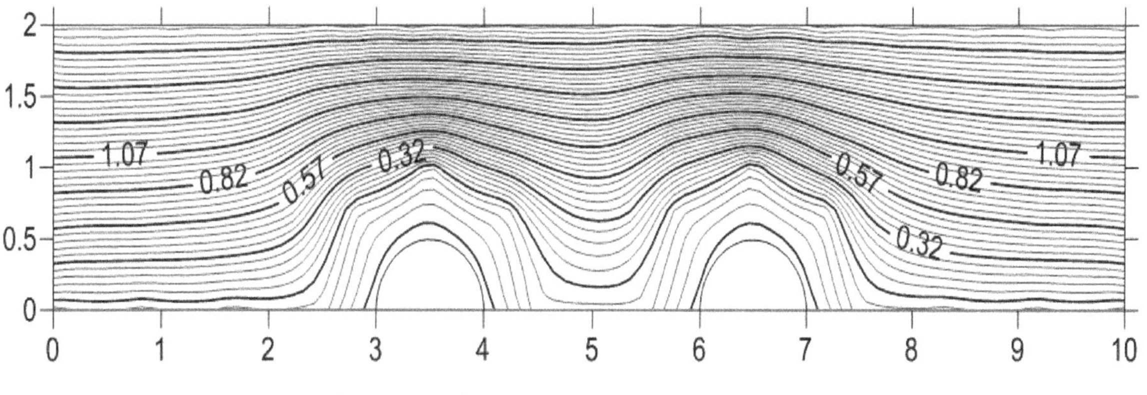

Figure 26. Stream lines contours from stream function formulation.

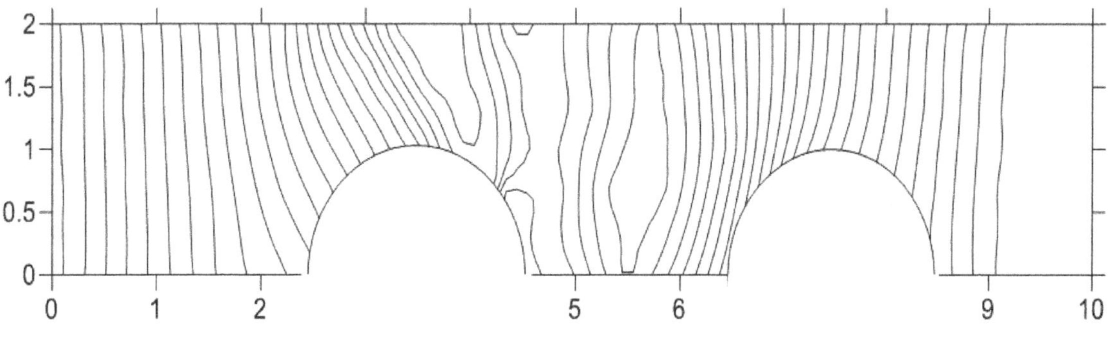

Figure 27. Velocity potential lines.

(iii) The calculated pressure co-efficient shows the pressure decreasing around the forwarded face from the initial total pressure at the stagnation point and reaching a minimum pressure at the top of the cylinder.
(iv) The calculated results depend to a certain extent on the discretization of the computational domain and accuracy increases with increase of number of elements.

Nomenclature: Ψ:Stream function, Φ:Velocity Potential function, N:Shape function, U:Free stream velocity, k^e:Element coefficient matrix, f^e: Element force vector.

Conflict of Interest

The authors have not declared any conflict of interest.

REFERENCES

Turner MJ, Clough RL, Martin HC, Topp LJ (1956). Stiffness and deflection analysis of complex structures. J. Aero. Sci. 23(9):805-825.
Chung TJ (1977). Finite Element Analysis in Fluid Dynamics". McGraw Hill, NY, USA. pp. 68-79.
Thom A (1933). "The flow past circular cylinders at low speeds". Proc. Royal Soc. 141(845):651-669.
Kawaguti M (1953). Discontinuous flow past a circular cylinder." J. Phys. Soc. Jap. 8:403-899.
Payne RB (1958). Calculation of Unsteady Viscous Flow Past Cylinder. J. Fluid Mech. 4:81.
Oden JT (1969). "A General Theory of Finite Elements". Int. J. Numer. Methods Eng. 1:247-259.
Dennis SCR, Chung GZ (1970). "Numerical Solutions for Steady Flow past a Circular Cylinder at Reynolds Numbers Up to 100". J. Fluid Mech. 42(3):471-489.
Tong P (1971). "The Finite Element Method for Fluid Flow". In: Gallagher RH, Oden JT, Yamada Y (eds.), Recent Advances in Matrix Method of Structural Analysis and Design. University of Alabama Press, Alabama. 904pp.
Olson MD (1974). "Variational-Finite Element Methods for Two-Dimensional and Axisymmetric Navier-Stokes Equations". Proceedings of the fourth International Symposium on Finite Element Methods in Flow Problems, Swansea, UK.
Hafez M (2004). "Inviscid Flows over a Cylinder". Comput. Methods Appl. Mech. Eng. 193:1981-1995.
Tarafder MS, Khalil GM, Islam MR (2010). "Analysis of potential flow around two-dimensional hydrofoil by source based lower and higher order panel method". J. Institut. Eng. Malaysia (71):2.

Permissions

All chapters in this book were first published in JMER, by Academic Journals; hereby published with permission under the Creative Commons Attribution License or equivalent. Every chapter published in this book has been scrutinized by our experts. Their significance has been extensively debated. The topics covered herein carry significant findings which will fuel the growth of the discipline. They may even be implemented as practical applications or may be referred to as a beginning point for another development.

The contributors of this book come from diverse backgrounds, making this book a truly international effort. This book will bring forth new frontiers with its revolutionizing research information and detailed analysis of the nascent developments around the world.

We would like to thank all the contributing authors for lending their expertise to make the book truly unique. They have played a crucial role in the development of this book. Without their invaluable contributions this book wouldn't have been possible. They have made vital efforts to compile up to date information on the varied aspects of this subject to make this book a valuable addition to the collection of many professionals and students.

This book was conceptualized with the vision of imparting up-to-date information and advanced data in this field. To ensure the same, a matchless editorial board was set up. Every individual on the board went through rigorous rounds of assessment to prove their worth. After which they invested a large part of their time researching and compiling the most relevant data for our readers.

The editorial board has been involved in producing this book since its inception. They have spent rigorous hours researching and exploring the diverse topics which have resulted in the successful publishing of this book. They have passed on their knowledge of decades through this book. To expedite this challenging task, the publisher supported the team at every step. A small team of assistant editors was also appointed to further simplify the editing procedure and attain best results for the readers.

Apart from the editorial board, the designing team has also invested a significant amount of their time in understanding the subject and creating the most relevant covers. They scrutinized every image to scout for the most suitable representation of the subject and create an appropriate cover for the book.

The publishing team has been an ardent support to the editorial, designing and production team. Their endless efforts to recruit the best for this project, has resulted in the accomplishment of this book. They are a veteran in the field of academics and their pool of knowledge is as vast as their experience in printing. Their expertise and guidance has proved useful at every step. Their uncompromising quality standards have made this book an exceptional effort. Their encouragement from time to time has been an inspiration for everyone.

The publisher and the editorial board hope that this book will prove to be a valuable piece of knowledge for researchers, students, practitioners and scholars across the globe.

List of Contributors

Marius-Constantin Popescu
Faculty of Electromechanical and Environmental Engineering, University of Craiova, B-dul Decebal, nr.107, 200440-Craiova, Dolj, România

Cristinel Popescu
Faculty of Electromechanical and Environmental Engineering, University of Craiova, B-dul Decebal, nr.107, 200440-Craiova, Dolj, România

Rakesh Singh Rajput
Department of Mechanical Engineering, Directorate of Technical Education Bhopal (M.P.), India

Sunil Kumar
Department of Mecanical Engineering, Rajeev Gandhi Technical University, Bhopal, M. P., India

Alok Chaube
Department of Mecanical Engineering, Rajeev Gandhi Technical University, Bhopal, M. P., India

J. P. Dwivedi
Department of Mechanical Engineering, IT-BHU, Varanasi, India

V. J. Bansod
Department of Mathematics, Technological University, Lonere India

B. Ambedkar
Department of Mathematics, Technological University, Lonere India

M. Emmanuel Adigio
Department of Mechanical Engineering, Niger Delta University, Wilberforce Island, Amassoma, Bayelsa State, Nigeria

B. K. C. Ganesh
Department of Mechanical Engineering, Narsaraopeta Engineering College, Narsaraopeta, A. P. India

N. Ramaniah
Department of Mechanical Engineering, College of Engineering (Autonomous), Andhra University, Visakhapatnam, India

P. V. Chandrasekhar Rao
Department of Mechanical Engineering, L. B. R. College of Engineering, Mylavaram. A. P. India

P. Maheshkumar
Department of Mechanical Engineering, National Institute of Technology, Calicut NIT Campus (P.O), Kerala, India 673601

C. Muraleedharan
Department of Mechanical Engineering, National Institute of Technology, Calicut NIT Campus (P.O), Kerala, India 673601

Mohd Nadeem Khan
Department of Mechanical Engineering, Krishna Institute of Engineering and Technology, Ghaziabad, India

Mohd Islam
Department of Mechanical Engineering, Jamia Millia Islamia, New Delhi, India

M. M. Hasan
Department of Mechanical Engineering, Jamia Millia Islamia, New Delhi, India

G. Jayaprakash
Department of Mechanical Engineering, Shivani Engineering College, Tiruchirapalli, India

K. Sivakumar
Department of Mechanical Engineering, Bannari Amman Institute of Technology, Sathiyamangalam, India

M. Thilak
Department of Mechanical Engineering, P.A.B.I.I.T, Tiruchirapalli, India

M. Mushtaq
Department of Mathematics, COMSATS Institute of Information Technology, Islamabad-Pakistan

M. Ashraf
Department of Mathematics, COMSATS Institute of Information Technology, Islamabad-Pakistan

S. Asghar
Department of Mathematics, COMSATS Institute of Information Technology, Islamabad-Pakistan
Department of Mathematics, King Abdul Aziz University, Jeddah, Saudi Arabia

M. A. Hossain
Department of Mathematics, University of Dhaka, Dhaka, Bangladesh

A. G. F. Alabi
Department of Mechanical Engineering, Faculty of Engineering, University of Ilorin, Kwara State, Nigeria

T. K. Ajiboye
Department of Mechanical Engineering, Faculty of Engineering, University of Ilorin, Kwara State, Nigeria

H.D. Olusegun
Federal University of Technology, Minna, Niger State, Nigeria

N. Shankar Ganesh
Vellore Institute of Technology, Vellore-632014, India

T. Srinivas
Vellore Institute of Technology, Vellore-632014, India

O. Obodeh
Mechanical Engineering Department, Ambrose Alli University, Ekpoma, Edo State, Nigeria

P. E. Ugwuoke
Mechanical Engineering Department, Petroleum Training Institute, Effurun, Delta State, Nigeria

Nasir .S. Hassen
Department of Thermo fluid, Faculty of Mechanical Engineering, University Technology Malaysia, UTM Johor Bahru, Malaysia

Nor Azwadi .C. Sidik
Department of Thermo fluid, Faculty of Mechanical Engineering, University Technology Malaysia, UTM Johor Bahru, Malaysia

Jamaludin .M. Sheriff
Department of Thermo fluid, Faculty of Mechanical Engineering, University Technology Malaysia, UTM Johor Bahru, Malaysia

R. M. Majumdar
Mechanical Engineering Department, G. H. Raisoni College of Engineering, Nagpur, 440016, India

V. M. Kriplani
Mechanical Engineering Department, G. H. Raisoni College of Engineering, Nagpur, 440016, India

A. T. Abdulrahim
Department of Mechanical Engineering, University of Maiduguri, Nigeria

I. S. Diso
Department of Mechanical Engineering, Bayero University Kano, Nigeria

A. S. Abdulraheem
Department of Mechanical Engineering, Bayero University Kano, Nigeria

S. N. Sapali
Department of Mechanical Engineering, Government College of Engineering, Shivaji Nagar, Pune, Maharashtra 411005, India

Pradeep A. Patil
Mechanical Engineering Department, Aissms College of Engineering, Pune University, Kennedy Road, Near R. T. O, Pune, Maharashtra, 411001 India

Md. Shahjada Tarafder
Department of Naval Architecture and Marine Engineering, Bangladesh University of Engineering and Technology, Dhaka-1000, Bangladesh

Nabila Naz
Department of Naval Architecture and Marine Engineering, Bangladesh University of Engineering and Technology, Dhaka-1000, Bangladesh

www.ingramcontent.com/pod-product-compliance
Lightning Source LLC
Chambersburg PA
CBHW080648200326
41458CB00013B/4779
9781682851098